PROTOPLANETARY DUST
Astrophysical and Cosmochemical Perspectives

Planet-formation studies uniquely benefit from three disciplines: astronomical observations of extrasolar planet-forming disks, analysis of material from the early Solar System, and laboratory astrophysics experiments. Pre-planetary solids, fine dust, and chondritic components are central elements linking these studies.

This book is the first comprehensive overview of planet formation, in which astronomers, cosmochemists, and laboratory astrophysicists jointly discuss the latest insights from the Spitzer and Hubble space telescopes, new interferometers, space missions including Stardust and Deep Impact, and laboratory techniques. Following the evolution of solids from their genesis through protoplanetary disks to rocky planets, the book discusses in detail how the latest results from these disciplines fit into a coherent picture. This volume provides a clear introduction and valuable reference for students and researchers in astronomy, cosmochemistry, laboratory astrophysics, and planetary sciences.

Dániel Apai is an Assistant Astronomer at the Space Telescope Science Institute. His research focuses on the observational characterization of the origins and properties of extrasolar planets and planetary systems.

Dante S. Lauretta is Associate Professor of Planetary Science and Cosmochemistry at the University of Arizona's Lunar and Planetary Laboratory. His research interests include the chemistry and mineralogy of asteroids and comets as determined by *in situ* laboratory analysis and spacecraft observations.

Cambridge Planetary Science

Series Editors: Fran Bagenal, David Jewitt, Carl Murray, Jim Bell, Ralph Lorenz, Francis Nimmo, Sara Russell

Books in the series

[†] Issued as a paperback

PROTOPLANETARY DUST

Astrophysical and Cosmochemical Perspectives

Edited by

DÁNIEL APAI
Space Telescope Science Institute

AND

DANTE S. LAURETTA
University of Arizona

CAMBRIDGE
UNIVERSITY PRESS

CAMBRIDGE
UNIVERSITY PRESS

The Edinburgh Building, Cambridge CB2 8RU, UK

Published in the United States of America by Cambridge University Press, New York

Cambridge University Press is part of the University of Cambridge.

It furthers the University's mission by disseminating knowledge in the pursuit of education, learning and research at the highest international levels of excellence.

www.cambridge.org
Information on this title: www.cambridge.org/9781107629424

© Cambridge University Press 2010

First published 2010
First paperback edition 2013

A catalogue record for this publication is available from the British Library

Library of Congress Cataloguing in Publication data

Protoplanetary dust : astrophysical and cosmochemical perspectives / edited by Dániel Apai and Dante S. Lauretta.
p.   cm. – (Cambridge planetary science ; 12)
ISBN 978-0-521-51772-0 (Hardback)
1. Cosmic dust.   2. Protoplanetary disks.   3. Cosmochemistry.   4. Solar system–Origin.
I. Apai, Dániel.   II. Lauretta, D. S. (Dante S.), 1970–   III. Title.   IV. Series.
QB791.P774 2010
523.2–dc22

2009036362

ISBN 978-0-521-51772-0 Hardback
ISBN 978-1-107-62942-4 Paperback

Contents

Contributing authors

Dániel Apai
Space Telescope Science Institute
3700 San Martin Drive
Baltimore, MD 21218, USA

Adrian J. Brearley
Department of Earth and Planetary Sciences
MSc03-2040
University of New Mexico
Albuquerque, NM 87131, USA

John R. Brucato
INAF – Osservatorio Astrofisico di Arcetri
L.go E. Fermi 5, 50125 Firenze, Italy

Subrata Chakraborty
University of California, San Diego
Department of Chemistry and Biochemistry
9500 Gilman Dr.
La Jolla CA 92093-0356, USA

John E. Chambers
Department of Terrestrial Magnetism
Carnegie Institution of Washington
5241 Broad Branch Road, NW
Washington, DC 20015-1305, USA

Fred J. Ciesla
Department of the Geophysical Sciences
University of Chicago
5734 South Ellis Avenue
Chicago, IL 60637–1433, USA

Harold C. Connolly Jr.
Earth and Planetary Sciences
Kingsborough College of the City University of New York
Department of Physical Sciences
2001 Oriental Blvd.
Brooklyn, NY 11235, USA
and
Lunar and Planetary Laboratory
University of Arizona
Tucson, AZ 85721, USA

Andrew M. Davis
Department of the Geophysical Sciences, Enrico Fermi Institute
and
Chicago Center for Cosmochemistry, University of Chicago
5734 South Ellis Avenue
Chicago, IL 60637-1433, USA

Cornelis P. Dullemond
Max Planck Institute for Astronomy
Koenigstuhl 17
D-69117 Heidelberg, Germany

George Flynn
316 Hudson Hall
SUNY Plattsburgh
101 Broad Str.
Plattsburgh, New York 12901, USA

Hans-Peter Gail
Center for Astronomy of the University of Heidelberg
Institute for Theoretical Astrophysics
Albert–Überle-Str. 2
D-69120 Heidelberg, Germany

Peter Hoppe
Particle Chemistry Department
Max Planck Institute for Chemistry
P.O. Box 3060, D–55020 Mainz, Germany

Dante S. Lauretta
Lunar and Planetary Laboratory
The University of Arizona

1629 E. University Blvd.
Tucson, AZ 85721, USA

Michiel Min
Astronomical Institute Anton Pannekoek
Kruislaan 403
1098 SJ Amsterdam, The Netherlands

Joseph A. Nuth III
Astrochemistry Laboratory, Code 691
NASA Goddard Space Flight Center
Greenbelt, MD 20771, USA

David P. O'Brien
Planetary Science Institute
1700 E. Ft. Lowell, Suite 106
Tucson, AZ 85719, USA

Ilaria Pascucci
Department of Physics & Astronomy
Johns Hopkins University
366 Bloomberg Center
3400 N. Charles Street
Baltimore, MD 21218, USA

Klaus M. Pontoppidan
Division of Geological and Planetary Sciences
California Institute of Technology, MS 150-21,
Pasadena, CA 91125, USA

Dmitry Semenov
Max Planck Institute for Astronomy
Koenigstuhl 17
D-69117 Heidelberg, Germany

Shogo Tachibana
Department of Earth and Planetary Science (Bldg.1)
The University of Tokyo
7-3-1 Hongo, Tokyo 113–0033, Japan

Mark Thiemens
University of California, San Diego
Department of Chemistry and Biochemistry
9500 Gilman Dr.
La Jolla, CA 92093-0356, USA

Preface

Some fundamental questions are surprisingly simple: Where did we come from? Are we alone in the Universe? These two simple questions have been pondered on and debated over by hundreds of generations. Yet, these questions proved to be very difficult to answer. Today, however, they have shifted from the realm of religious and philosophical discussions to the lecture rooms and laboratories of hard sciences: they are, indeed, among the drivers of modern astrophysics and planetary sciences.

Fortunately, and perhaps surprisingly, the Universe provides a means to address these important questions. Today we are witnessing as the answers emerge to these age-old questions. We now know that asteroids and comets of the Solar System have preserved a detailed record of the dramatic events that four billion years ago gave birth to our planetary system in only a few million years. Gravity and radiation pressure conspire to deliver almost pristine samples of the early Solar System to Earth in the form of meteorites and interplanetary dust particles. We have also taken this process one step further with the successful return of particles from the coma of comet Wild 2 by NASA's Stardust mission. Detailed chemical and mineralogical analyses of these materials allow for the reconstruction of the history of our planetary system.

We can address the questions of the ubiquity of planetary systems in our galaxy by comparing the conditions and events of the early Solar System to circumstellar disks in star-forming regions. Technological wonders, such as the Hubble and Spitzer space telescopes, have allowed direct imaging of disks in which planetary systems are thought to form and enable comparative mineralogy of dust grains hundreds of light years away.

Over the past decade these exciting advances have transformed our understanding of the origins of planetary systems. Astronomers provide exquisite observations of nascent planetary systems. Cosmochemists reconstruct the detailed history of the first ten million years of the Solar System. Circumstellar disks and, in particular, the evolution of dust grains play a pivotal role in the formation and early evolution of planetary systems, including our own.

The chance collisions and sticking of a few tiny dust grains around a young star: these are the first steps in a long and fascinating journey that a few million years later culminate in violent, catastrophic collisions of hot, molten protoplanets as a new planetary system is born. The evolution of these dust grains and the dust disk itself is the best-studied and most-constrained phase of planet formation. We can observe dust grains as they form during the death throes of a previous generation of stars and as they are injected into interstellar space. We know that these grains are then altered by the harsh radiation fields and shock waves that propagate through the interstellar medium. Dust, concentrated into giant molecular clouds, is entrained in the gas that dominates the mass of these systems. We can identify evolutionary snapshots as some of the densest parts of clouds become unstable, collapse, and form stars surrounded by accretion disks. The dynamic and turbulent conditions in these disks lead to the evaporation, melting, crystallization, amorphization, and agglomeration of primordial and newly formed dust grains. The dust particles accrete into planetesimals, many of which persist throughout the stellar lifetime. These small bodies collide with each other, producing more dust but also, in some cases, growing to planetary bodies. This book is an attempt to synthesize our current state of knowledge of the history of this dust, from the interstellar medium where stars and planets are born to the final stages of planetary accretion using both astronomical and cosmochemical perspectives.

Astronomers study the evolution of protoplanetary disks on large scales, measuring simple, general properties of hundreds of disks. Planetary scientists, in contrast, unravel the detailed history of our Solar System by meticulous characterization of the solid remnants of the earliest epochs combined with dynamical simulation of the formation and accretion of particles from dust grains to planets. However, there has long been a disconnection between specialists in these two allied disciplines. Although they study the same processes and address the same questions, communication has been difficult because of differences in methods, concepts, terminology, instrumentation, analytical techniques, and the scientific forums where cutting-edge results are presented. This problem is not new. Twenty-seven years ago Tom Gehrels in his Introduction to *Protostars and Planets* noted the "growing separation between astronomers and planetary scientists." Although the problem persists, we believe that today astronomy and planetary science are intersecting in many places; questions where the two disciplines overlap benefit from a diversity of constraints and allow the transport of ideas and concepts. In particular, there appears to be an important convergence in the study of the origins of planetary systems.

This book builds bridges between astronomy and planetary sciences. It does so to capitalize from the value of the common questions and the different approaches.

Therefore, in designing this volume we decided from day one to merge diverse perspectives in each topic. The authors for each chapter were selected to represent distinct disciplines focused on the same question. The long, heated, and constructive discussions that ensued from pairing specialist authors with different backgrounds brought a real novel value to these chapters. This mix was further enriched by the referees' work – typically three or four for each chapter – that added diverse perspectives. They worked very hard to check the emerging text repeatedly and their essential help made this book truly a community effort.

We are immensely satisfied with the results. In the course of this work we have learned an enormous amount, from the contributing authors and also from each other. This volume presents the comprehensive history of the birth and early development of planetary systems – it provides a complex and fascinating story to partly answer a simple, yet fundamental question.

We hope you enjoy reading the book as much as we enjoyed compiling it.

Dániel Apai and Dante S. Lauretta

Acknowledgments

We are grateful to the following colleagues for motivating discussions or for reviewing chapter manuscripts: Anja Andersen, Phil Armitage, Ted Bergin, Roy van Boekel, Jade Bond, Bill Bottke, Fred Ciesla, Cathie Clarke, Jeff Cuzzi, Ann Dutrey, Ian Franchi, Lee Hartmann, Louis d'Hendecourt, Frank Hersant, Shigeru Ida, Lindsay Keller, Thorsten Kleine, Guy Libourel, Casey Lisse, Gary Lofgren, Harry Y. McSween, Jr., Scott Messenger, Knut Metzler, James Muzerolle, Larry Nittler, Ilaria Pascucci, Matt Pasek, Mike Sitko, Mario Triloff, Gerhard Wurm, Hisayoshi Yurimoto, Thomas Henning, and Michael R. Meyer. We thank Linda L. Mamassian for compiling the index for this volume.

1

Planet formation and protoplanetary dust

Dániel Apai and Dante S. Lauretta

Abstract Planet formation is a very complex process through which initially submicron-sized dust grains evolve into rocky, icy, and giant planets. The physical growth is accompanied by chemical, isotopic, and thermal evolution of the disk material, processes important to understanding how the initial conditions determine the properties of the forming planetary systems. Here we review the principal stages of planet formation and briefly introduce key concepts and evidence types available to constrain these.

Tiny solid cosmic particles – often referred to as "dust" – are the ultimate source of solids from which rocky planets, planetesimals, moons, and everything on them form. The study of the dust particles' genesis and their evolution from interstellar space through protoplanetary disks into forming planetesimals provides us with a bottom-up picture on planet formation. These studies are essential to understand what determines the bulk composition of rocky planets and, ultimately, to decipher the formation history of the Solar System. Dust in many astrophysical settings is readily observable and recent ground- and space-based observations have transformed our understanding on the physics and chemistry of these tiny particles.

Dust, however, also obscures the astronomical view of forming planetary systems, limiting our knowledge. Astronomy, restricted to observe far-away systems, can only probe some disk sections and only on relatively large scales: the behavior of particles must be constrained from the observations of the whole disk.

However, planet formation is a uniquely fortunate problem, as our extensive meteorite collections abound with primitive materials left over from the young Solar System, almost as providing a perfect sample-return mission from a protoplanetary disk. A remarkable achievement of geochronology is that many of these samples can be dated and the story of the Solar System's formation reconstructed.

A thorough and quantitative understanding of planet formation is impossible without using the puzzle pieces from both astronomy and cosmochemistry. With chapters chronologically ordered and co-written by experts from both fields, this book attempts to lay out the pieces available for the first 10 Myr of planet formation and arrange them in a meaningful pattern. We focus on a common large-scale context for astronomical and meteoritical findings, rather than providing specialized reviews on specific details: the goal is developing a grander new picture rather than scrutinizing evidence. The book identifies controversial questions, but aims to remain impartial in debates.

In this chapter we first introduce the types of evidence, basic concepts, and planet-formation timeline that are used throughout the book and briefly review the constraints available for different epochs of planet formation within the first 10 Myr. Table 1.1 provides a summary of the types of constraints on the different stages of planet formation and the chapters in which they are discussed in the book.

1.1 Types of extraterrestrial material available

Meteorites are fragments of planetary material that survive passage through the Earth's atmosphere and land on the surface of the Earth. To date all known meteorites are pieces of either asteroids, the Moon, or Mars, with the former dominating the flux of material. Asteroidal meteorites show an amazing diversity in their texture and mineralogy and illustrate the geologic diversity of the small bodies in our Solar System. They are uniformly ancient, dating from the first 10 Myr of Solar System history. These samples are invaluable in providing a detailed, albeit biased, history of planetary evolution. Table 1.1 summarizes the types of extraterrestrial material and astronomical observations available for the key stages of planet formation. Figure 1.1 illustrates the classification of the primitive materials most relevant to planet formation.

Meteorites are divided into two broad categories: chondrites, which retain some record of processes in the solar nebula; and achondrites, which experienced melting and planetary differentiation. The nebular record of all chondritic meteorites is obscured to varying degrees by alteration processes on their parent asteroids. Some meteorites, such as the CI, CM, and CR chondrites, experienced aqueous alteration when ice particles that co-accreted with the silicate and metallic material melted and altered the primary nebular phases. Other samples, such as the ordinary and enstatite chondrites, experienced dry thermal metamorphism, reaching temperatures ranging from about 570 to 1200 K. In order to understand the processes that occurred in the protoplanetary disk, we seek out the least-altered samples that best preserve the record of processes in the solar nebula. The CV, CO,

Table 1.1 *The astronomical and cosmochemical evidence available on the key stages of the evolution of protoplanetary disks and the chapters in which they are discussed.*

	Chapters	Meteoritical evidence	Astronomical evidence	Laboratory experiments
Interstellar medium	2	Presolar grains	Radio: cold gas; optical/infrared extinction: dust	Condensation experiments
Protostellar collapse	2	Organic residues on presolar grains	Radio: gas lines, near-infrared extinction maps	
Disk formation, dust condensation	3, 4, 5	Oxygen isotopes, noble gases, volatility trends in chondrites	Spectral energy distributions, scattered light images, disk silhouettes	Heating experiments, photochemistry
Dust coagulation	6, 7	"Pre-chondrule aggregates," AOAs, fine-grained matrix, fine-grained CAIs	Spectroscopy (8–30 micron) mm-interferometry scattered light	Zero-G or micro-G experiments (space, parabolic flights, drop tower, sounding rocket)
Thermal processing	5, 8, 9	Chondrite components: chondrules, compact CAIs	Spectroscopy (8–30 micron)	Condensation and heating experiments
Planetesimals	10	Chondrites, achondrites, iron meteorites	Debris around white dwarfs, debris disks	
Planets	10	Lunar and martian meteorites, planetary bulk composition	Exoplanets	

and CH carbonaceous chondrites along with the unequilibrated ordinary chondrites offer the best record of early Solar System evolution and are the subject of intense investigation.

The most primitive chondrites consist of coarse-grained (mm-sized) mineral assemblages embedded in fine-grained (10 nm–5 μm) matrix material (see Fig. 1.2). The coarse-grained chondritic components are diverse in their composition and mineralogy and include calcium–aluminum-rich inclusions (CAIs), amoeboid olivine aggregates (AOAs), Al-rich chondrules, Fe–Mg chondrules, Fe-rich metals, and iron sulfides. The CAIs are composed largely of calcium, aluminum, and titanium

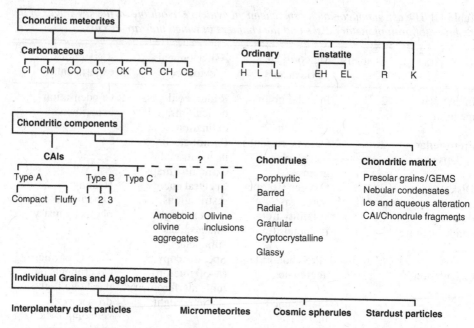

Figure 1.1 Types of primitive, unprocessed materials available for studies.

oxides. The AOAs contain CAI nuggets surrounded by magnesium-rich olivine. Most chondrules have porphyritic textures (large crystals surrounded by the fine-grained mesostasis). Other textural types include barred-olivine, radial-pyroxene, granular, cryptocrystalline, and glassy (see Fig. 1.1). Aluminum-rich chondrules contain Al–Ti-rich pyroxene and olivine crystals in glassy, calcium-rich mesostasis. Ferromagnesian chondrules are composed largely of olivine, pyroxene, metal, sulfide, and glassy mesostasis. Matrix material is an aggregate of mineral grains that surrounds the coarse components and fills in the interstices between them. It is made largely of forsterite and enstatite grains, and amorphous silicate particles. Matrix also contains metal sulfide grains, refractory oxides, carbon-rich material, and a few parts per million of presolar silicate, carbide, and oxide grains. Appendix 1 provides a summary of the minerals common in astrophysical settings and Appendix 2 describes high-resolution analytic techniques important for studying them.

In addition to meteorites, three other important types of extraterrestrial material are available for analysis: interplanetary dust particles (IDPs), micrometeorites, and Stardust samples. Interplanetary dust particles are collected in the stratosphere by high-altitude research aircrafts. Most of these samples are smaller than 20 μm in diameter, although some of the highly porous cluster particles probably exceeded

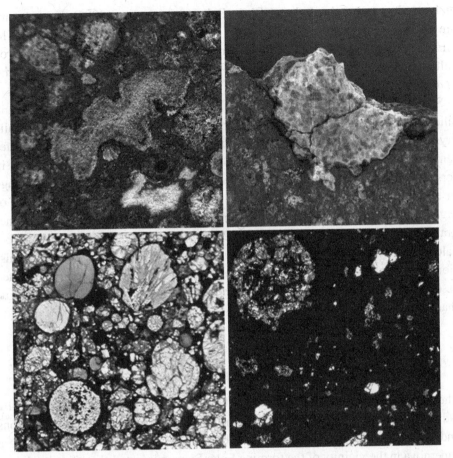

Figure 1.2 Components of chondritic meteorites: fluffy CAI (upper left), compact CAI (upper right), chondrule (lower left), and matrix (lower right).

100 μm before they fragmented on the collection surface. The IDPs include samples of both asteroids and comets. Micrometeorites are much more massive than typical IDPs. They can be collected in vast numbers, and include particles near the mass flux peak at 200 μm size that dominates the bulk of cosmic matter accreted by Earth. Micrometeorites exhibit a diversity of compositions and structures, with the majority being dominated by fine-grained anhydrous minerals. Stardust samples were collected in the coma of comet Wild 2 by high-velocity aerogel capture and returned to Earth for detailed analysis.

From a mineralogy viewpoint, IDPs are aggregates of mostly sub-micron-sized crystalline silicates (olivine and pyroxene), amorphous silicates, sulfides, and minor refractory minerals, held together by an organic-rich, carbonaceous matrix. Large fractions, 30–60 wt%, of these IDPs are amorphous silicates, known as glass with embedded metals and sulfides (GEMS, Keller & Messenger 2007). These grains are roughly spherical and range from 0.1 μm to 1 μm in size (Bradley 1994). The GEMS particles also contain finely dispersed nanocrystals of Fe–Ni alloy and iron sulfide and organic carbon molecules ($\sim 12\,\mathrm{wt}\%$, Schramm *et al.* 1989; Thomas *et al.* 1994; Flynn *et al.* 2004).

Stardust grains are composed of olivine, low-Ca pyroxene, sulfides, sodium silicates, and refractory minerals similar to CAIs (Zolensky *et al.* 2006). While crystalline grains are abundant, the intrinsic abundance of amorphous silicates remains unknown. Olivine grains in the Stardust samples span a large range in forsterite abundance and the low-Ca pyroxene grains exhibit a similarly large range in enstatite abundance. The Wild 2 grains also contain a significant amount of organic matter, which in many ways resembles the organic material observed in fine-grained, anhydrous IDPs (Sandford *et al.* 2006). Carbonates are rare in P/Wild 2, but calcite, dolomite, and ferroan magnesite grains occur (Flynn *et al.* 2008). This mission shows that cometary dust is heterogeneous and represents an un-equilibrated assortment of mostly solar materials resembling chondritic material, far less pristine than anticipated (Brownlee *et al.* 2006).

1.2 Chronology of planet formation

The events that lead to the formation of the Solar System can be reconstructed by radioisotopic dating of extraterrestrial samples originating from different locations and epochs of the proto-solar nebula. The isotopic dating is possible because a supernova in the vicinity of the forming Solar System injected short-lived radionuclides (e.g. ^{26}Al, ^{60}Fe, ^{41}Ca, ^{36}Cl, ^{53}Mn) into the proto-solar cloud; the decay of these short-lived nuclides provides high-resolution chronology which, in combination with the decay of long-lived isotopes (mostly U and Th), provides today accurate clocks for dating critical events in the early Solar System.

In contrast, astronomical constraints on the evolution of protoplanetary disks are provided by studies of nearby groups of young stars with ages $< 1\,\mathrm{Myr}$ to $> 100\,\mathrm{Myr}$ (Fig. 1.3). Stars in the these co-eval groups provide snapshots of the disk evolution at different evolutionary stages. The diversity observed at any given age reveals a large spread in the possible evolutionary paths of disks. Although stellar clusters and co-moving stellar groups can be dated with several different methods, typical age uncertainties for young clusters (< 20 Myr) remain 50–100%.

In the remainder of the chapter we review the major stages and key open questions of planet formation, drawing on the detailed discussions presented in the subsequent chapters.

Chronology of planet formation

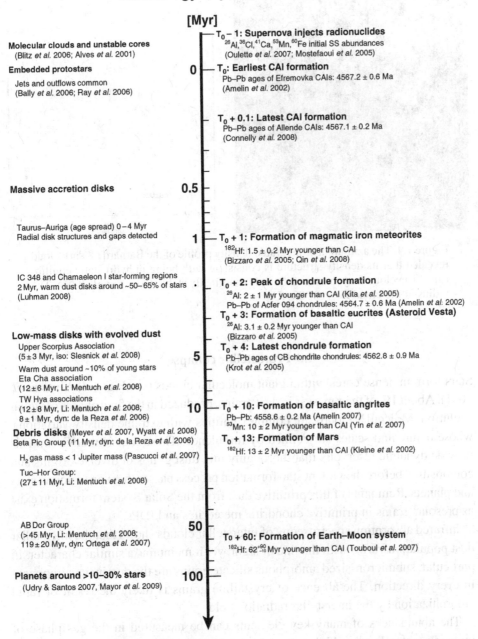

[Myr]

T₀ – 1: Supernova injects radionuclides
^{26}Al, ^{36}Cl, ^{41}Ca, ^{53}Mn, ^{60}Fe initial SS abundances
(Oulette *et al.* 2007; Mostefaoui *et al.* 2005)

Molecular clouds and unstable cores
(Blitz *et al.* 2006; Alves *et al.* 2001)

Embedded protostars

Jets and outflows common
(Bally *et al.* 2006; Ray *et al.* 2006)

0 — **T₀: Earliest CAI formation**
Pb–Pb ages of Efremovka CAIs: 4567.2 ± 0.6 Ma
(Amelin *et al.* 2002)

T₀ + 0.1: Latest CAI formation
Pb–Pb ages of Allende CAIs: 4567.1 ± 0.2 Ma
(Connelly *et al.* 2008)

Massive accretion disks **0.5**

Taurus–Auriga (age spread) 0 – 4 Myr
Radial disk structures and gaps detected

1 — **T₀ + 1: Formation of magmatic iron meteorites**
^{182}Hf: 1.5 ± 0.2 Myr younger than CAI
(Bizzaro *et al.* 2005; Qin *et al.* 2008)

IC 348 and Chamaeleon I star-forming regions
2 Myr, warm dust disks around ~50– 65% of stars
(Luhman 2008)

T₀ + 2: Peak of chondrule formation
^{26}Al: 2 ± 1 Myr younger than CAI (Kita *et al.* 2005)
Pb–Pb of Acfer 094 chondrules: 4564.7 ± 0.6 Ma (Amelin *et al.* 2002)

T₀ + 3: Formation of basaltic eucrites (Asteroid Vesta)
^{26}Al: 3.1 ± 0.2 Myr younger than CAI

Low-mass disks with evolved dust
Upper Scorpius Association
(5 ± 3 Myr, iso: Slesnick *et al.* 2008)

(Bizzaro *et al.* 2005)

5 **T₀ + 4: Latest chondrule formation**
Pb–Pb ages of CB chondrite chondrules: 4562.8 ± 0.9 Ma
(Krot *et al.* 2005)

Warm dust around ~10% of young stars
Eta Cha association
(12 ± 6 Myr, Li: Mentuch *et al.* 2008)

TW Hya associations
(12 ± 8 Myr, Li: Mentuch *et al.* 2008;
8 ± 1 Myr, dyn: de la Reza *et al.* 2006)

10 — **T₀ + 10: Formation of basaltic angrites**
Pb–Pb: 4558.6 ± 0.2 Ma (Amelin 2007)
^{53}Mn: 10 ± 2 Myr younger than CAI (Yin *et al.* 2007)

Debris disks (Meyer *et al.* 2007, Wyatt *et al.* 2008)
Beta Pic Group (11 Myr, dyn: de la Reza *et al.* 2006)

T₀ + 13: Formation of Mars
^{182}Hf: 13 ± 2 Myr younger than CAI (Kleine *et al.* 2002)

H₂ gas mass < 1 Jupiter mass (Pascucci *et al.* 2007)

Tuc–Hor Group:
(27 ± 11 Myr, Li: Mentuch *et al.* 2008)

AB Dor Group **50**
(> 45 Myr, Li: Mentuch *et al.* 2008;
119 ± 20 Myr, dyn: Ortega *et al.* 2007)

T₀ + 60: Formation of Earth–Moon system
^{182}Hf: 62^{+90}_{-10} Myr younger than CAI (Touboul *et al.* 2007)

Planets around >10–30% stars **100**
(Udry & Santos 2007, Mayor *et al.* 2009)

Figure 1.3 Chronology of the planet formation in the Solar System and astronomical analogs. The isotopes given identify the radioisotope systems that served as a basis for the dating. For the astronomical ages, Li refers to ages derived from stellar atmospheric Li abundances, dyn refers to dynamically derived ages, iso refers to ages derived through stellar isochrone fitting. Note that the zero points of the two systems were assumed here to coincide.

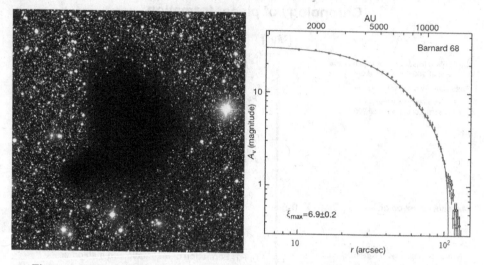

Figure 1.4 The azimuthally integrated density profile of the Barnard 68 dark cloud revealed that its density structure is consistent with being in hydrostatic equilibrium. The cloud may, however, be only marginally stable (Alves *et al.* 2001). The radius of the cloud is $\sim 12\,500$ AU.

1.3 Protostellar collapse

Stars form in dense cores within giant molecular clouds (see Fig. 1.4, Alves *et al.* 2001). About 1% of their mass is in dust grains, produced in the final phases of stellar evolution. Molecular clouds are complex entities with extreme density variations, whose nature and scales are defined by turbulence. These transient environments provide dynamic reservoirs that thoroughly mix dust grains of diverse origins and composition before the violent star-formation process passes them on to young stars and planets. Remnants of this primitive dust from the Solar System formation exist as presolar grains in primitive chondritic meteorites and IDPs.

Infrared absorption spectroscopy of interstellar clouds shows that the interstellar dust population varies with the line of sight, yet it maintains a similar character. In particular, submicron-sized amorphous silicate grains are the dominant component in every direction. The absence of crystalline grains is likely the result of rapid amorphization by the interstellar radiation field.

The abundances of many key elements can be measured in the gas phase of interstellar clouds using high-resolution spectroscopy (Savage & Sembach 1996). The composition of the solid phase – dust grains and ice mantles – can be calculated by subtracting the gas composition from the assumed bulk composition. In addition, X-ray spectroscopy probes not only the elements in the gas phase, but also those in the dust grains. Using bright X-ray binaries as background sources Ueda *et al.* (2006) have been able to determine the total abundances of the elements in the interstellar medium (ISM) and find that most of them are approximately solar, with the exception of oxygen. The silicates are predominantly rich in magnesium.

At the low temperatures characteristic of these dense clouds volatile molecular species (H_2O, CO, CO_2, HCO, H_2CO, CH_3OH, NH_3, and CH_4) condense onto the dust grains as icy mantles (Sandford & Allamandola 1993; Bergin *et al.* 2002; Walmsley *et al.* 2004). Ultraviolet photolysis of the mantles converts some of this material into coatings of refractory organic matter. Interestingly, some presolar grains are partially embedded in carbonaceous matter with isotopic ratios that reflect fractionation at extremely low temperatures ($\sim 20\,K$) expected in molecular cloud environments (Messenger *et al.* 2009).

Rising temperatures in the collapsing molecular cloud cores lead to the sublimation of first the icy grain mantles, and then, in the innermost regions of the newly formed protoplanetary disk, the more refractory dust grains. Star formation converts 10–30% of the molecular cloud core mass to stars. During the collapse of a cloud core its mass, initially distributed over parsec scales, is concentrated to $\sim AU$ scales, leading to a factor of $\sim 10^{10}$ decrease in its moment of inertia. To allow this compression, angular momentum must be redistributed, resulting in the formation of a viscous accretion disk. The small fraction of the disk mass that moves outward carries with it a substantial angular momentum, allowing the inner disk to lower its angular momentum and fall onto the protostar.

The extent to which the dust from the ISM survives planet formation intimately depends on the details of the core collapse and the formation of the accretion disk.

1.4 Structural evolution of protoplanetary disks

The collapse of rotating molecular cloud cores leads to the formation of massive accretion disks that evolve to more tenuous protoplanetary disks. Disk evolution is driven by a combination of viscous evolution, grain coagulation, photoevaporation, and accretion to the star. The pace of disk evolution can vary substantially, but massive accretion disks are thought to be typical for stars with ages < 1 Myr and lower-mass protoplanetary disks with reduced or no accretion rates are usually 1–8 Myr old. Disks older than 10 Myr are almost exclusively non-accreting debris disks (see Figs. 1.3 and 1.5).

The fundamental initial parameters of protoplanetary disk evolution are the masses and sizes of the disks. Optical silhouettes of disks in the Orion Nebula Cluster (McCaughrean & O'Dell 1996), scattered light imagery (e.g. Grady *et al.* 1999), interferometric maps in millimeter continuum or line emission (e.g. Rodmann *et al.* 2006; Dutrey *et al.* 2007), and disk spectral energy distributions

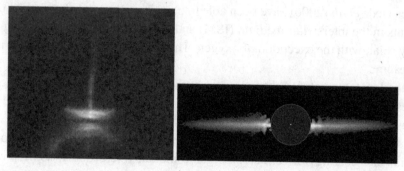

Figure 1.5 *Left panel:* the young, accreting star–disk system HH30 seen edge-on at visible wavelengths. The optically thick disk occults the star and the scattered light image shows the flaring disk surface. The system also drives a powerful jet (NASA/Space Telescope Science Institute, Burrows *et al.* 1996). *Right panel:* debris disk around the 12 Myr-old low-mass star AU Mic. The disk is geometrically flat, optically thin and depleted in gas (NASA/ESA/STScI).

(SEDs) demonstrate that massive disks often extend to hundreds of astronomical units. A lower estimate for the initial mass distribution of our Solar System is provided by the minimum mass solar nebula (MMSN) model, which is the minimum mass required to produce the observed distribution of solids from a disk with solar composition. This analysis predicts a disk mass between 0.01 and 0.07 $M_\odot$ extending out to 40 AU. Mass estimates for circumstellar disks derived from submillimeter and longer-wavelength observations are consistent with the range estimated for the MMSN (e.g. Beckwith *et al.* 1990; Williams *et al.* 2005).

The structure of disks can be probed through multiple techniques, including direct imaging of light scattered by dust and models of the SEDs. These measurements show that most young disks (< 3–5 Myr) around Sun-like stars display a flared disk structure, in which the disk opening angle increases with the radius. Some disks, especially those around very low-mass stars, often show reduced flaring or flat disks (e.g. Apai *et al.* 2005).

The flaring geometry naturally arises from the combination of turbulent gas and micron-sized dust grains that can efficiently couple to it. However, models demonstrate that with grain sizes increasing through random collisions and coagulation the dynamical coupling weakens, leading to dust settling and the overall flattening of the disk structure (e.g. Dullemond & Dominik 2005; Meyer *et al.* 2007).

The thermal structure of the disks plays a central role in determining the chemistry and the observable spectrum. The thermal structure, in turn, is set by the disk geometry and accretion rate, an important heat source. As a function of these parameters the mid-plane temperature of the disk can vary between the mild $T \sim r^{-1/2}$ for a flared disk to the rapidly declining of $T \sim r^{-3/4}$ for a flat disk. The highest temperatures in the static disk are reached at its innermost edge, directly exposed to the star.

1.5 Chemical evolution of the gas disks

The intricate interplay between the viscous evolution of the disk, large-scale flows, small-scale turbulence, dust settling, and growth provides a complex and evolving environment for the chemical evolution of the primitive planet-building materials. Telescopic observations of the gas-phase chemistry, as well as the isotopic and mineralogic studies of primitive Solar System material, provide insights into the chemical evolution during the first few million years of planet formation.

Models of irradiated disks predict four chemically distinct zones (see Fig. 4.1). (I) Zone of ices in the cold mid-plane opaque to incoming radiation. Chemistry in this region is dominated by cold gas-phase and grain-surface reactions. Here Infrared Space Observatory (ISO) and Spitzer observations confirmed the existence of ices, various silicates and PAHs (polycyclic aromatic hydrocarbons e.g. van den Ancker *et al.* 2000; van Dishoeck 2004; Bouwman *et al.* 2008). (II) Zone of molecules, a warm molecular layer adjacent to the mid-plane, dominated by ultraviolet/X-ray-driven photochemistry; (III) the heavily irradiated zone of radicals, a hot dilute disk atmosphere deficient in molecules; and (IV) the inner zone, inside of the ice line where terrestrial planets form.

Mid- and far-infrared, submillimeter, and radio-wavelength observations allow probing the presence and abundance of simple molecules in these zones and provide constraints and boundary conditions for coupled disk evolution and chemical network models. The observed abundances and predictions from the disk models can be directly compared to constraints derived from the early Solar System.

In the inner Solar System, the bulk composition of bodies shows a marked depletion of volatile elements compared to solar composition, which is thought to match the composition of the initial protostellar cloud. These volatile depletions are best explained as a result of the condensation of solids from hot gas ($< 1800\,K$, e.g. Davis 2006). This gas must have been well mixed and nearly homogeneous since refractory elements have relative abundances within 10% of solar composition and are isotopically uniform within 0.1% across all classes of primitive meteorites. Correspondingly, observations of hot CO gas show that gas is present in some of the innermost regions of disks where temperatures are too high ($> 2000\,K$) even for the most refractory elements to exist in solid state. These hot inner-disk regions readily mix and homogenize the elemental and isotopic composition of the gas-phase material.

In the solar nebula oxygen isotopes provide a particularly important and enigmatic tracer of physical and chemical processes (Clayton 2007; Thiemens 2006). Simultaneously present in both gaseous and solid phases, oxygen is the only rock-forming element that shows a wide range of isotopic heterogeneity at the bulk level. The complicated structure of meteoritic oxygen isotopes is difficult to reproduce simply by mixing of different reservoirs. Self-shielding of CO from photodissociation has been proposed as a possible solution (Lyons & Young 2005).

Isotopic tracers, together with bulk compositions of the Solar System bodies and remote-sensing observations of gas-phase molecules in disks, reveal the complexity of the chemical evolution of planet-forming disks. The chemical coupling of the gas to the dust grains, through grain-surface chemistry and evaporation/condensation processes, necessitates a thorough understanding of the dust behavior beyond what can be achieved by remote sensing of natural systems.

1.6 Laboratory dust analogs

Due to the complexity, temporal and physical scales, and uncontrollable nature of astrophysical processes remote sensing can only provide limited insights. Laboratory studies, in contrast, are well suited for studying the evolution of dust grains from their formation through their journey in the ISM to their reprocessing in the protoplanetary disks via a series of condensation, crystallization, amorphization, and grain-surface catalysis experiments. These studies are also crucial for the correct interpretation of astronomical observations.

Synthesizing astrophysical condensate grain analogs in a variety of chemical systems (e.g. Mg–Fe–SiO, Mg–SiO, Al_2O_3–SiO_2, and Al_2O_3–Fe_3O_4–SiO_2) is a key step in understanding grain formation and the effects of violent thermal reprocessing (e.g. Rietmeijer *et al.* 1999; Rietmeijer & Nuth 2000). The experiments reveal complex chemical processes during the extremely short duration of the vapor-growth phase. Surprisingly, condensation produces separate populations of amorphous grains with distinct compositions. Most intriguing is the fact that separate populations of iron silicate and magnesium silicate grains form from mixed Fe–Mg–SiO vapors.

Laboratory investigations confirm that crystalline silicates form in stellar outflows and in protoplanetary disks. In contrast, dust grains in the ISM are dominated by amorphous materials; less than 2.2% of the grains are crystalline silicates (Kemper *et al.* 2005). Laboratory simulations of the harsh interstellar radiation fields demonstrate that ion irradiation of crystalline silicates quickly leads to their amorphization (e.g. Jäger *et al.* 2003; Brucato *et al.* 2004).

In outflows and disks, grains may also be annealed either by absorption of energetic photons or by contact with the hot gas. Magnesium silicates thermally anneal more rapidly than iron silicates at the same temperature (Hallenbeck *et al.* 1998). Thus, the crystalline magnesium silicate minerals observed in protoplanetary disks may result from thermal annealing of amorphous magnesium and iron silicate condensates. In contrast, shock annealing produces both crystalline magnesium and iron silicates. In addition, experiments demonstrate that the spectral properties of annealing silicate smoke grains are highly dependent on their thermal history. Variations in the mid-infrared spectra occur in stages as crystals grow first on the surface, and then in the inner volume, until they merge into a single crystal.

Surface catalysis on dust grains is another important mechanism for the formation of many simple gas-phase molecules (H_2, H_2O, CO_2, etc.), more complex molecules (e.g. CH_3OH), radicals, and organic refractory material. In particular, catalytic reactions can convert CO and H_2 into hydrocarbons or N_2 and H_2 into reduced nitrogen compounds such as NH_3. Experiments have been performed to test the relative efficiency of various common dust materials and follow the catalytic properties as a function of time and temperature (Nuth *et al.* 2008). A surprising result of these experiments is that the macromolecular carbonaceous coating produced on the grains is a better catalyst than inorganic dust grains. Such a self-perpetuating catalyst that forms naturally on every grain surface may result in giant organic chemical factories that turn abundant CO, N_2, and H_2 into complex hydrocarbons.

Armed with the results of laboratory studies of astrophysical dust processing, we are able to interpret the complex and varied history of dust in protoplanetary disks. This information is complemented by the detailed analysis of the solid material that remains from the earliest epochs of Solar System formation.

1.7 Dust composition in protoplanetary disks

The composition of protoplanetary dust is key to understanding the bulk composition of planetesimals and planets and also provides insight into the planet formation processes. Several lines of evidence demonstrate that primitive dust underwent dramatic transformations in the early Solar System. These findings underline the need for understanding the links between disk evolution and the evolution of dust composition. The variety of dust species that can be identified via remote sensing include silicates, carbonaceous grains and carbonates, and sulfide-bearing grains. The abundances of these species, as derived from observations of disks, can be directly compared to meteoritic samples, IDPs, GEMS particles, and Stardust grains.

In the Solar System the bulk elemental composition of the most volatile-rich CI chondrites resembles closely that of the solar photosphere. Indeed, models that follow the condensation of a solar-composition hot gas reproduce many of the minerals and abundance trends observed in the Solar System. These are also consistent with some of the astronomical observations of dust in protoplanetary disks. Stardust grains also show approximately solar bulk composition in the measurable elements, albeit with some variations (Flynn *et al.* 2006). Some IDPs – mainly the fine-grained, porous, and anhydrous particles – match the solar elemental abundances, while some other grains or IDP subunits show non-solar isotopic ratios, often indicating presolar origin.

Dust mineralogy offers insights complementary to bulk elemental compositions. In disks, silicate grains are the best-studied dust component. While the composition of amorphous silicates remains difficult to constrain, crystalline grains display mid-infrared spectral features sensitive to their composition. The most abundant crystalline silicates are pyroxene and olivine, the latter highly magnesium-rich ($> Fo_{90}$). The observed radial gradients in the relative amount of amorphous-over-crystalline dust can provide constraints on the thermal history and mixing of protoplanetary dust (van Boekel *et al.* 2004). In addition to silicates, an abundance of carbonaceous grains is expected in protoplanetary disks. Much more difficult to observe directly, their presence is indirectly deduced through modeling of the observed spectra (e.g. Min *et al.* 2005; Lisse *et al.* 2006). Crystalline carbon grains – nano-diamonds and graphite – also display identifiable spectral features and have been detected in a few young disks.

1.8 Dust coagulation

Collisions at moderate velocities between micron-sized dust grains and their subsequent sticking leads to dramatic changes in the grain size distribution in disks. With increasing sizes the grains decouple from the turbulent gas and sink toward the disk mid-plane, also enhancing their relative velocities to other grains and increasing the frequency of collisions. Because larger grains have larger geometrical cross-sections, they will accrete faster, as long as the relative velocities of the particles remain moderate (Blum & Wurm 2008).

With the contributions of grain formation, destruction, and fragmentation to grain growth, the grain size evolution in disks is a complex process, which also leads to the key first step toward planet formation. Characterizing the grain size evolution and developing a predictive picture requires combining astronomical observations with studies of materials from the early Solar System and dust coagulation experiments. The comparison, however, is not straightforward: while astronomical observations often probe the disk surface or the cold outer regions, chondrites are thought to probe a narrow annulus ($< 1\,AU$) in the disk mid-plane, centered on 3.5–4.0 AU.

The presence of grains in the micron-to-centimeter size range in protoplanetary disks can be remotely probed by infrared, submillimeter, and radio observations, with the largest grains detectable at the longest wavelengths. While the interstellar medium is dominated by submicron-sized grains, larger grains have been detected in many young disks. Infrared spectroscopy reveals that disk surfaces of young, massive disks are often abundant in micron-sized grains, although clear correlations between the presence of these grains and fundamental disk properties remain elusive. Millimeter-wavelength interferometric observations identified millimeter-to-centimeter-sized grains in a handful of outer disks around young stars, the largest such grains yet observed (1–8 Myr, e.g. Calvet *et al.* 2002; Testi *et al.* 2003; Rodmann *et al.* 2006).

Meteoritic evidence shows that dust coagulation continued in the disk mid-plane for an extended period during the early disk evolution. Chondritic meteorites contain a mixture of coarse-grained materials, such as chondrules and CAIs, embedded within a fine-grained ($< 1\,\mu m$) matrix. The majority of the primitive dust preserved in the least-processed chondrites is in the submicron size range, also detectable in protoplanetary disks via spectroscopy or scattered light imaging. Chondritic matrices also contain mineral grains of $1–10\,\mu m$ in size, albeit at low concentrations (< 1 vol%); some matrices show subunits with distinct mineralogical and compositional properties. The observed large diversity on submicron scales demonstrates an extremely thorough mixing of grains that formed in different locations within the disk. Many, or perhaps most, of these grains have been altered and processed during chondrule-forming events.

The ubiquitous presence of silicate emission features in young protoplanetary disks is evidence that a population of small (a few micron) particles persists on million year timescales, much longer than the grain coagulation timescales (e.g. Dullemond & Dominik 2005; Brauer *et al.* 2008). This demonstrates that an efficient mechanism must operate that replenishes particles in the $1–10\,\mu m$ size range, at least in the upper layers of protoplanetary disks.

The presence of interstellar (presolar) grains intimately embedded in diverse meteoritic materials of nebular origin shows that mixing of dust must have occurred after the violent chondrule-forming events began to wane in either frequency or intensity. Thus, the processes of dust formation and dust coagulation must have continued for millions of years, throughout the period of chondrule formation. Chondrule-forming events themselves may also lead to the repeated evaporation and condensation of small grains, a plausible way to continually replenish the population of fine dust in protoplanetary disks. It remains to be seen if astronomical studies of these objects can identify such mechanisms at work. In order to accomplish this task, it is important to review the nature of transient heating events in protoplanetary disks.

1.9 Thermal processing of the pre-planetary material

Cold disk regions ($< 300\,K$), both in the Solar System and in protoplanetary disks, have been found to be very rich in crystalline silicates and once-molten solids. The evidence for such widespread thermal processing – considering that heating and melting fundamentally transforms the pre-planetary material – motivates detailed studies of the heating processes.

Spectroscopic observations of protoplanetary disks offer snapshots of dust processing through time. In particular, the characteristic mid-infrared spectra of submicron-sized crystalline silicate grains reveals their presence in many disks, with diverse disk geometries and in various stages of disk evolution. Thermal annealing is a strong function of temperature even for small grains: timescales for crystallization range from 5 Myr at 630 K to a few seconds at 1200 K. The crystals observed in young disks (< 1 Myr) must have been exposed to high temperatures (> 1000 K), perhaps repeatedly. The detection of these crystals in disk regions too cold for annealing ($T < 300$ K) thus challenges disk physics, including disk evolution, mixing, and shock-wave propagation.

Sensitive observations enable comparative surveys of silicate emission features from disks around low-mass, intermediate-mass, and Sun-like stars. While no strong correlations have been found with disk properties, flatter disks and disks around the coolest stars more often show crystalline silicate features. Cool stars and very low-mass disks display prominent crystalline silicate emission peaks (Apai *et al.* 2005; Merín *et al.* 2007; Pascucci *et al.* 2009). Thus, whatever processes are responsible for the presence of crystals around Sun-like stars must be capable of very efficiently producing crystals around low-mass stars, too. Interferometric measurements suggest that the amorphous/crystalline dust mass fraction is higher in the inner disk than at medium separations (van Boekel *et al.* 2004; Ratzka *et al.* 2007). The surveys also show that amorphous silicate grains frequently have similar magnesium and iron abundances in protoplanetary disks. In contrast, those with crystalline silicates are always dominated by Mg-rich grains (e.g. Malfait *et al.* 1998; Bouwman *et al.* 2008).

Similarly to the protoplanetary disks observed, primitive material from the Solar System underwent dramatic heating and cooling events prior to its incorporation into planetesimals. Much of the primitive planetary materials, such as igneous CAIs and chondrules, have been melted, slowly cooled, and crystallized. The precursors of these igneous objects, thought to have been free-floating in the solar nebula, were heated up to 2200 K for seconds to minutes and then slowly cooled (tens to hundreds of Kelvin per hour). In addition, fine-grained matrix in chondrites is in part a collection of amorphous and crystalline material that likely condensed from the vapor during chondrule formation.

While the amount of dust and small particles that underwent thermal processing remains difficult to constrain both in the entire proto-solar nebula and in protoplanetary disks around other stars, in the Asteroid Belt over 80% of the pre-chondritic components have been melted. These heating events may play a crucial role in defining the bulk composition of planetesimals and planets by reprocessing much or all of the pre-planetary material. The heating mechanism or mechanisms that produced the transient heating responsible for chondrule and igneous CAI formation is still unknown, but may have also been able to produce crystalline silicate dust. The energetic mechanism(s) that melted these objects operated multiple times over millions of years, with highly variable intensities. Flash heating by shock fronts is not only the leading explanation for chondrule formation, but also with the most quantitative framework to date (Ciesla & Hood 2002; Desch & Connolly 2002; Desch *et al.* 2005). However, the source and nature of the shock waves is yet ambiguous. Alternative mechanisms include shocks induced by X-ray flares, processing in the X-wind, and lightning. Astronomical observations may be able to play a key role in identifying heating events in other disks.

1.10 Dispersal of protoplanetary disks

The lifetime of protoplanetary disks determines the time available for planet formation; with the loss of the dusty gas disks no raw material is left to form planetesimals or giant planets. Thus, disk mass as a function of time is perhaps the single most important constraint on the formation of both the rocky and the giant planets. The most readily observable, albeit imperfect, indicator of disks is the presence of excess emission above the stellar photosphere, emerging from small, warm dust grains.

Observations of near-infrared excess emission from hundreds of disks with ages covering the first 10 Myr demonstrate fundamental structural evolution and the eventual loss of the fine dust from the inner disk ($< 1\,$AU). The declining fraction of stars with dust disks suggests a disk half-life of 3 to 5 Myr (see Chapter 9, e.g. Hernández *et al.* 2007). Longer-wavelength infrared observations, primarily from the Spitzer Space Telescope, show a similar picture for the intermediate disk radii (1–$5\,$AU). The combination of these lines of evidence is interpreted as a rapid (< 1–3 Myr) dispersal of the fine dust in most systems, probably progressing inside-out.

Although the disk mass is dominated by hydrogen, much less is known about its dispersal. Tracers of hot gas in the innermost disk regions show a one-to-one correspondence to the presence of hot dust (Hartigan *et al.* 1995) and gas accretion to the stars declines at the same rate as hot dust disperses. Spitzer studies of mid-infrared ro-vibrational lines probe warm gas on orbits similar to Jupiter's and demonstrate the loss of gas in few tens of millions of years (Pascucci *et al.* 2007). Gas in the coldest disk regions can be traced through CO rotational lines; such studies also suggest a gas depletion by 10 Myr. The combined astronomical evidence shows that: (1) dust disks dissipate in 3$-$8 Myr via rapid inside-out dispersal; (2) gas dissipates in a similar, or perhaps even shorter timescale.

The Solar System, in comparison, offers two lines of evidence to constrain the timescale for the lifetime of the proto-solar nebula and the epoch of planetesimal formation. On one side, relatively unaltered chondritic components preserve traces of their chemical and thermal history; on the other side, dynamical information is imprinted on the hierarchy of the Solar System.

The observation that bulk chondrites are isotopically homogeneous – with the exception of H, C, N, and O – is evidence for a very thorough mixing in an early phase in the hot nebula (> 2000 K). Chondrules themselves provide a variety of constraints on the dust and gas content of their natal environment. Their chemical and isotopic compositions, size, and shape distributions all suggest dusty gas reservoirs during their formation epoch, which lasted 1–3 Myr after CAI formation, possibly with a peak at 2 Myr.

Minor planets and satellite systems of giant planets also provide constraints on the presence of gas during their formation. For example, Asuka 881394 – a fragment probably derived from the asteroid 4 Vesta – dates to ∼0.5 Myr after CAI formation. This shows that large planetesimals formed early, while gas was still present in the proto-solar nebula. Similarly, the irregular shape and the equatorial ridge of Iapetus, the Saturnian satellite, is well explained by excess heating from ^{26}Al, if formed within 2.5–5 Myr after CAI formation (Castillo-Rogez et al. 2007). The fact that Saturn must have formed prior to Iapetus provides further evidence for a gas-rich disk before 2.5–5 Myr.

The key uncertainty in relating the astronomical observations to the Solar System constraints is at the zero point. However, the different constraints seem to line up well if CAIs formed less than 1 Myr after the protostellar collapse. If so, chondrules would have formed within 3 Myr, consistent with the presence of fine dust in many astronomical analogs. The presence of millimeter- and centimeter-sized objects at a few million years after CAI formation is also broadly consistent with the astronomical constraints, as is the timing for planetesimal collisions indicated by the freed planetary debris. This phase is likely to have started by 3–5 Myr after CAI formation and would have lasted for tens to hundreds of million years, until the final planetary architecture was reached.

1.11 Accretion of planetesimals and rocky planets

Planets, satellites, and small bodies provide a wide range of dynamical and chemical constraints on the building of the Solar System from planetesimals. In addition to the primary parameters of planets, the planet mass and semi-major axis distributions, the relative masses of the cores (exceptionally large for Mercury and low for the Moon) provide further constraints. In addition, the Asteroid Belt seems to be depleted in mass by three to four orders of magnitude and its medium- to small-sized members show a size distribution characteristic of collisional erosion. The surface compositions of main-belt asteroids show a correlation with semi-major axis, possibly resulting from compositional gradients present at the time of their formation. In addition, the volatile-element depletions appear to increase from carbonaceous chondrites through ordinary chondrites and terrestrial planets to some differentiated meteorites. Complementing these compositional constraints is the chronology of key events of planet formation in the Solar System (see Fig. 1.3), providing a detailed constraint set on the assembly of planets.

Initial conditions adopted from disk studies allow the development of a model that traces the evolution from planetesimals to planets and that is, in most parts, consistent with the key constraints observed in the Solar System. In these models kilometer-sized planetesimals serve as the basic building blocks of planets and their pairwise collisions build bodies of increasing sizes. The mutual dynamical interactions of the largest bodies and their interactions with the planetesimal disk regulate accretion processes and define three phases: the initial runaway growth (leading to > 100 kilometer-sized bodies at 1 AU in 10^4 yr), the slower oligarchic growth (forming lunar- to Mars-sized bodies in $\sim 10^6$ yr), and the stochastic postoligarchic growth (defining the final hierarchy of the planetary systems). This latter stage is characterized by large-scale, stochastic mixing of the planetary embryos and their catastrophic collisions. One such collision likely formed the Moon. Simulations show that, in the case of the Solar System, the oligarchic and postoligarchic growth stages in the inner few astronomical units are strongly influenced by the presence of Jupiter and Saturn.

Planet formation unfolds differently beyond the snowline, where water condensation enhances the surface density. Here massive cores (> 5–10 M_{Earth}) may form rapid enough to accrete directly and retain nebular gas. These massive cores, if formed prior to the dispersal of the gas disk, rapidly reach Jupiter masses, forming giant planets. An alternative mechanism that may be responsible for the formation of some giant planets is gravitational instability in a massive, marginally unstable disk (e.g. Boss 2007; Mayer *et al.* 2007).

1.12 Key challenges and perspectives

1.12.1 How typical is the Solar System?

Whether the Solar System can serve as a template for the typical planetary systems in general remains one of the fundamental questions of astronomy. Radial velocity surveys have been very successful in finding planets unlike the ones in our Solar System and are now reaching sensitivity and temporal coverage to detect Jupiter-like planets on Jupiter-like orbits. Giant planets on orbits < 4 AU are found around $\sim 6\%$

of the Sun-like stars (e.g. Udry & Santos 2007) and $\sim$20–40% of the Sun–like stars may harbor Neptune-mass planets (Mayor *et al.* 2009). Because eccentric, close-in and massive planets induce the largest radial velocity signal, it is no surprise that such planets dominate the current census of exoplanets. In the coming years the COROT and Kepler missions are expected to establish a large sample of rocky and giant planets that will help to place our inner Solar System in the context of the extrasolar planet population.

1.12.2 How to pass the meter-sized boundary?

The accretion of rocky and icy planetesimals (the asteroids and comets of today's Solar System) is one of the least understood phases of Solar System history. A key unknown factor, which dominates the evolution of particles in the pre-planetesimal stage, is the degree to which the nebula was turbulent. Surface forces help small dust grains stick to each other, forming macroscopic fractal aggregates, which are presumably made progressively more compact by collisions. How far planetesimals can grow in this way is unclear. If the nebula is turbulent, collisions may become disruptive as particles grow larger and relative velocities increase, stalling accretion at around a meter in size (Blum & Wurm 2008). An additional severe problem is the drift of the growing particles towards the Sun, due to gas drag. Bodies with sizes of order of a meter are removed from a region faster than they can grow. This combination of problems is usually called the *meter-size barrier*. While making sticking more difficult, turbulence helps us understand observations that suggest widespread mixing in the early nebula, such as the recent finding of high-temperature, crystalline minerals in samples of comet P/Wild 2 returned by Stardust.

To overcome the meter-size barrier and avoid uncertainties about sticking, it has been suggested that planetesimals might form quickly by gravitational instability, in a dense particle layer close to the nebula mid-plane. However, these dense layers themselves generate local turbulence which disperses the particles and prevents gravitational instability. Although the idea of classical gravitational instability has been recently resurrected in the context of very small particles, this latter scenario is hampered by even tiny amounts of nebular turbulence.

1.12.3 Transition from protoplanetary disks to debris disks

Observations of the first 10 Myr of disk evolution have shown more dramatic changes than during the entire remaining lifetime of the systems. Not only is the disk material lost and the disk structure flattens, but the dust grain population, too, changes fundamentally. Older debris disks are almost always devoid of micron-sized silicate grains, crystalline or amorphous. It is not yet understood in what way or ways the massive gas-rich protoplanetary disks may evolve into debris disks. The formation of massive giant planets and the collisions of planetesimals are expected to shape and replenish the disk, but the exact mechanisms remain unexplored. Specific open questions include the following. How does the fraction of collisionally replenished dust vary with time? What disk structures will forming planets with different masses introduce? Tying disk properties to the properties of the forming planetary systems is essential to fully exploit disks as diagnostic tools for the diversity of planetary systems.

1.12.4 Disk properties as a function of stellar mass

Surprisingly, detailed Spitzer studies show that disks around cool stars and brown dwarfs are markedly different from those around more massive stars. First, the disk lifetimes around the lowest-mass stars appear to be significantly longer than around their higher-mass counterparts (Lada *et al.* 2006; Carpenter *et al.* 2006). Second, the disk structures appear to be significantly flatter even at young ages (e.g. Testi *et al.* 2003). Third, the growth and thermal processing of the warm ($100-300$ K) dust in the inner disks is more prominent (Apai *et al.* 2005; Kessler-Silacci *et al.* 2007). Finally, the accretion rate appears to scale with the square of the stellar mass ($\dot{M} \propto M_*^2$), resulting in very small accretion rates in the lowest-mass objects (Natta *et al.* 2004; Muzerolle *et al.* 2005). These observed differences demonstrate that the initial and boundary conditions for planet formation around low-mass stars differ from those around Sun-like stars, suggesting that planet formation may unfold very differently (Laughlin *et al.* 2004; Ida & Lin 2005; Payne & Lodato 2007; Kennedy *et al.* 2007).

1.12.5 The supernova explosion and the Sun's natal environment

Multiple lines of evidence exist for a population of live radionuclides, such as ^{26}Al or ^{60}Fe, which were injected into the proto-solar cloud or disk prior to the formation of CAIs (e.g. Tachibana *et al.* 2006). Isotopic abundances suggest that the isotopes have originated in a supernova, possibly with a very massive star progenitor that also underwent a Wolf–Rayet phase (Bizzarro *et al.* 2007). If this interpretation is correct, the Sun must have formed in a very rich and dense stellar cluster, such as the Carina Nebula, very much unlike the Taurus or other low-mass star-forming regions. Luminous massive stars in such clusters may truncate or fully evaporate protoplanetary disks around other cluster members. Two key questions remain open. How close in time and space did the supernova explode to the proto-solar nebula? Did the supernova or the other neighboring massive stars trigger the collapse, truncate the disk, or influence the proto-solar nebula in a significant way? Answering these questions will help understanding whether the supernova explosion (or the nearby massive stars) could have made the Solar System different from typical planetary systems.

1.12.6 *Heterogeneity of the disks: oxidation states and isotopes*

The CI chondrites are generally regarded as being representative of the bulk elemental composition of the Solar System, with the exception of the highly volatile elements hydrogen, carbon, nitrogen, oxygen, and the noble gases. Relative to CI chondrites, all other groups of meteorites, as well as bulk terrestrial planets, are depleted in volatile elements. These depletions differ from group to group, but for the most part have the common characteristic of being reasonably smooth functions of volatility (expressed as 50% condensation temperatures). Thus, equilibrium condensation sequences capture many first-order observations of the volatility dependent fractionations of the elements, the identities of their host phases in meteorites, and even the chemical structure of the Solar System. However, a variety of oxidation states are exhibited by the most primitive (Type-3) chondritic meteorites. These samples suggest that the nebular redox state varied from several orders of magnitude more oxidizing than a solar gas to several orders of magnitude more reducing. These wide variations are thought to result from large variations in the relative abundances of gas, dust, ice, and carbonaceous material within the nebular mid-plane. To date, no comprehensive model of the dynamics and chemistry of the solar nebula is able to reproduce the wide range of oxidation states recorded in the meteorite record.

1.12.7 *Chondrule formation*

Transient heating events were important in the formation of the Solar System and provided the energy to produce chondrules and refractory inclusions. These objects are not predicted in astrophysical models for the formation of planetary systems. They comprise 50–80% of the mass of many primitive meteorites. However, the mechanism that produced the transient heating events is still unknown. Future work must focus on putting the details of the petrographic and chemical analysis of these rocks into an astrophysical and cosmochemical framework of Solar System formation.

Leading models for transient heating mechanisms include shock waves, X-ray flares, the X-wind, and partially molten planetesimals. Shock waves have been shown to be able to reproduce the deduced thermal histories of chondrules, yet their driving mechanism remains elusive. The X-wind model, in contrast, needs to provide detailed thermal histories that can be compared to existing data. Similarly, a detailed quantitative model for chondrule and igneous CAI formation by interactions between planetary bodies has not yet been developed. Understanding the relationship between the mechanism of transient heating and the environment of CAI and chondrule formation would be a critical step toward identifying the heating mechanism. Finally, we need to understand why CAI formation was limited to 0.1 Myr while chondrule formation persisted for 4 Myr and how the heating events fit into the collapse of the molecular cloud and formation of the disk.

1.12.8 Asteroid heating source

Meteoritic and asteroidal studies both reveal the presence of an intense, selective, and short-lived heat source in the early Solar System. Systematic trends in the spatial distribution of asteroid taxonomic types, as defined by their reflectance spectrum, are thought to represent variations in their primordial thermal history. Complementary studies of meteorites show that all of their asteroidal parent bodies have been thermally altered by internal heating to some degree, ranging from 400 to > 1500 K. Discerning the heat source(s) responsible for thermal processing of asteroids is one of the great mysteries of planetary science. Of the myriad proposed heat sources, three are still actively discussed today: decay of ^{26}Al and other short-lived radionuclides (e.g. Bennett & McSween 1996), impact heating (e.g. Rubin 1995), and electromagnetic induction. The role of induction heating in the early Solar System has not been explored in light of recent results, such as the role of magneto-rotational instability and X-ray emissions from pre-main-sequence young stellar objects. Based on these developments and more recent observations of star-forming regions, the interaction between protostellar magnetic fields and planetesimal bodies must be re-evaluated.

1.12.9 Origin of the bulk planet composition: volatile delivery and terrestrial Mg/Si ratio

The Earth's primitive upper mantle has atomic (e.g. Mg/Si and Mg/Al) and isotopic (e.g. ^{187}Os/^{188}Os) ratios that are distinctly different from those in the CI chondrites. In fact, no primitive material similar to the Earth's mantle is currently represented in our meteorite collections. The "building blocks" of the Earth must instead be composed of a yet unsampled meteoritic (chondritic or differentiated) material. Some of the elemental variation is potentially due to its partitioning into the core. It is unlikely, however, that this can explain all of the observed variations. It is also possible that cosmochemical processes in the inner nebula led to depletions in the moderately volatile and volatile elements, including silicon and magnesium. In addition, it may be that the lower mantle of the Earth has a different composition from the upper mantle and that the bulk composition of the silicate Earth is poorly constrained. The relative and absolute abundances of these elements need to be examined in detail in order to better understand the accretionary processes and initial composition of nebular material in the feeding zone of the growing proto-Earth.

References

Alves, J. F., Lada, C. J., & Lada, E. A. 2001, *Nature*, **409**, 159.

Amelin, Y. 2007, *Lunar and Planetary Science Conference Abstracts*, **38**, 1669.

Amelin, Y., Krot, A. N., Hutcheon, I. D., & Ulyanov, A. A. 2002, *Science*, **297**, 1678.

Apai, D., Pascucci, I., Bouwman, J., *et al.* 2005, *Science*, **310**, 834.

Bally, J., Reipurth, B., & Davis, C.J. 2007, in *Protostars and Planets V*, ed. B. Reipurth, D. Jewitt, & K. Keil, University of Arizona Press, 215–230.

Beckwith, S. V. W., Sargent, A. I., Chini, R. S., & Guesten, R. 1990, *Astronomical Journal*, **99**, 924.

Bennett, III, M. E. & McSween, Jr., H. Y. 1996, *Meteoritics and Planetary Science*, **31**, 783.

Bergin, E. A., Alves, J., Huard, T., & Lada, C. J. 2002, *Astrophysical Journal*, **570**, L101.

Bizzarro, M., Baker, J. A., Haack, H., & Lundgaard, K. L. 2005, *Astrophysical Journal*, **632**, L41.

Bizzarro, M., Ulfbeck, D., Trinquier, A., *et al.* 2007, *Science*, **316**, 1178.

Blitz, L., Fukui, Y., Kawamura, A., *et al.* 2007, in *Protostars and Planets V*, ed. B. Reipurth, D. Jewitt, & K. Keil, University of Arizona Press, 81–96.

Blum, J. & Wurm, G. 2008, *Annual Review of Astronomy and Astrophysics*, **46**, 21.

Boss, A. P. 2007, *Astrophysical Journal*, **661**, L73.

Bouwman, J., Henning, T., Hillenbrand, L. A., *et al.* 2008, *Astrophysical Journal*, **683**, 479.

Bradley, J. P. 1994, *Science*, **265**, 925.

Brauer, F., Dullemond, C. P., & Henning, T. 2008, *Astronomy and Astrophysics*, **480**, 859.

Brownlee, D., Tsou, P., Aléon, J., *et al.* 2006, *Science*, **314**, 1711.

Brucato, J. R., Strazzulla, G., Baratta, G., & Colangeli, L. 2004, *Astronomy and Astrophysics*, **413**, 395.

Burrows, C. J., Stapelfeldt, K. R., Watson, A. B., *et al.* 1996, *Astrophysical Journal*, **473**, 437.

Calvet, N., D'Alessio, P., Hartmann, L., *et al.* 2002, *Astrophysical Journal*, **568**, 1008.

Carpenter, J. M., Mamajek, E. E., Hillenbrand, L. A., & Meyer, M. R. 2006, *Astrophysical Journal*, **651**, L49.

Castillo-Rogez, J. C., Matson, D. L., Sotin, C., *et al.* 2007, *Icarus*, **190**, 179.

Ciesla, F. J. & Hood, L. L. 2002, *Icarus*, **158**, 281.

Clayton, R. N. 2007, *Annual Review of Earth and Planetary Sciences*, **35**, 1.

Connelly, J. N., Amelin, Y., Krot, A. N., & Bizzarro, M. 2008, *Astrophysical Journal*, **675**, L121.

Davis, A. M. 2006, *Meteorites and the Early Solar System II*, ed. D. S. Lauretta & H. V. McSween, University of Arizona Press, 295–307.

de la Reza, R., Jilinski, E., & Ortega, V. G. 2006, *Astronomical Journal*, **131**, 2609.

Desch, S. J. & Connolly, Jr., H. C. 2002, *Meteoritics and Planetary Science*, **37**, 183.

Desch, S. J., Ciesla, F. J., Hood, L. L., & Nakamoto, T. 2005, *Astronomical Society of the Pacific Conference Series*, **341**, 849.

Dullemond, C. P. & Dominik, C. 2005, *Astronomy and Astrophysics*, **434**, 971.
Dutrey, A., Guilloteau, S., & Ho, P. 2007, in *Protostars and Planets V*, ed. B. Reipurth, D. Jewitt, & K. Keil, University of Arizona Press, 495–506.
Flynn, G. J., Bleuet, P., Borg, J., *et al.* 2006, *Science*, **314**, 1731.
Flynn, G. J., Keller, L. P., Jacobsen, C., & Wirick, S. 2004, *Advances in Space Research*, **33**, 57.
Flynn, G. J., Leroux, H., Tomeoka, K., *et al.* 2008, *Lunar and Planetary Science Conference Abstracts*, **39**, 1979.
Grady, C. A., Woodgate, B., Bruhweiler, F. C., *et al.* 1999, *Astrophysical Journal*, **523**, L151.
Hallenbeck, S. L., Nuth, J. A., & Daukantas, P. L. 1998, *Icarus*, **131**, 198.
Hartigan, P., Edwards, S., & Ghandour, L. 1995, *Astrophysical Journal*, **452**, 736.
Hernández, J., Hartmann, L., Megeath, T., *et al.* 2007, *Astrophysical Journal*, **662**, 1067.
Ida, S. & Lin, D. N. C. 2005, *Astrophysical Journal*, **626**, 1045.
Jäger, C., Fabian, D., Schrempel, F., *et al.* 2003, *Astronomy and Astrophysics*, **401**, 57.
Keller, L. P. & Messenger, S. 2007, in Proceedings of the Chronology of Meteorites and the Early Solar System, Kauai, Hawaii, November 5–7, Lunar and Planetary Institute Contribution No. 1374, 88.
Kemper, F., Vriend, W. J., & Tielens, A. G. G. M. 2005, *Astrophysical Journal*, **633**, 534.
Kennedy, G. M., Kenyon, S. J., & Bromley, B. C. 2007, *Astrophysics and Space Science*, **311**, 9.
Kessler-Silacci, J. E., Dullemond, C. P., Augereau, J.-C., *et al.* 2007, *Astrophysical Journal*, **659**, 680.
Kita, N. T., Huss, G. R., Tachibana, S., *et al.* 2005, *Astronomical Society of the Pacific Conference Series*, **341**, 558.
Kleine, T., Münker, C., Mezger, K., & Palme, H. 2002, *Nature*, **418**, 952.
Kleine, T., Palme, H., Mezger, K., & Halliday, A. N. 2005, *Science*, **310**, 1671.
Krot, A. N., Amelin, Y., Cassen, P., & Meibom, A. 2005, *Nature*, **436**, 989.
Lada, C. J., Muench, A. A., Luhman, K. L., *et al.* 2006, *Astronomical Journal*, **131**, 1574.
Laughlin, G., Bodenheimer, P., & Adams, F. C. 2004, *Astrophysical Journal*, **612**, L73.
Lisse, C. M., VanCleve, J., Adams, A. C., *et al.* 2006, *Science*, **313**, 635.
Luhman, K. L., Allen, L. E., Allen, P. R. 2008, *Astrophysical Journal*, **675**, 1375.
Lyons, J. R. & Young, E. D. 2005, *Nature*, **435**, 317.
Malfait, K., Waelkens, C., Waters, L. B. F. M., *et al.* 1998, *Astronomy and Astrophysics*, **332**, L25.
Mayer, L., Lufkin, G., Quinn, T., & Wadsley, J. 2007, *Astrophysical Journal*, **661**, L77.
Mayor, M., Udry, S., Lovis, C., *et al.* 2009, *Astronomy and Astrophysics*, **493**, 639.
McCaughrean, M. J. & O'dell, C. R. 1996, *Astronomical Journal*, **111**, 1977.
Merín, B., Augereau, J.-C., van Dishoeck, E. F., *et al.* 2007, *Astrophysical Journal*, **661**, 361.
Messenger, S., Keller, L. P., & Lauretta, D. S. 2009, *Science*, **309**, 737.
Meyer, M. R., Backman, D. E., Weinberger, A. J., & Wyatt, M. C. 2007, in *Protostars and Planets V*, ed. B. Reipurth, D. Jewitt, & K. Keil, University of Arizona Press, 573–588.
Min, M., Hovenier, J. W., de Koter, A., Waters, L. B. F. M., & Dominik, C. 2005, *Icarus*, **179**, 158.
Mostefaoui, S., Lugmair, G. W., & Hoppe, P. 2005, *Astrophysical Journal*, **625**, 271.
Muzerolle, J., Luhman, K. L., Briceño, C., Hartmann, L., & Calvet, N. 2005, *Astrophysical Journal*, **625**, 906.
Natta, A., Testi, L., Muzerolle, J., *et al.* 2004, *Astronomy and Astrophysics*, **424**, 603.

Nuth, J. A., III, Johnson, N. M., & Manning, S. 2008, *Astrophysical Journal*, **673**, L225.

Ortega, V. G., Jilinski, E., de la Reza, R., & Bazzanella, B. 2007, *Monthly Notices of the Royal Astronomical Society*, **377**, 441.

Ouellet, C., Vockenhuber, C., The L. S., *et al.* 2007, *American Physical Society Meeting Abstracts*, October, DNP.EE0020.

Pascucci, I., Apai, D., Luhman, K., *et al.* 2009, *Astrophysical Journal*, **696**, 143.

Pascucci, I., Hollenbach, D., Najita, J., *et al.* 2007, *Astrophysical Journal*, **663**, 383.

Payne, M. J. & Lodato, G. 2007, *Monthly Notices of the Royal Astronomical Society*, **381**, 1597.

Qin, L., Dauphas, N., Wadhwa, M., Masarik, J., & Janney, P. E. 2008, *Earth and Planetary Science Letters*, **273**, 94.

Ratzka, T., Leinert, C., Henning, T., *et al.* 2007, *Astronomy and Astrophysics*, **471**, 173.

Ray, T., Dougados, C., Bacciotti, F., Eislöffel, J., & Chrysostomou, A. 2007, in *Protostars and Planets V*, ed. B. Reipurth, D. Jewitt, & K. Keil, University of Arizona Press, 231–244.

Rietmeijer, F. J. M. & Nuth, J. A., III 2000, *Earth, Moon and Planets*, **82**, 325.

Rietmeijer, F. J. M., Nuth, J. A., III, & Karner, J. M. 1999, *Astrophysical Journal*, **527**, 395.

Rodmann, J., Henning, T., Chandler, C. J., Mundy, L. G., & Wilner, D. J. 2006, *Astronomy and Astrophysics*, **446**, 211.

Rubin, A. E. 1995, *Icarus*, 113, 156.

Sandford, S. A., Aléon, J., Alexander, C. M. O., *et al.* 2006, *Science*, **314**, 1720.

Sandford, S. A. & Allamandola, L. J. 1993, *Astrophysical Journal*, **417**, 815.

Savage, B. D. & Sembach, K. R. 1996, *Annual Review of Astronomy and Astrophysics*, **34**, 279.

Schramm, L. S., Brownlee, D. E., & Wheelock, M. M. 1989, *Meteoritics*, **24**, 99.

Slesnick, C. L., Hillenbrand, L. A., & Carpenter, J. M. 2008, *Astrophysical Journal*, **688**, 377.

Tachibana, S., Huss, G. R., Kita, N. T., Shimoda, G., & Morishita, Y. 2006, *Astrophysical Journal*, **639**, L87.

Testi, L., Natta, A., Shepherd, D. S., & Wilner, D. J. 2003, *Astronomy and Astrophysics*, **403**, 323.

Thiemens, M. H. 2006, *Annual Review of Earth and Planetary Sciences*, **34**, 217.

Thomas, K. L., Clemett, S. J., Flynn, G. J., *et al.* 1994, *Lunar and Planetary Science Conference Abstracts*, **25**, 1391.

Touboul, M., Kleine, T., Bourdon, B., Palme, H., & Wieler, R. 2007, *Nature*, 450, 1206.

Udry, S. & Santos, N. C. 2007, *Annual Review of Astronomy and Astrophysics*, **45**, 397.

Ueda, Y., Mitsuda, K., Murakami, H., & Matsushita, K. 2006, *Astrophysical Journal*, **652**, 845.

van Boekel, R., Min, M., Leinert, C., *et al.* 2004, *Nature*, **432**, 479.

van den Ancker, M. E., Bouwman, J., Wesselius, P. R., *et al.* 2000, *Astronomy and Astrophysics*, **357**, 325.

van Dishoeck, E. F. 2004, *Annual Review of Astronomy and Astrophysics*, **42**, 119.

Walmsley, C. M., Flower, D. R., & Pineau des Forêts, G. 2004, *Astronomy and Astrophysics*, **418**, 1035.

Williams, J. P., Andrews, S. M., & Wilner, D. J. 2005, *Astrophysical Journal*, **634**, 495.

Yin, Q. Z., Jacobsen, B., Moynier, F., & Hutcheon, I. D. 2007, *Astrophysical Journal*, **662**, L43.

Wyatt, M. C. 2008, *Annual Review of Astronomy and Astrophysics*, **46**, 339.

Zolensky, M. E., Zega, T. J., Yano, H., *et al.* 2006, *Science*, **314**, 1735.

2

The origins of protoplanetary dust and the formation of accretion disks

Hans-Peter Gail and Peter Hoppe

Abstract Dust is an important constituent in the Universe. About 1% of the mass of the interstellar matter is in dust. This dust is either stardust that condensed in the winds of evolved stars and in the ejecta of supernova and nova explosions or dust that formed in dense interstellar clouds. Here, we will discuss the cycle of matter from stars to the interstellar medium and how interstellar clouds evolve to protostars and protostellar disks. We will discuss the nature and origin of interstellar dust and how it entered the Solar System. A small fraction of the stardust grains survived the earliest stages of Solar System formation and can be recognized by highly anomalous isotopic compositions as presolar grains in meteorites, interplanetary dust particles, and cometary matter, with concentrations at the subpermil level. Imprints of likely interstellar chemistry are seen as D and ^{15}N enrichments in organic matter in primitive Solar System materials.

Dust is an important constituent in the Universe and its meaning for astrophysics is manifold. In the interstellar medium (ISM) about 1% of the mass is in dust. A major fraction of the refractory elements in the ISM is locked up in dust leading to a depletion of these elements in the gas phase. Dust is responsible for interstellar extinction (absorption and scattering of light). It was this feature that led to the first firm identification of dust in the ISM in the twentieth century. Detailed studies of interstellar extinction imply the presence of solid particles with sizes of the wavelength of visible light, i.e. in the submicrometer range. Dust affects interstellar chemistry in many ways and it also plays an important role in the different stages of stellar evolution, from stellar birth to stellar death.

Stars form from collapsing interstellar gas and dust clouds. Interstellar dust also provides the raw material from which planets form. Under favorable conditions, i.e. low temperatures, the interstellar dust may survive processing in protoplanetary disks and later planetary metamorphism, e.g. in asteroids and comets. By far the largest fraction of the starting material of a newly forming stellar system goes into the protostar. Its chemical and isotopic composition will be altered by nuclear fusion reactions in the star's interior during later stages of stellar evolution. At the end of the stellar lifetime large amounts of nucleosynthetically processed matter are ejected by stellar winds or explosions. This leads to a progressive enrichment of the ISM with heavier elements. Dust forms in the stellar ejecta, an important source for the dust in the ISM.

Here we will focus on the origins of protoplanetary dust, on the earliest stages of star formation, and on the formation of accretion disks around young stars. Section 2.1 deals with the nature and origin of dust in the ISM and how it entered the Solar System. Some of the interstellar dust that went into the making of our Solar System can be recognized as "presolar grains" in primitive Solar System materials, namely, in meteorites, micrometeorites, comets, and interplanetary dust particles (IDPs). In Section 2.2 we describe how presolar grains (and organics) can be identified in the laboratory and discuss their mineralogy and (stellar) sources. Presolar grains are only a minor component of dust in protoplanetary disks. But since they constitute the only refractory high-temperature materials with proven extra-solar origin, they will be discussed in great detail. Section 2.3 deals with various aspects of the earliest stages of star formation. We discuss how interstellar clouds and cores evolve to protostars and protostellar disks and how these stages can be modeled.

2.1 Dust in the interstellar medium

That starlight is somehow weakened in the space between stars was already recognized by a number of astronomers during the nineteenth and the beginning twentieth century, but it was not before 1930 that the origin of this obscuration could unequivocally be unraveled. In that year Trümpler (1930) detected the reddening of starlight and could show that the presence of very fine dust material distributed in the space between stars can explain the observation of increasing weakening and reddening of starlight with increasing distance. It was also rapidly recognized that scattering by such interstellar grains also explained the already known existence of reflection nebulae and the diffuse galactic emission. The wavelength dependence of absorption was determined in the following years and found to show a λ^{-1} variation in the 4000–6300 Å and the 1–3 μm region. This observation of a λ^{-1} variation excludes Rayleigh-scattering by extremely small particles as the origin of the absorption. At the same time, the observed strength of the absorption could be shown to exclude atoms, molecules or free electrons as sources of the interstellar extinction. The first explicit application of Mie theory of extinction by small particles to the problem of interstellar extinction was then given by Schalén (1934), who considered small iron grains and showed that they can reproduce the observations. Since about 1935 the existence and the fundamental properties of interstellar grains (sizes small compared to optical wavelengths, grains exhibiting strong forward scattering and high albedos) were firmly established. In the following period, a number of theoretical studies tried to develop models for the composition and the origin of this dust (e.g. Lindblad 1935; Oort & van de Hulst 1946). It needed nearly three further decades until observations in the far-ultraviolet and infrared spectral region became possible, that the composition of this dust became gradually clearer. The details of this historical development of our knowledge on cosmic dust are described, from two very different points of view, by Dorschner (2003) and by Li (2005).

2.1.1 Nature and origin of the dust in the ISM

The information on the nature and properties of the interstellar dust are presently obtained through the following observations:

   (i) extinction of starlight in the optical and ultraviolet spectral region
  (ii) scattering of starlight
 (iii) polarization of starlight
 (iv) infrared absorption and emission features
  (v) re-radiation of absorbed starlight at infrared and millimeter wavelengths
 (vi) haloes around X-ray sources
(vii) depletion of the refractory elements
(viii) collection and laboratory analysis of presolar grains, regardless of where they are collected from.

The basic information on the nature of the dust has been obtained from analysis of interstellar extinction from the near-infrared to the far-ultraviolet spectral region, using ground-based telescopes and the first space-borne ultraviolet telescopes. The derived interstellar extinction curve is in most parts rather smooth and shows only one broad and strong absorption feature centered around 220 nm (cf. Fitzpatrick & Massa 2007). This feature is explained by carbonaceous dust grains (Stecher & Donn 1965) with a wide distribution of sizes. The true nature of the carbonaceous dust material remains still somewhat unclear, but seems to be some kind of amorphous carbon (cf. Draine 2003, 2004, for a detailed discussion).

From theoretical considerations it was concluded in the 1960s that carbon (graphite) or silicate dust can condense in carbon-rich or oxygen-rich[1] cool stars and that this dust can be driven out from the stars by radiation pressure (Hoyle &

---

[1] Carbon- or oxygen-rich means that either carbon is more abundant than oxygen or oxygen more abundant than carbon. The exceptionally high bond energy of the CO molecule has the consequence that the less abundant of the two elements C and O is completely blocked in CO, whenever efficient molecule formation is possible. As a result, in carbon-rich environments the excess carbon forms soot and the rich zoo of hydrocarbon compounds, in oxygen-rich environments the excess oxygen forms with the abundant refractory elements oxides and silicates and a small number of O-bearing molecular species.

Wickramasinghe 1962; Wickramasinghe *et al.* 1966). Condensation of some other solid phases was also proposed, e.g. of SiC in C-stars (Friedemann 1969), and iron and other solids in supernova (SN) ejecta as they cool by expansion (Hoyle & Wickramasinghe 1970).

Already the first infrared observations of late-type giant stars have revealed that many of them are indeed surrounded by thick dust shells (Woolf & Ney 1969). These were rapidly found to consist of carbonaceous dust (some kind of soot) if the stellar spectrum indicates the star to be carbon-rich, and to be silicate dust (olivine, pyroxene) if the star is oxygen-rich (Gilman 1969). Since this dust is mixed into the interstellar medium due to mass loss by stellar winds, it was then assumed that silicate and carbon particles are abundant dust components in the interstellar medium.

This motivated a number of attempts, starting around 1970 with the models published by Hoyle & Wickramasinghe (1969), Wickramasinghe (1970), and Wickramasinghe & Nandy (1970) to reproduce the interstellar extinction curve with mixtures of silicate and carbon grains, and, occasionally, additional components. These models provided already successful fits to the observed extinction curve. This established silicate and carbon dust as the primary dust components of interstellar dust. In most of these studies it was assumed that interstellar dust is stardust, i.e. dust born in stellar ejecta.

With improved possibilities for infrared spectroscopy, broad extinction bands around 9.7 μm and 18 μm have been detected, which were ascribed to the stretching (Woolf & Ney 1969) and bending (Treffers & Cohen 1974) modes in the SiO_4 tetrahedron forming the building block of silicates, because they correspond to known absorption bands seen in all terrestrial silicates. These bands are also seen in the emission from dust shells around O-rich stars. This gave the first observational hints on the mineralogy of the silicate dust. The smooth, structureless nature of the bands indicated that the silicates in the ISM and in circumstellar dust shells are amorphous.

The detection of a strong depletion of refractory elements from the gas phase of the interstellar medium relative to stellar abundances (Morton *et al.* 1973; for a review see Savage & Sembach 1996) provided further clues to the composition of the dust. This observation enforced the conclusion that silicates and carbon dust dominate the interstellar dust because of the strong deficiency of Mg, Fe, and Si from the gas phase and a significant deficiency of carbon. The only acceptable explanation is that these elements are bound in dust particles. The finding of Field (1974) one year later that the degree of depletion increases rather smoothly with condensation temperature of the refractory grains was generally taken as an indication that the dust was formed in a gradually cooling environment, in stellar outflows, or in expanding shells after explosions (though this type of correlation may also have other explanations, e.g. Snow 1975).

The efficiency of scattering relative to absorption depends strongly on particle size in the size regime where particle sizes are of the order of the wavelength of the scattered light (see Appendix 3). The wavelength dependence of scattering of starlight by interstellar dust particles therefore contains information on their size distribution. The same holds for polarization of starlight by non-spherical grains (e.g. Wilking *et al.* 1980). The increasing contribution of scattering to the total extinction toward the blue and ultraviolet spectral regions has already shown that interstellar dust grains are $\ll 1\,\mu m$. The first model that tried to reproduce observed extinction, scattering, and polarization properties of interstellar dust and to observe at the same time the conditions resulting from depletion observations, i.e. that Mg, Si, Fe completely condensed into dust and that carbon partially condensed, was the model by Mathis, Rumpl, & Nordsieck (1977), generally known as the MRN model. The best fit in this model was obtained with magnesium iron silicate with composition $MgFeSiO_4$, a 60% condensation of carbon in graphite and a distribution $f(a)$ of grain radii a according to a power law

$$f(a) \propto a^{-3.5} \tag{2.1}$$

between $a = 5\,nm$ and $a = 250\,nm$. This has been improved by Draine & Lee (1984) by deriving significantly improved optical constants and became the standard model for a long time; it is still used as a first approximation.

Several further improvements have been added by many people; the last and most complex version of the model is that of Zubko *et al.* (2004) that provides a different form of the size distribution. In particular, scattering observations show that there is an additional component of very small grains, not included in the MRN model. The existence of a population of very small grains was already detected earlier by Witt & Lillie (1973) from investigations of the diffuse galactic radiation field in the far-ultraviolet and even earlier it was proposed by Donn (1968) that very small carbon grains may be responsible for strong ultraviolet extinction. After the Infrared Astronomical Satellite (IRAS) made far-infrared observations possible in 1984, it was found by analyzing the newly detected "cirrus" emission at $60\,\mu m$ and $120\,\mu m$ from the ISM, that the emission of very cold dust in the ISM shows an excess at $12\,\mu m$. This can only be explained by emission from extremely small particles that suffer high-temperature excursions following absorption of energetic photons (Boulanger & Pérault 1988). The detection of strong emission features at 3.3, 6.2, 7.7, 8.6, and $11.3\,\mu m$ (Gillett *et al.* 1973), originating from polycyclic aromatic hydrocarbons (PAHs), showed that this subpopulation of small grains are PAHs (Léger & Puget 1984). Such particles are now included in dust models (Weingartner & Draine 2001; Zubko *et al.* 2004). Modern models predict about 10% of the total carbon to be condensed in the small PAHs and about 50% in amorphous carbon.

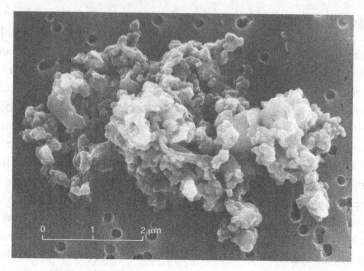

Figure 2.1 An IDP composed of very small <0.2 μm subunits. The IDPs show the fluffy structure of particles that is expected for dust agglomerates in protoplanetary accretion disks. The IDPs are suspected to be composed partly from particles of interstellar origin. A small fraction of the subunits can be identified as presolar dust grains. (Courtesy of NASA, http://stardust.jpl.nasa.gov/science/sci2.html.)

Sometimes, dust models have been considered that contain coated particles, e.g. silicate grains with a mantle of carbonaceous material. Such models can probably be excluded (Zubko *et al.* 2004); the best fits to observations are obtained for bare grains.

A potential source of information on the nature and composition of interstellar dust grains comes from IDPs, which are collected in the Earth's stratosphere (e.g. Bradley 2003; Alexander *et al.* 2007). The IDPs with bulk compositions in the range of chondritic meteorites can be divided into two broad categories, compact hydrous particles and porous anhydrous particles. The latter are aggregates of submicrometer building blocks (see Fig. 2.1). They represent a disequilibrium assemblage of minerals and amorphous silicates and it has long been suspected that some of them are dust particles released during the flyby of comets at the Sun. If true, these grains are particles from the formation time of the Solar System that have survived for 4.6 Gyr inside extremely cold (approximately 20 K) cometary bodies. Although modified in several important ways (Stern 2003, see also Section 2.3.4), these are some of the most pristine bodies known. Since some material of the protoplanetary disk in the region of comet formation is interstellar material from the parent molecular cloud (see Section 2.3.4), cometary IDPs should contain interstellar dust grains. In particular, one of the components forming IDPs is the so-called GEMS (glass with embedded metals and sulfides), a glassy material with inclusions of iron

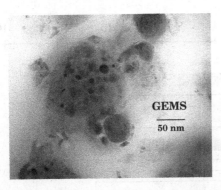

Figure 2.2 A GEMS particle found in an IDP, consisting of a glassy matrix with silicate-like composition and embedded nanometer-sized inclusions of iron metal and iron sulfide (dark patches). Some of them are possibly interstellar silicate dust grains. (From NASA http://stardust.jpl.nasa.gov/science/sci2.html.)

metal and iron sulfide (see Fig. 2.2). Such GEMS particles have been considered to be true interstellar silicate dust grains (Bradley 2003), which may have formed by accreting thoroughly mixed refractory elements from the interstellar gas phase that is isotopically normal. However, more recent investigations show that most GEMS grains with solar O-isotopic compositions have chemical compositions that appear to be incompatible with inferred compositions of interstellar dust. The origins of GEMS with solar O-isotopic composition are still a matter of debate and currently it is not possible to draw definite conclusions. There are, however, a few GEMS particles with O-isotopic anomalies (Messenger *et al.* 2003). These grains are undoubtedly stardust grains (see Section 2.2).

2.1.2 Cycle of matter from stars to the ISM to the Solar System

The visible matter of galaxies is concentrated in mainly three components: stars, interstellar matter, and stellar remnants. Since the early days of galaxy formation there is a vivid exchange of matter between the stellar component and the interstellar matter. Stars are formed in local concentrations of the ISM, the molecular clouds; they live for a certain period of time while burning their nuclear fuels; and they die

- by massive mass loss by a stellar wind in the case of low- and intermediate-mass stars (initial masses $M_0 < 8\,M_\odot$), leaving a white dwarf remnant,
- or by collapse of the core region of a massive star to a neutron star ($8\,M_\odot \lesssim M_0 \lesssim 40\,M_\odot$), associated with an ejection of most of their mass at that instant in a SN event,
- or by collapse to a black hole ($M_0 \gtrsim 40\,M_\odot$) without mass ejection,
- or by detonation of white dwarfs as supernovae of Type Ia, caused by mass transfer in certain binary systems.

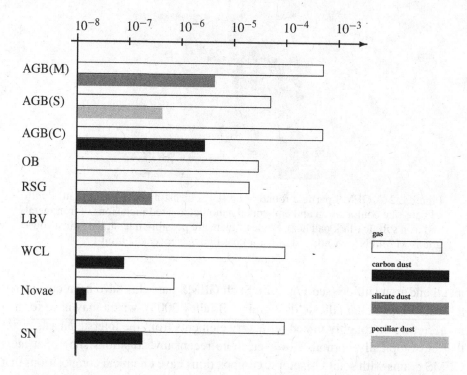

Figure 2.3 Dust production and gas mass return rate by different stellar types in solar masses per year and kpc^{-2} in the galaxy at the solar cycle. Stars produce mainly silicate or carbon dust; only in some cases is a different kind of dust material formed, probably iron or some iron alloy (peculiar dust). Many additional dust components with much smaller abundance are formed in most cases (Data from Tielens 1999; Zhukovska *et al.* 2008). Abbreviations of stellar types: AGB = asymptotic giant branch stars of spectral types M, S, or C; OB = massive stars of spectral types O and B on or close to the main sequence; RGB = massive stars on the red giant branch; LBV = luminous blue variables; WCL = Wolf–Rayet stars from the lower temperature range; Novae = mass ejecta from novae; SN = mass ejecta from supernovae.

In effect, stars return between about 50% and 90% of their initial mass by winds or by explosive mass ejection. Figure 2.3 gives an overview of the relative contributions of the different stellar types to the total mass replenishment. The mass returned by the stars becomes part of the ISM and serves as raw material for the formation of the next stellar generations. In this way part of the baryonic matter in a galaxy is continuously cycled between stars and the interstellar matter. Only the very-low-mass stars (initial masses $\lesssim 0.8\,M_\odot$) are not involved in this matter cycle because they have lifetimes exceeding the present age of the Universe and have not yet evolved very much. Some fraction of the matter therefore accumulates in very-low-mass stars and in stellar remnants, but at least part of this loss from the matter cycle seems to be replenished by continued mass infall of fresh matter from intergalactic space (see Matteucci 2007, for more details on the matter cycle). Presently the turnaround time of matter between stars and the ISM in the solar vicinity of the Milky Way is of the order of 2.5 Gyr, as follows from a local surface density of the ISM of $10 \pm 3\,M_\odot\,pc^{-2}$ (e.g. Dickey 1993) and a local star formation rate of $4 \pm 1.5\,M_\odot\,Gyr^{-1}\,pc^{-2}$ (e.g. Rana 1991).

The returned mass from the many different stellar sources is mixed on rather short timescales with the interstellar matter by turbulent flows. The result is an interstellar material with rather homogeneous elemental composition, which forms the raw material for the formation of the next stellar generation. The small internal scatter of stellar elemental abundances[2] of significantly less than ± 0.15 dex (e.g. Pont & Eyer 2004) observed for G dwarfs from the solar vicinity with ages comparable to the age of the Sun is an observational confirmation of a well-mixed interstellar medium. The elemental composition of the matter from which stars are formed, therefore, shows no substantial variation at any given instant and radial zone within the Milky Way, only a slow increase of the heavy-element content of successive stellar generations. There exist, however, moderate radial abundance gradients in the Milky Way: the metallicity of stars decreases significantly between the central region and the outermost parts.

The density and temperature distribution of interstellar matter, contrary to its elemental composition, is strongly inhomogeneous. At least three different phases exist (e.g. Tielens 2005): (i) extended low-density bubbles of hot ionized gas (hot interstellar medium or HIM, mass fraction ≈ 0.003, volume fraction ≈ 0.5), resulting from series of SN explosions in mass-rich stellar clusters; (ii) cold and dense clouds of neutral gas (cold and neutral interstellar medium or CNM, mass fraction ≈ 0.3, volume fraction ≈ 0.01), resulting from sweeping up of warm gas; and (iii) a warm, either ionized or neutral, medium in between (warm interstellar medium or WIM, mass fraction ≈ 0.5, volume fraction ≈ 0.5). The essential properties of the three phases are indicated in Fig. 2.4. The coolest and most massive of the clouds are the molecular clouds (MC, mass fraction ≈ 0.2, volume fraction ≈ 0.0005), a separate component, that are the places of star formation, where new stars are formed as stellar clusters with total masses between about 200 and several $10^6\,M_\odot$.

The interstellar matter continuously cycles through these phases and the molecular clouds: HIM $\leftrightarrow$ WIM $\leftrightarrow$ CNM $\leftrightarrow$ MC $\rightarrow$ HIM, WIM, and stars. The molecular clouds live typically for 20 Myr (Blitz *et al.* 2007) and convert about 25% of their mass into stars (10% to 30% according to Lada & Lada 2003). Since they contain $\approx 20\%$ of total mass, $4 \times 5 = 20$ cycles are required for 100% conversion of interstellar matter into stars. Since the cycle time between stars and the ISM is about

---

[2] Elemental abundance from stars other than the Sun can be determined with an accuracy usually not much better than ± 0.15 dex.

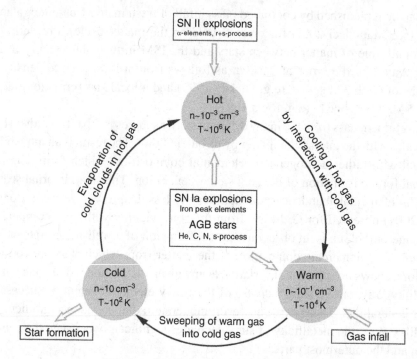

Figure 2.4 Matter cycle in the Milky Way.

2.5 Gyr, a cycle of the interstellar matter through the phases requires about 125 Myr. All these numbers are, of course, only crude estimates.

Massive stars return mass into the HIM, because almost all massive stars in rich stellar clusters end their life within the super-bubble driven into the ISM by the SN explosions (e.g. Higdon & Lingenfelter 2005). The expansion of the bubble is also largely responsible for the dispersion of the remaining part of the MC within which the cluster has formed. Due to their long lifetimes the mass return by low- and intermediate-mass stars occurs well after the dissolution of their parent stellar clusters. By that time they have become randomly distributed field stars. Because of the large volume-filling factors of HIM and WIM the mass return of these stars is to HIM and WIM, but not to CNM.

2.1.3 Sources and destruction of dust in the ISM

Low- and intermediate-mass stars ultimately become asymptotic giant branch (AGB) stars and start to form dust in the cool stellar wind by which they eject their envelopes (see also Chapter 5). This dust is also mixed into the ISM like the returned gaseous material. About one half of the mass returned by AGB stars is oxygen-rich and injects mainly silicate dust to the ISM; the other half is carbon-rich and injects mainly carbon and silicon carbide dust to the ISM (see Fig. 2.3). Additionally, some minor dust components are formed in both cases. The details of the dust production by AGB stars are discussed in Ferrarotti & Gail (2006), the observations in Molster & Waters (2003).

Cameron (1973) speculated that grains from stellar sources survive in the interstellar medium, become incorporated into bodies of the Solar System, and may be found in meteorites, because some meteorites represent nearly unprocessed material from the time of Solar System formation. These grains may be identified by unusual isotopic abundance ratios of some elements, since material from nuclear burning zones is mixed at the end of the life of stars into the matter from which dust is formed. Indeed, these "presolar" dust grains[3] were found in the late 1980s in meteorites (and later also in other types of primitive Solar System matter) and they contain rich information on their formation conditions and on nucleosynthetic processes in stars (see Section 2.2). By identifying such grains in primitive Solar System matter it is possible to study the nature and composition of at least some components of the interstellar dust mixture in the laboratory.

Laboratory studies of presolar dust grains also show that dust is formed in the mass ejected after an SN explosion, as will also be discussed in Section 2.2. Observations show that the ejected mass shells occasionally do form dust some time after the SN explosions (e.g. Bianchi & Schneider 2007), but generally the efficiency of dust production seems to be rather low (Bianchi & Schneider 2007; Zhukovska *et al.* 2008). Other important sources of stardust are red supergiants (mostly silicate dust). Most of the dust from red supergiants, however, is not expected to survive the shock wave from the subsequent SN explosion of the star (Zhukovska *et al.* 2008). Some dust is also formed by novae (Amari *et al.* 2001b), Wolf–Rayet stars (WRs, Crowther 2007), and luminous blue variables (LBVs, Voors *et al.* 2000), but the dust quantities formed by these are very small. Stardust – i.e. dust that is formed in stellar outflows or ejecta – in the interstellar medium is dominated by dust from AGB stars.

Figure 2.3 gives an overview of the relative contributions of the different stellar types to the total dust injection rate by stars.

Dust particles in the ISM are subject to a number of destruction processes (Draine & Salpeter 1979), e.g. sputtering of dust particles by energetic ions in hot ($T > 5 \times 10^5$ K) supernova bubbles, evaporation of dust grains in high-velocity grain–grain collisions behind shock waves, or photo-sputtering of dust due to irradiation by hard ultraviolet photons. Such processes return part of the dust material

---

[3] Presolar means that the particles were formed before the formation of the Solar System. In contrast to "stardust" this specifically means that such grains are identified by laboratory investigations unequivocally as to be of stellar origin by anomalous isotopic abundances of at least one element.

to the gas phase and therefore reduce the dust content of the ISM. Dust particles may also be shattered into smaller pieces by low-velocity grain–grain collisions behind shocks (Tielens *et al.* 1994; Jones *et al.* 1994). This changes the size distribution of the interstellar dust particles, but not the total dust content of the ISM. The shattering of grains is thought to be responsible for a power-law type size distribution of interstellar dust grains like the popular MRN-size distribution (Mathis *et al.* 1977).

The most important process for dust destruction is sputtering behind supernova shocks that are driven into the WIM (Jones *et al.* 1994, 1996). Shocks in the HIM are ineffective due to its low density, while shocks in the CNM are ineffective owing to the rapid shock deceleration in this component. For the WIM, an average lifetime of grains against shock destruction of 0.6 Gyr for carbon dust and 0.4 Gyr for silicate dust is estimated by Jones *et al.* (1996). Recent determinations of residence times in the ISM for some presolar SiC grains (Gyngard *et al.* 2007; Heck *et al.* 2008), estimate ranges from a few million years to more than one billion years, are consistent with lifetimes of this order of magnitude.

The average lifetimes of dust grains in the ISM of about 0.5 Gyr have to be compared with a turnaround time of about 2.5 Gyr for the matter cycle between stars and the ISM, which would result in a small depletion $\delta \approx 0.8$ of the refractory elements in the ISM into dust, if depletion of the refractory elements in the returned mass from stars was strong and if no accretion of refractory elements onto dust occurred in the ISM. This clearly contradicts the high observed depletion in the ISM. Hence, most of the interstellar dust is formed in the ISM and is not stardust (Draine 1995; Zhukovska *et al.* 2008). The most likely place for dust growth in the ISM is in the dense molecular clouds (Draine 1990), but the processes responsible for growth are presently unknown.

One of the unexpected results of the far-infrared observations of dust shells around stars by the infrared space observatory (ISO) satellite in 1996 was the finding that oxygen-rich objects show a rich structure from crystalline silicates in their spectra beyond a wavelength of 25 μm. Most of the observed features can be identified with absorption bands of olivine (with composition $Mg_{2x}Fe_{2(1-x)}SiO_4$) and pyroxene (with composition $Mg_x Fe_{1-x}SiO_3$) as measured in the laboratory (Molster & Waters 2003). The crystalline material seems to be iron-poor ($x > 0.9$) and represents about 10–20 % of the condensed silicate phase in outflows (Molster & Waters 2003), if mass-loss rates are very high. Crystalline grains are absent in lower mass-loss rate outflows. This finding is in sharp contrast with the finding that interstellar dust does not show any indication of the presence of crystalline silicates, only the bell-shaped broad profiles of the 9.7 and 18 μm silicate bands that are typical for amorphous material; upper limits on the degree of crystallinity in the ISM between 1% and 2% have been given by different authors (Kemper *et al.*

2005; Min *et al.* 2007). It is assumed now that interstellar dust grains are subject to amorphization processes in the ISM by bombardment with energetic ions (e.g. H^+, He^+) that destroy a regular lattice structure (Demyk *et al.* 2001; see also Chapters 5 and 8), but probably there is in any case no discrepancy, since ISM dust is mostly not stardust (Draine 1995; Jones *et al.* 1996).

2.1.4 Element abundances

The primordial composition of baryonic matter in the early Universe at the time of the onset of star formation was rather simple: H, He, and traces of Li. All heavier elements from carbon on, up to the actinides, that today form a mass fraction of about 2% of the baryonic matter content of spiral galaxies, are products of nucleosynthetic processes in stars. Elements up to the iron group result mainly from burning processes in massive stars. Heavier nuclei are products of neutron-capture processes during the final evolution of low- and intermediate-mass stars on the thermal pulsing AGB (slow neutron capture, the s-process elements) or during the explosive burning processes during SN events (rapid neutron capture, the r-process elements). Part of the products of the nucleosynthetic processes are contained in that fraction of the stellar mass that is returned to the ISM. This gradually increases the heavy-element content of the interstellar matter, its metallicity (= the mass fraction of all elements heavier than He).

Of particular interest are the abundant refractory elements that form the dust material in space and the mantles and cores of planets. These are the following 10 of the 83 stable and long-lived radioactive elements: C, O, Si, Mg, Fe, S, Al, Ca, Na, and Ni. Only four elements, O, C, Fe, and Ne, already account for 80% of the mass fraction of all elements heavier than He, 16% is contributed by Si, Mg, N, and S, and the final 4% shared between all other elements heavier than He. Almost all of the abundant elements heavier than He, except for C, N, Fe, and Ni, are mainly or exclusively formed by massive stars. They have been continuously produced since the onset of the matter cycle in the galaxy about 13 Gyr ago. Because the element production of massive stars is not strongly metallicity dependent, the relative abundances of the rock-forming elements O, Si, Mg, S, Al, Ca, and Na do not change much over time, only their absolute abundances. The production of C and N by AGB-stars and of Fe and Ni by exploding white dwarfs in binary stars commences with a delay of ≈ 1.5 Gyr because of the much longer lifetime of intermediate-mass stars. After an initial period of rapid increase that lasted until the metallicity of the ISM reached about $\frac{1}{10} Z_\odot$ (which happened to occur in the Milky Way about 10 Gyr ago), the abundance of these elements has been continuously increasing nearly in pace with that of the other rock-forming elements.

Since $\frac{1}{10}Z_\odot$ seems to set the lower limit for efficient dust formation,[4] planetary systems are only expected to form around stars with at least this metallicity. The relative abundances of rock-forming elements in such systems are essentially the same as in the Solar System. Only the total amount of such elements may vary considerably, depending on the birthplace and birth-time of such systems. Planetary systems with unusual elemental compositions, e.g. like those observed in many highly evolved stars, are not expected to exist; in particular, oxygen is always more abundant than carbon such that for planetary systems there is no counterpart to the carbon-rich circumstellar dust shells.

2.2 Presolar grains in primitive Solar System materials

Until the 1960s it was believed that the material that went into the making of the Solar System had been thoroughly homogenized in a hot solar nebula, resulting in uniform isotope ratios. In fact, the vast majority of meteoritic minerals and components have rather uniform isotopic compositions in the rock-forming elements. However, first hints of the survival of trace levels of presolar matter came from isotopic anomalies (i.e. deviations from average Solar System matter) in hydrogen and noble gases in primitive meteorites. With the discovery of isotope anomalies in other elements such as C, N, O, Mg, Ca, and Ti in the following two decades the idea of the presence of relict presolar material in meteorites found wide acceptance. The search for the carriers of exotic Ne and Xe components led finally to the isolation of presolar grains at the University of Chicago in the late 1980s, samples of stardust that formed long before our Solar System was born (Anders & Zinner 1989). In general, the term "presolar grain/mineral" is used only for dust samples from other stars. A secure identification of presolar minerals relies on large isotopic anomalies in the major elements in individual grains. Dust grains that formed in the ISM may not necessarily be characterized by isotope anomalies and are thus difficult to detect, but imprints of likely interstellar chemistry are seen as D and ^{15}N enrichments in organic matter in primitive Solar System materials (see Section 2.2.5).

Presolar grains exhibit large isotopic anomalies not only in their major elements, but also in many minor elements. Isotopic ratios vary over many orders of magnitude, indicative of contributions from different types of stellar sources, namely evolved stars, novae, and SN explosions. Isotope anomalies are also seen in objects with Solar System origin, which, however, are much smaller than those in presolar grains. For example, the calcium–aluminum-rich inclusions (CAIs), the earliest

---

[4] Formation of dust from elements different from C, for instance formation of SiC, in stellar outflows ceases at about the metallicity of the Small Magellanic Cloud ($Z = \frac{1}{8}Z_\odot$), see Sloan *et al.* (2006), while in the Large Magellanic Cloud ($Z = \frac{1}{4}Z_\odot$) dust production is observed, see Zijlstra *et al.* (2007).

Table 2.1 *Presolar minerals with identified stellar origin from primitive meteorites*

Mineral	Abundance[1] (ppm)	Stellar sources	Relative contribution
SiC	30	1.5–3 $M_\odot$ AGB stars	>90%
		J-type C-stars, post-AGB (?)	5%
		Type II supernovae	1%
		Novae	0.1%
Graphite	10	Type II supernovae	<80%
		AGB stars	>10%
		J-type C-stars, post-AGB (?)	<10%
		Novae	2%
Si_3N_4	0.002	Type II supernovae	100%
Oxides	50	1–2.5 $M_\odot$ AGB stars	≈90%
		Type II supernovae	≈10%
		Novae	<1%
Silicates	200[2]	1–2.5 $M_\odot$ AGB stars	≈90%
		Type II supernovae	≈10%

[1] Reported maximum values (normalized to volume of meteorite matrix) are given.
[2] Higher abundances (≈375 ppm on average) are observed in isotopically primitive IDPs.

condensates in the solar nebula, show enrichments in ^{16}O of up to 5 % relative to terrestrial O. They also exhibit isotope anomalies (typically on the order of several percent relative to average Solar System material) in a large number of other elements (e.g. Mg, Ca, Ti, Cr, Fe, Ni, Sr, Ba, Nd, and Sm). These anomalies are likely the result of incorporation of incompletely mixed, but heavily processed presolar material. The CAIs and other minerals that formed in the earliest stages of Solar System formation also show excesses in certain isotopes that can be attributed to the decay of now-extinct radioactive isotopes that were part of the solar nebula and still alive when these solids formed (e.g. ^{10}Be, ^{26}Al, ^{41}Ca, ^{53}Mn, ^{60}Fe).

Presolar grains are found in small quantities (with concentrations of ppb to several 100 ppm, see Table 2.1) in all types of primitive Solar System materials (Lodders & Amari 2005; Zinner 2007). This includes primitive meteorites (the chondrites), IDPs, some of which might originate from comets, Antarctic micrometeorites (AMMs), and samples from comet Wild 2 collected by NASA's Stardust mission. Presolar grains are nanometer to micrometer in size. The isotopic compositions, chemistry, and mineralogy of individual grains with sizes >100 nm can be studied in the laboratory. Important analysis techniques are secondary ion mass spectrometry (SIMS) and resonance ionization mass spectrometry (RIMS)

that permit isotope measurements of a large number of elements at the micrometer and submicrometer scale (see Appendix 2) as well as electron microscopy
(scanning electron microscopy/energy dispersive X-ray spectroscopy (SEM/EDX),
transmission electron spectroscopy (TEM), Auger spectroscopy). These studies
have provided a wealth of information on stellar nucleosynthesis and evolution,
on mixing in nova and SN ejecta, on dust formation in stellar environments, dust
chemistry in the ISM, and on the nature of the starting material from which our
Solar System formed some 4.6 Gyr ago.

2.2.1 Types of presolar grains

The first identified presolar minerals were diamond (Lewis *et al.* 1987), silicon carbide (SiC, Bernatowicz *et al.* 1987; Tang & Anders 1988), and graphite (Amari *et al.*
1990). Refractory presolar oxides and a nitride were found a few years later, first
corundum and other forms of Al_2O_3 (Hutcheon *et al.* 1994; Nittler *et al.* 1994) and
later also spinel ($MgAl_2O_4$), hibonite ($CaAl_{12}O_{19}$), titanium oxide, silicon nitride
(Si_3N_4) as well as wüstite (FeO; for a list of the common minerals in the Solar System see Appendix 1). Presolar silicates were found only recently (Messenger *et al.*
2003). Among the silicates are amorphous grains, GEMS, olivines [$(Mg,Fe)_2SiO_4$],
and a $MgSiO_3$ with a perovskite- (high-pressure-) like structure. The structure of
the latter has been explained to be likely the result of shock-transformation of a
silicate precursor, either in the ISM or induced by the grain's parent star. Some of the
silicates exhibit a complex, i.e. heterogeneous chemistry. One of them consists of
an Al-rich core, probably Al_2O_3, surrounded by a silicate mantle and another grain
comprises at least three different Ca–Al-rich minerals and can thus be regarded
as a presolar CAI. A complex internal structure is also evident for many presolar
graphite grains which host tiny subgrains. The first identified subgrains were TiC
(Bernatowicz *et al.* 1991). Later also Zr-, Mo-, and Ru-rich carbides, kamacite
(Fe–Ni), and cohenite [$(Fe,Ni)_3C$] were found (see Table A1.1 in the appendix for
a list of relevant mineral compositions).

Diamonds are found in comparatively large quantities (>1000 ppm) in primitive
meteorites. However, they are only 2–3 nm in size, which makes it impossible to
measure isotopic compositions of single grains and only ensembles consisting of
thousands or millions of grains can be studied. A presolar origin is indicated by
exotic Xe and Te isotopic compositions. Although their nitrogen isotope ratio is
anomalous it is compatible with what was inferred for Jupiter. On the other hand, the
carbon isotopic ratio of the diamonds is essentially solar (a signature not expected
for stardust, see below). Most of them might have formed in the ISM or in the Solar
System and it remains an open question what fraction of the diamonds actually
represents stardust.

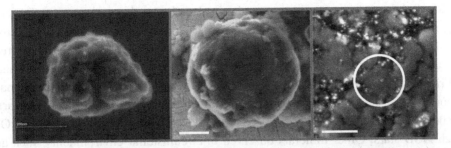

Figure 2.5 Presolar grains from primitive meteorites. Left: presolar SiC from the Murchison meteorite scale bar is 200 nm. Middle: presolar graphite from Murchison; scale bar is 1 μm. Right: presolar silicate grain (within the white circle) in the matrix of the Acfer 094 meteorite; scale bar is 500 nm. Photo credit: Max Planck Institute for Chemistry.

Apart from diamonds, the most abundant presolar minerals are the silicates, with concentrations of several 100 ppm in the most primitive meteorites and IDPs, followed by the refractory oxides, SiC, and graphite with abundances of 10–50 ppm (Table 2.1). Silicon nitride is very rare and is found only at the ppb level. Surprisingly, matter from comet Wild 2 returned by the Stardust mission contains only low abundances of presolar grains (McKeegan *et al.* 2006). To date only four presolar grains have been identified (three oxygen-rich grains and one SiC grain). A current estimate for the abundance of O-rich presolar dust is 11 ppm, which, however, still bears large uncertainties. Most presolar SiC (Fig. 2.5, left) and oxide grains are submicrometer in size (diameter); a small fraction of these grains, however, have sizes of several micrometers. Presolar graphite grains are round. They are larger than most SiC and oxide grains with typical sizes in the micrometer range (Fig. 2.5, middle). Presolar silicates have sizes between 150 and 800 nm with most grains having sizes between 200 and 400 nm (Fig. 2.5, right).

While most of the presolar minerals (e.g. SiC, graphite, corundum, spinel) can be chemically separated from meteorites, this does not hold for presolar silicates. For this reason the presolar silicates remained unrecognized for a long time and only the application of advanced analytical techniques (see Appendix 2) led to their discovery, first in an IDP and later also in primitive meteorites and AMMs. The GEMS particles with solar isotopic compositions are an abundant component of porous anhydrous IDPs. It was pointed out in Section 2.1.1 that the question on the origin of GEMS particles is not yet settled. Interstellar and/or Solar System origins have been proposed. In any case, even if some of them did have an interstellar origin, their solar O-isotopic ratios would be the result of irradiation and/or cycling of material in the ISM, i.e. they would not represent unaltered stardust (in contrast to the few identified GEMS grains with large O-isotopic anomalies).

Most of the presolar stardust minerals are also observed in various astrophysical environments. Oxygen- and carbon-rich dust are important constituents of circumstellar and interstellar matter (see Section 2.1 and Molster & Waters 2003). Features of amorphous and crystalline silicates are seen in the spectra of young stars, C- and O-rich AGB stars, post-AGB stars, planetary nebulae, and massive stars. Carbonaceous dust is seen in the spectra of C-rich AGB stars, post-AGB stars, and planetary nebulae. Refractory oxides such as Al_2O_3 and $MgAl_2O_4$ are observed around O-rich AGB stars. Other minerals observed in circumstellar environments, such as sulfides (MgS, FeS), on the other hand, are still missing from the inventory of presolar grains. But it should be noted that the most diagnostic (with respect to a circumstellar origin) isotope ^{36}S is very rare, making a definite identification of presolar sulfides very difficult. Stardust is only a minor component of dust in protoplanetary disks. The topic of the overall dust composition in protoplanetary disks is discussed in detail in Chapter 6.

2.2.2 Carbonaceous dust

The noble gases played a key role in the isolation of carbonaceous presolar grains from meteorites as they served as beacons in the development of the chemical separation procedure. Presolar diamonds are tagged with the noble gas component Xe-HL (enhanced abundances of the light and heavy Xe isotopes), SiC with Ne-E(H), almost pure ^{22}Ne that releases at high temperatures, and Xe-S (imprints of s-process nucleosynthesis), and graphite with Ne-E(L), essentially pure ^{22}Ne that releases at lower temperatures than Ne-E(H) in SiC. These samples turned out to be a treasure trove for astrophysics and a wealth of information has been obtained from the laboratory study of individual SiC and graphite grains in the last two decades (Lodders & Amari 2005; Zinner 2007).

Silicon carbide is by far the best-characterized presolar mineral. The reason for this is that it can be isolated from primitive meteorites in almost pure form, is available as a large number of micrometer-sized grains, and that it contains high concentrations of trace elements such as N, Al, noble gases, and refractory intermediate-mass and heavy elements. The C- and nitrogen-isotopic compositions of individual presolar SiC grains are displayed in Fig. 2.6. Carbon- and nitrogen-isotopic ratios vary over almost four orders of magnitude with $^{12}C/^{13}C$ ratios between 2 and 7000 and $^{14}N/^{15}N$ ratios between 5 and 20000. Based on C-, N-, and Si-isotopic compositions, SiC has been divided into distinct populations (Hoppe *et al.* 1994): mainstream, Type A, B, X, Y, Z, and nova grains. Most abundant are the mainstream grains ($\approx 90\%$), most of which show enhanced ^{13}C and ^{14}N with respect to solar isotopic abundances, the signature of H burning by the CNO cycle. Silicon-isotopic anomalies are smaller than those in C and N. Most mainstream

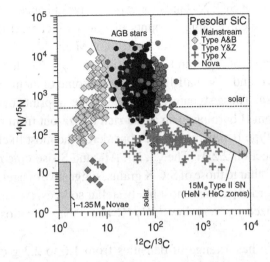

Figure 2.6 Carbon- and nitrogen-isotopic compositions of presolar SiC grains. Predictions from stellar models are shown for comparison. Solar metallicity AGB star models: Nollett *et al.* (2003), Type II SN: Rauscher *et al.* (2002), novae: José *et al.* (2004). For data sources see Lodders & Amari (2005); Zinner (2007). Note that for the solar $^{14}N/^{15}N$ ratio the value inferred for Jupiter's atmosphere is shown.

grains show enrichments in the neutron-rich isotopes ^{29}Si and ^{30}Si of up to 20% and in an Si-three-isotope representation they plot along a characteristic correlation line ($\delta^{29}Si = 1.37 \times \delta^{30}Si - 20$; $\delta^i Si = [(^i Si/^{28}Si)_{Grain}/(^i Si/^{28}Si)_\odot - 1] \times 1000$), the "SiC mainstream line." For historical reasons, it is also interesting to note that there are clear hints for the initial presence of ^{99}Tc (half-life 210 000 yr; Savina *et al.* 2004), an isotope, whose observation in the spectra of AGB stars has provided the first direct evidence that the chemical elements are produced in the interior of stars (Merrill 1952). The isotopic patterns of the intermediate-mass and heavy elements (e.g. Kr, Sr, Zr, Mo, Ru, Xe, Ba, Nd, Sm, and Dy) show the characteristic imprints of s-process nucleosynthesis (slow neutron-capture reactions). The Y and Z grains (several percent of SiC) differ from mainstream grains in that they plot to the ^{30}Si-rich side of the SiC mainstream line. While the C-isotopic ratios of the Type-Z grains are not different from those of mainstream grains, the Y grains have $^{12}C/^{13}C$ ratios of higher than 100, which is outside the range of mainstream grains. Carbon-rich AGB stars are the most likely stellar sources for the mainstream, Y, and Z grains (for a detailed discussion of stellar sources see Section 2.2.4). The A and B grains (several percent of SiC) are generally treated as a single population. They have $^{12}C/^{13}C$ ratios of <10 and many of them exhibit enrichments in ^{15}N relative to solar. Their Si-isotopic signature is essentially the same as that of mainstream grains. The origin of the A and B grains is still not understood.

The X grains ($\approx 1\%$ of SiC; X stands for exotic) have enhanced ^{12}C (most grains), ^{15}N, and ^{28}Si. They carry large amounts of now-extinct ^{26}Al (half-life 700 000 yr), with inferred ^{26}Al/^{27}Al ratios of up to 0.6. Many X grains show large isotopic overabundances of ^{44}Ca and ^{49}Ti, resulting from the decay of radioactive ^{44}Ti (half-life 60 yr) and ^{49}V (half-life 330 d) and some of them exhibit an unusual Mo isotopic pattern. These signatures are clearly the imprints of different nuclear burning stages, from H burning to an alpha-rich freeze-out from nuclear statistical equilibrium, and Type II supernovae are considered the most likely stellar sources of X grains (see Section 2.2.4). The C-, N-, Al-, and Si-isotopic ratios of presolar Si$_3$N$_4$ grains are similar to those of SiC X grains, suggestive of a related origin. Nova grains, named according to their most likely stellar sources, are very rare ($\approx 0.1\%$) and are characterized by very low ^{12}C/^{13}C and ^{14}N/^{15}N ratios, and large ^{30}Si excesses.

Presolar graphite has a range of densities from 1.6 to 2.2 g cm^{-3}. Two basic morphologies are evident: "onions," grains with smooth or shell-like platy surfaces (Fig. 2.5) and "cauliflower," dense aggregates of smaller particles. Presolar graphite grains have a similar range of ^{12}C/^{13}C ratios as presolar SiC. The distribution of C-isotopic ratios, however, is clearly different (Fig. 2.7). Only few graphite grains have the C-isotopic characteristics as typically observed for presolar SiC. Some presolar graphite grains have C-isotopic ratios in the range of the SiC A and B grains but most graphite grains have higher than solar ^{12}C/^{13}C as similarly observed for the rare SiC X grains. Grains of lower density have higher trace-element concentrations and have thus been studied in more detail for their isotopic compositions. Many of these grains have excesses in ^{15}N, ^{18}O, and ^{28}Si and they carry the decay products of radioactive ^{26}Al, ^{41}Ca (half-life 105 000 yr), ^{44}Ti, and ^{49}V. This implies a related origin of low-density graphite grains and the supernova SiC X grains. Imprints of s-process nucleosynthesis are seen in the Si-, Zr- and Mo-isotopic patterns measured in a few high-density graphite grains as well as in the Kr-isotopic pattern of graphite bulk samples. These grains are likely to come from C-rich AGB stars (see Section 2.2.4).

2.2.3 Oxygen-rich dust

Compared to carbonaceous presolar grains much less isotope information, mostly for the major elements, is available for O-rich presolar dust (Lodders & Amari 2005; Zinner 2007). This has several reasons. Other than presolar SiC and graphite, presolar oxides have only low trace-element concentrations. Moreover, it is not possible to produce chemical separates that consist mostly of presolar oxide grains. Presolar silicates can be found only *in situ* by ion imaging (see Appendix 2), a time-consuming task. The O-isotopic data of refractory oxides (corundum and other

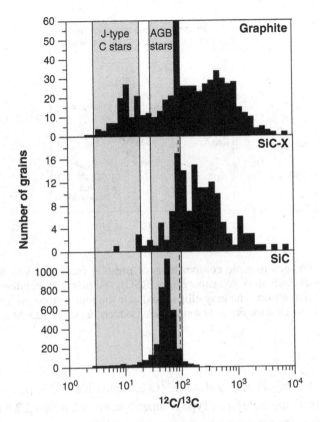

Figure 2.7 Histograms of $^{12}C/^{13}C$ ratios of presolar SiC (all grain types), SiC X grains, and graphite. The solar ratio is indicated by the dashed line. Observed ranges of $^{12}C/^{13}C$ ratios in AGB stars and J-type carbon stars are shown for comparison. For data sources see Lodders & Amari (2005) and Zinner (2007).

forms of Al_2O_3, spinel, hibonite) and silicates are shown in Fig. 2.8. Similar to C- and N-isotopic ratios in presolar SiC, the O-isotopic ratios range over more than three orders of magnitude with $^{17}O/^{16}O$ ratios between 0.08 and 40 times the solar ratio and $^{18}O/^{16}O$ ratios from 0.01 to 13 times the solar ratio. Based on the O-isotopic systematics, the O-rich presolar grains were divided into four isotopically distinct groups (Nittler *et al.* 1997). Most abundant are Group 1 grains (>70%), which are characterized by enrichments in ^{17}O and slightly lower than or close to solar $^{18}O/^{16}O$ ratios, qualitatively the signature of H burning. Group 2 grains ($\approx$15% of all grains) have enrichments in ^{17}O and strong depletions in ^{18}O. Oxygen-rich AGB stars are the most likely sources of the Group 1 and 2 grains (see Section 2.2.4). Group 3 grains (several percent of all grains) have slightly lower than solar $^{17}O/^{16}O$ and $^{18}O/^{16}O$ ratios. The origin of those grains is still not clear. Group 4 grains ($\approx$10% of all grains) are characterized by moderate

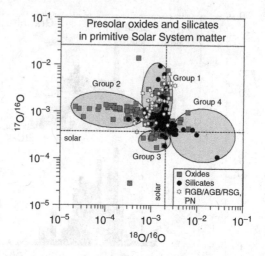

Figure 2.8 Oxygen-isotopic compositions of presolar oxide and silicate grains. Data for RGB/AGB stars, red supergiants (RSG), and planetary nebulae (PN) are shown for comparison. The gray ellipses indicate the four isotope groups defined for presolar O-rich dust. For data sources see Lodders & Amari (2005) and Zinner (2007).

to strong enrichments in ^{18}O and both ^{17}O enrichments and depletions. Group 4 grains are likely to originate from Type II supernovae (see Section 2.2.4). There is no systematic difference between refractory oxides and silicates except for the fact that the Group 2 grains appear to be less abundant among the presolar silicates. However, this may be an experimental bias because presolar silicates are identified *in situ* and measured isotope ratios may be diluted by contributions from surrounding matrix material, resulting in less extreme isotopic ratios. Two of the known oxide grains are outside this classification scheme, one with strong enrichment in ^{16}O and one with very high ^{17}O/^{16}O and moderate depletion in ^{18}O.

Many (but not all) grains carry radiogenic ^{26}Mg from the decay of radioactive ^{26}Al. Inferred ^{26}Al/^{27}Al ratios are generally highest in Group 2 grains in which ratios of larger than 0.1 can be reached. A few hibonite grains carry radiogenic ^{41}K from the decay of radioactive ^{41}Ca. Silicon-isotopic compositions of presolar silicate grains from Groups 1 and 2 are similar to those of SiC mainstream grains. The extremely ^{17}O-rich Group 1 grains, however, are comparably rich in ^{30}Si. The Si-isotopic signatures of the Group 4 silicate grains are distinct from those of the other grains. Most grains show lower than solar ^{30}Si/^{28}Si and ^{29}Si/^{28}Si ratios. This is qualitatively the signature of the supernova SiC X grains, pointing to a close relationship between these presolar grain populations.

2.2.4 Stellar sources

As it was pointed out above, AGB stars are the most prolific suppliers of stardust into the ISM (see Fig. 2.3). An example for dust injection by an AGB star into its surroundings is shown in Fig. 2.9. Oxygen-rich dust particles, mainly silicates, are expected to form at an early stage on the AGB, namely, when $C/O < 1$ in the stellar wind. Once C exceeds O in the wind, carbonaceous dust can form which occurs at a later stage on the AGB (see below).

The isotopic compositions of low- to intermediate-mass ($1-8\,M_\odot$) star envelopes are changed in three dredge-up episodes when matter from the star's interior is mixed outward. The first dredge-up occurs after the exhaustion of core H burning, i.e. when a main sequence star reaches the red giant branch (RGB), and matter that experienced H burning by the CNO cycle is mixed into the star's envelope. Following the core He flash, He burns gently in the core and H in a thin shell. After exhaustion of core He burning the AGB is reached and stars more massive than $\approx 3-4\,M_\odot$ experience a second dredge-up, again enriching the star's envelope with the ashes of H burning. Subsequently, He burns in a thin shell around the CO core, alternating with quiescent H burning in an overlying shell. As a result of thermal instabilities in the He shell, the base of the convective envelope moves inward, mixing matter that experienced H burning (^{26}Al), He burning (mostly ^{12}C, but also ^{22}Ne), and the s-process (intermediate-mass to heavy elements) to the stellar surface (Busso *et al.* 1999). This is called the third dredge-up (TDU). The C-isotopic ratio in the envelope is affected by all three dredge-up episodes, those

Figure 2.9 Hubble Space Telescope image of the Cat's Eye Nebula. The evolved AGB star (center) ejected its mass in a series of pulses at 1500 year intervals which created concentric dust shells. Photo credit: ESA, NASA, the Hubble European Information Centre and the Hubble Heritage Team, STScI/AURA.

of N and O essentially only in the first and second dredge-up. After a number of TDU events, when the C/O ratio exceeds unity, the star turns into a carbon star. Astronomical observations and presolar grain data indicate that there must be an additional episode, "cool bottom processing" (CBP), during the RGB and/or AGB phases, in which the envelope CNO isotope composition is modified by partial H burning (Nollett *et al.* 2003).

The SiC mainstream grains most likely originate from 1.5–3 $M_\odot$ AGB stars of roughly solar metallicity (Lugaro *et al.* 2003). This is supported by several facts; their C-isotopic compositions are compatible with those of C-rich AGB stars inferred from astronomical observations (Fig. 2.7); model predictions for such stars can account for the observed isotopic compositions of C and N (if CBP is considered, see Fig. 2.6); and the large overabundances of ^{22}Ne and the s-process isotopic signatures of the intermediate-mass to heavy elements in the grains. Although Si is affected by the s-process, the TDU changes the initial Si-isotopic composition of the envelope in solar metallicity AGB stars only marginally. Thus, the spread in Si-isotopic compositions represents mostly a range of the initial Si compositions of the parent stars, which is the result of the galactic chemical evolution of the Si isotopes, both in time and space. Similarly, the SiC Y and Z grains are believed to come from 1.5–3 $M_\odot$ AGB stars, but of lower metallicity than the parent stars of the mainstream grains (Hoppe *et al.* 1997; Amari *et al.* 2001a). The origin of the ^{13}C-rich A and B grains (Fig. 2.7) is still not clear. Among the proposed stellar sources are J-type carbon stars which have ^{12}C/^{13}C ratios compatible to those of A and B grains and post-AGB stars that experienced a very late thermal pulse (e.g. Sakurai's object; Amari *et al.* 2001c). The same holds for some graphite grains with ^{12}C/^{13}C ratios in the range of the SiC A and B grains. Some (unknown) fraction of presolar graphite appears to come from AGB stars (Croat *et al.* 2005; Jadhav *et al.* 2006), as indicated by the chemical composition of internal refractory carbide grains and the imprints of s-process nucleosynthesis.

Most presolar oxide/silicate grains (Group 1) have O-isotopic compositions as observed for RGB and AGB stars, red supergiants, and planetary nebulae (Fig. 2.8). The O-isotopic data are well explained by model predictions for the first and second dredge-up in 1–2.5 $M_\odot$ RGB and AGB stars (Boothroyd & Sackmann 1999). While the ^{17}O/^{16}O ratio is largely determined by the initial stellar mass, the ^{18}O/^{16}O ratio is only slightly lowered from its initial value and the observed spread in ^{18}O/^{16}O requires a range of initial compositions, i.e. a range of metallicities around solar. The very low ^{18}O/^{16}O ratios of the Group 2 grains can not be explained by distinctly lower metallicities of the parent stars, but require the operation of CBP (Nittler *et al.* 1997, 2008). The Group 3 grains may either come from stars of lower metallicity and lower mass than the parent stars of most Group 1 and 2 grains and/or from Type II supernovae (Nittler *et al.* 2008).

Grains with evidence for ^{26}Al apparently formed during the AGB phase when the TDU was operating; those without ^{26}Al during the AGB phase did, too, but prior to the TDU.

Novae apparently made only a small contribution to the presolar grain inventory. Nova models (José *et al.* 2004) predict low ^{12}C/^{13}C and ^{14}N/^{15}N ratios (Fig. 2.6) and large excesses in ^{30}Si. These characteristics are qualitatively shown by a handful of presolar SiC grains, the nova grains. Novae are also the most likely sources for a few graphite grains with low ^{12}C/^{13}C and ^{20}Ne/^{22}Ne ratios and possibly for the oxide grain with the largest ^{17}O enrichment (Fig. 2.8).

There is clear evidence that many presolar grains originate from Type II supernovae. All massive stars heavier than $\approx 8\,M_\odot$ are believed to explode as Type II supernovae. Before the explosion such stars consist of concentric layers that experienced different stages of nuclear burning (Rauscher *et al.* 2002). The presence of now-extinct ^{44}Ti and ^{49}V in SiC X grains can be considered as evidence for a SN origin because these isotopes are produced only in the deep interior of massive stars. The SN origin of SiC X grains is also supported by the large overabundance of ^{28}Si. The isotopic compositions of a large number of elements in SiC X grains can be satisfactorily explained by mixing matter from different SN layers in appropriate proportions (Hoppe *et al.* 2000), although there are problems in accounting quantitatively for some of the observed isotopic signatures (e.g. the ^{15}N enrichments; see Fig. 2.6). The close isotopic relationship of the presolar Si_3N_4 grains and of many low-density graphite grains with the SiC X grains suggests a SN origin also for these grains (Nittler *et al.* 1995; Travaglio *et al.* 1999). The same holds for presolar oxide grains with a large excess in ^{16}O. In addition, most (if not all) Group 4 oxide and silicate grains are probably from Type II supernovae as indicated by the enrichments in ^{18}O and Si-isotopic signatures of silicate grains (Vollmer *et al.* 2008) and by multi-element (O, Mg, Ca, K) isotope data from a couple of oxide grains (Nittler *et al.* 2008).

The most likely stellar sources of presolar grains and their relative contributions are summarized in Table 2.1. About 90% of the presolar grains appear to come from AGB stars, in agreement with the relative contribution of dust injection into the ISM from these stars (see Fig. 2.3). The abundance of carbonaceous grains (c. 10–20% not considering diamonds) is lower than expected (about one third) but this may be the result of the preferential destruction of C-rich dust in the oxidizing environment of the solar nebula and the meteorite and IDP parent bodies. Only stars in the mass range 1–3 $M_\odot$ apparently have contributed to the population of presolar dust grains from AGB stars. This contrasts with the expectation that AGB stars in the mass range 4–8 $M_\odot$ should inject similar amounts of dust into the ISM (Gail *et al.* 2009). A solution to this apparent puzzle is still outstanding.

2.2.5 Presolar organics

Organic matter is an important constituent of IDPs and carbonaceous chondrites. Most of this is insoluble high-molecular-weight macromolecular material (insoluble organic matter, "IOM"). Hydrogen and nitrogen isotope studies of bulk organic matter revealed enrichments in D and ^{15}N, especially in the IOM. Ion microprobe imaging of H- and N-isotopic ratios in IDPs and meteoritic IOM finally showed hotspots of very large D and ^{15}N enrichments (D/H up to 25 times the terrestrial ratio, ^{15}N/^{14}N up to 4 times the terrestrial ratio) at the micrometer scale (Busemann *et al.* 2006; Messenger 2000). Specific carriers could not be determined in these studies, but the absence of larger C-isotopic anomalies excludes stardust as a carrier. Deuterium and ^{15}N hotspots are spatially not correlated, indicating a variety of formation processes.

Large D enrichments are observed in cold interstellar clouds and large ^{15}N enrichments are predicted for these environments. The observed D and ^{15}N anomalies in primitive Solar System material suggest that at least some fraction of the organic matter has an interstellar origin or is material that formed in the cold outer regions of the protoplanetary disk. Possible mechanisms to produce large D and ^{15}N enrichments in either environment are low-temperature (10 K) ion–molecule reactions in the gas phase and catalytic processes on dust grains.

Enrichments in D and ^{15}N are also evident for organic globules, mostly sub-micrometer in size, in the Tagish Lake meteorite (Nakamura–Messenger *et al.* 2006). It has been proposed that these globules originated as organic ice coatings on pre-existing grains that were photochemically processed into refractory organic matter. That stardust grains may be covered by coatings of organic material in the ISM is also demonstrated by the discovery of a presolar SN olivine which is partly encased by ^{15}N-rich carbonaceous matter (Messenger *et al.* 2005).

2.3 Star formation

2.3.1 Protostars and protostellar collapse

Planetary systems are now generally believed to be by-products of the process of star formation. Star formation, therefore, is the natural starting point for discussions of planet formation. Almost all stars are born as members of stellar clusters that, in turn, are born in molecular clouds. Formation of isolated stars seems to be possible according to observations, but this is a rare process. Whether the Sun and its associated planetary system formed in isolation or as member of a cluster is not known; some indications hint to formation in a cluster (see Hester & Desch 2005; Gounelle & Meibom 2008; and Chapter 9, this volume).

How star formation is initiated is still not completely clear, but seems to be related to the highly supersonic turbulent flows in a molecular cloud (e.g. McKee & Ostriker 2007). Local density enhancements due to compression of finite volumes of material within the flow become gravitationally unstable, if their size λ exceeds the critical Jeans-length $\lambda_J = (\pi c_s^2 / G\rho)^{1/2}$, where $c_s^2 = kT/\mu$ is the isothermal sound velocity, T the gas temperature in a molecular cloud, typically $10\,K$, μ mean molecular weight ($\approx 7/3m_H$), G is the gravitational constant, and ρ the local density. If $\lambda > \lambda_J$, thermal pressure cannot resist self-gravity and a runaway collapse ensues. This results in an isothermal near free-fall contraction of the material, during which its density ρ increases by at least 15 orders of magnitude. The contraction continues until the central region becomes opaque and compressional heating starts to exceed radiative cooling of the gas–dust mixture, causing the temperature to increase. The free-fall timescale during this phase is $t_{ff} = (3\pi/32G\rho)^{1/2}$ for a homogeneous initial mass distribution, which amounts to

$$t_{ff} = 6.4 \times 10^4 \left(\frac{M}{M_\odot}\right)\left(\frac{10\,K}{T}\right)^{3/2} \text{[yr]} \tag{2.2}$$

For more concentrated initial distributions, the collapse time is even shorter.

During this initial collapse phase the rotational degrees of freedom of H_2 are not excited ($\Delta E_{exc} = 512\,K$ for the $J = 2-0$ transition). If the temperature approaches $100\,K$ the ratio $\gamma = c_p/c_v$ of specific heats changes from $\gamma = 5/3$ (rotation of H_2 not excited), to $\gamma = 7/5$ (rotation excited). The critical γ for a self-gravitating adiabatic gaseous sphere to be stable is $\gamma = 4/3$. The pressure in the central core then supports the matter against further collapse and the inflow steepens to a shock at the border of the opaque core. The "first core" has formed.

The continued addition of matter increases the density and temperature of the core until H_2 begins to dissociate. The dissociation consumes heat, which holds temperature approximately constant, i.e. the heat capacity becomes very high and $\gamma \rightarrow 1$. The stability condition $\gamma > 4/3$ becomes violated and a new collapse of the core ensues. The core collapses until all H_2 is dissociated and the H finally becomes ionized. The temperature then increases again with further contraction and the "second core" is formed that approaches stellar density. The second collapse phase is short and lasts for a solar-type star of the order of 10^3 years. By this event a protostellar embryo is born, which continues to grow in mass by collecting the remaining material from its environment.

Table 2.2 shows timescales required to increase the mass by given mass increments and average mass-infall rates during these periods for three different final masses of low-mass stars. The formation of the initial stellar embryo requires about one free-fall time and is very short. The subsequent addition of 80% of the final mass requires about four free-fall times, i.e. is rather rapid. The mass accretion rate

Table 2.2 *Time intervals* Δt *required to increase the mass of a protostar from the lower to the upper end of the indicated intervals of fractions of the final mass, and average accretion rate during that period*

Mass fraction	0.1 $M_\odot$		0.5 $M_\odot$		1.0 $M_\odot$	
	Δt kyr	$\dot{M}_{av}$ $M_\odot\,yr^{-1}$	Δt kyr	$\dot{M}_{av}$ $M_\odot\,yr^{-1}$	Δt kyr	$\dot{M}_{av}$ $M_\odot\,yr^{-1}$
0–0.8	30	2.7×10^{-6}	126	3.2×10^{-6}	189	4.2×10^{-6}
0.8–0.9	11	8.8×10^{-7}	52	9.8×10^{-7}	85	1.2×10^{-6}
0.9–0.99	32	2.7×10^{-7}	144	3.0×10^{-7}	280	3.2×10^{-7}
0.99–0.999	28	3.0×10^{-8}	139	3.2×10^{-8}	264	3.4×10^{-8}

Note: Results of Wuchterl & Tscharnuter (2003) for three different final masses and for spherically symmetrical collapse.

during this phase is very high, with a peak value as high as $10^{-4}\,M_\odot\,yr^{-1}$ during the very initial build-up of the star, but rapidly declining. The average mass-infall rate during the initial collapse phase is several times $10^{-6}\,M_\odot\,yr^{-1}$. The collapse of the last 20% of mass towards the mass center takes place on a much longer timescale, about 0.5 Myr for a solar mass star, with a rapidly declining rate.

This describes the final steps of stellar evolution of single stars. Since stars are formed in clusters, the initially unstable region in the molecular cloud has to fragment into many collapsing subunits. How this works is a yet unsolved problem. Additionally, a large fraction of stars are formed with at least one companion star; this process is also not well understood. Both problems remain important challenges in our understanding of star formation (McKee & Ostriker 2007).

2.3.2 Disk formation

The material of a gravitationally unstable region of a molecular cloud carries inevitably some angular momentum. This prevents the material from collapsing exactly towards the mass center of the local mass condensation. Conservation of angular momentum requires an infalling gas parcel to move on an elliptic orbit around the center, the distance of closest approach being more distant from the center the higher the angular momentum is (for details, see Cassen & Moosman 1981, and Chapter 3). Since matter is falling in from all sides, symmetry requires the gas parcels to collide at a plane through the mass center and orthogonal to the total angular momentum vector of the cloud. They are stopped there and a flat rotating gas disk around the growing star starts to form. For a uniformly rotating cloud, material starting infall close to the symmetry axis, i.e. having low angular momentum, falls directly onto the star, while material starting at some distance

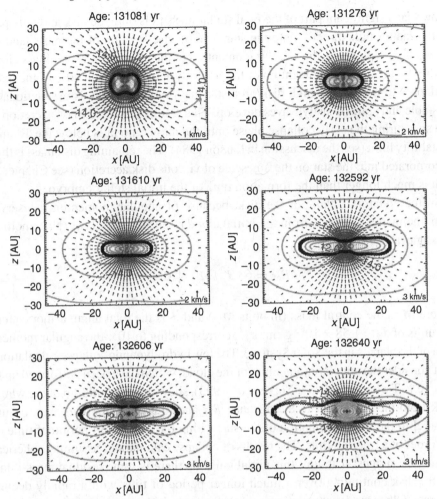

Figure 2.10 Cylindrically symmetric hydrodynamical model of accretion flow with rotation during the early collapse phase, showing the inflow of matter in the meridional plane and the build-up of a flat rotating disk structure after about 1.05 free-fall times. Arrows indicate matter flow direction and velocity, gray lines indicate cuts of isodensity surfaces with meridional plane. Dark crosses outline locations of supersonic to subsonic transition of inflow velocity; this corresponds to the position of the accretion shock. Matter falling along the polar axis and within the equatorial plane arrive within 1600 yr almost simultaneously, which results in an almost instantaneous formation of an extended initial accretion disk [new model calculation following the methods in Tscharnuter (1987), figure kindly contributed by W. M. Tscharnuter].

from the symmetry axis, i.e. having significant angular momentum, misses the star and falls onto the gas disk. Fig. 2.10 shows the initial stages of the build-up of a star and the associated gas-disk, as it is found from two-dimensional cylindrically symmetric hydrodynamic model calculations (Tscharnuter 1987). Model calculations show that a significant part of the final stellar mass falls onto the disk and only part of the matter falls directly onto the star. The precise fraction of the mass landing on the disk depends on the angular momentum of the cloud. For a solar mass cloud with $10\times$ the angular momentum of the Solar System, more than 90% of the mass collapses to a disk. For high disk/protostar mass ratios the disk is gravitationally unstable and spiral density waves develop. This results in a rapid mass accretion to the star during the early collapse phase until the mass ratio falls below the Toomre instability limit (see the discussion in Larson 1984). The remaining disk mass is then incorporated into the star on the timescale of viscous disk accretion (see Chapter 3) that is much longer than the formation time of the initial stellar embryo.

In the equatorial plane of the disk, because of angular momentum conservation, the matter contracts from the initial cloud radius to a radius of (Nakamoto & Nakagawa 1995)

$$R_{cf} = 63\, J^2/M^3 \quad [AU] \tag{2.3}$$

where M is the central mass (in units $M_\odot$) and J is the total angular momentum (in units of $J_0 = 1.58 \times 10^{53}\,\mathrm{g\,cm^2\,s^{-1}}$, corresponding to $10\times$ the angular momentum of the planets in the Solar System). The few hydrodynamic collapse calculations that consider the angular momentum of the cloud, which have been performed up to now (e.g. Tscharnuter 1987; Yorke *et al.* 1993), show that the build-up of the whole disk from the center out to the distance R_{cf} and to most of its final mass occurs parallel to the initial formation of the star within the same short period of time of the order of a free-fall time. The results of collapse calculations assuming spherical symmetry suggest that after this initial build-up there is some residual mass-infall to this disk continuing over a much longer period of time with a rapidly declining rate. Other model calculations, e.g. Nakamoto & Nakagawa (1995), consider a slow inside-out build-up of the accretion disk from the star to R_{cf} over much longer timescales (6.4×10^5 yr for a solar-type star), based on the analytic collapse model of Shu (1977). Such models likely do not represent the real mode of formation of accretion disks.

The fraction of mass that failed to fall on the stellar embryo but formed an accretion disk cannot directly be incorporated into the star. Its high angular momentum has to be preserved and this prevents it from spiraling in toward the star. An inward migration requires that angular momentum transport processes within the disk transfer some fraction of the angular momentum to matter lying further outward in order that part of the matter can migrate inward. The processes of matter transport in a protostellar disk are slow and for this reason part of the collapsed matter is stored for a long period of time in an accretion disk (see Chapter 3). This inefficiency of accretion makes planet formation possible.

2.3.3 Young stellar objects

Young stellar objects are frequently classified into four classes that characterize four important evolutionary stages:

Class 0. Sources with a very faint protostar in the optical and near-infrared spectral region, but with considerable luminosity in the submillimeter spectral region. During this phase the molecular cloud material starts to collapse and to build up a stellar embryo and its associated disk.

Class I. Sources with average spectral index

$$\alpha_{IR} = \frac{d \log \lambda F_\lambda}{d \log \lambda} > 0 \tag{2.4}$$

in the wavelength region 2.5 μm $< \lambda <$ 10–25 μm. These are protostars with circumstellar disks and envelopes. During this phase stars accrete most of their final mass from the disk and the surrounding envelope.

Class II. Sources with spectral index $-1.5 < \alpha_{IR} < 0$. These are pre-main-sequence stars with observable accretion discs (classical T Tauri stars).

Class III. Sources with spectral index $\alpha_{IR} < -1.5$. These are pre-main-sequence stars without detectable accretion (weak-lined T Tauri stars).

The initial evolutionary stages (Classes 0 and I) are optically hidden by the dust of the collapsing envelope. These phases can be only observed in the far-infrared. Optically visible are the Class II and III objects, that form the long-known class of T Tauri stars. They have reached almost their final mass ($\approx 90\%$; see Beckwith *et al.* 1990) at the transition from Class I to Class II, which occurs for solar-mass stars approximately 2×10^5 yr after the onset of collapse (see Table 2.2).

After a few million years of evolution, where most of the remaining disk mass is accreted to the T Tauri star, the residual disk is dispersed (see Chapter 8) and the star continues its further evolution to the main sequence as a Class III object. The processes going on in these disks, that are usually called *protoplanetary* during this final phase of protostar evolution, are the subject of the following chapters of this book.

The observed properties of some Class II objects and their accretion disks are shown in Table 2.3. The properties and evolution of accretion disks are discussed in detail in Chapter 3.

All young stellar objects show indications for stellar winds and outflows. These phenomena are always observed to occur in systems that undergo mass accretion that interacts with magnetic fields and rotation. They are not limited to star formation but are also observed in other cases, e.g. during accretion onto central black holes in galaxies.

Table 2.3 *Observed properties of some young stellar objects and their accretion disks: spectral type, effective temperature T_{eff}, luminosity L_*, estimated stellar mass M_*, stellar radius R_*, accretion rate $\dot{M}$, disk radius R_{disk} as observed by dust emission, inclination of disk with respect to sight line, and disk mass M_{disk} estimated from submillimeter dust emission*

Object	Spect. type	T_{eff} [K]	L_* [$L_\odot$]	M_* [$M_\odot$]	R_* [$R_\odot$]	$\dot{M}$ [$M_\odot$ yr^{-1}]	R_{disk} [AU]	Incl. [degr.]	M_{disk} [$M_\odot$]
TW Hya	K7	4009	0.25	0.7	1.8	$5.0 \cdot 10^{-10}$	448	0	0.03
BP Tau	K7	4060	0.83	0.77	1.9	$1.3 \cdot 10^{-8}$	108	28	0.0012
DG Tau	K3	4775	3.63	2.2	2.8	$7.4 \cdot 10^{-7}$	550	75	0.03
GM Aur	K3	4730	1.01	1.22	1.5	$6.6 \cdot 10^{-9}$	525	54	0.07
LkCa 15	K2	4350	1.05	1.05	1.5	$1.5 \cdot 10^{-9}$	190	57	0.01

Data from: Akeson *et al.* 2005, Dutrey *et al.* 2003, Muzerolle *et al.* 2000, Sargent *et al.* 2006, and Qi *et al.* 2003.

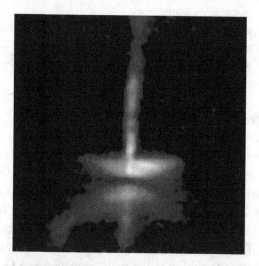

Figure 2.11 The HH-30 object, a young stellar object showing two thin jets flowing out from the central region of an accretion disk. The outflow velocity in the jets is 90–270 km s^{-1}. The two bowl-shaped regions are starlight scattered by the dust in the uppermost layers of the disk. The dark lane in between is the accretion disk seen side-on. The radial optical depth in the disk is too high for starlight to penetrate in this direction. The radial extension of the disk is 425 AU. (Photo credit: Hubble Space Telescope, NASA/ESA and STScI.)

In heavily obscured regions with ongoing star formation one observes the so-called Herbig–Haro (HH) objects: thin collimated jets of matter rapidly flowing (up to several hundred kilometers per second) out from young stellar objects. An example is shown in Fig. 2.11. These jets are mainly associated with Class 0 and I objects but sometimes are also observed for T Tauri stars. The outflows interact with the surrounding matter and previously ejected matter, forming shocks that result in a complex and rapidly time-varying structure. The jets are finally stopped by interactions with the ISM matter in a bow-shock, where the ISM material is pushed aside to pave the way for the following jet material. The shock-heated material emits a rich spectrum of emission lines that can easily be observed and serves to identify HH objects.

The jets seem to be launched in the central region of the objects, either by the star itself or by the disk surface close to the star and it is thought that magnetic fields are responsible for the collimation of the thin jets. The detailed processes underlying the phenomenon are not well understood because of the inherent difficulties of treating magneto-hydrodynamic problems.

Herbig–Haro objects are often found to be associated with molecular outflows observed in rotational lines of cosmically abundant molecules (e.g. CO, NH_3) that come from regions enveloping the jet. The outflow velocity is generally much lower. The origin of these molecular outflows is not clear; either they result from winds from the accretion disk or from the interaction between the jet and its surroundings.

Some further discussions on outflows are given in Chapters 3 and 8. A detailed overview on the present state of theory and observation of jets and bipolar outflows is found in Königl & Ruden (1993), Königl & Pudritz (2000), and Pudritz *et al.* (2007).

2.3.4 Dust composition and chemistry

The material falling from the parent molecular cloud directly on the forming star is rapidly dissociated and ionized and solids are vaporized, once the matter passes the accretion shock on the stellar surface. The material falling on the inner parts of the accretion disk suffers the same fate, since the matter has to pass through a standing shockwave on the surface of the disk. At this shock the infall of matter is stopped and the flow characteristics change from infall of envelope material to nearly Keplerian rotation of disk material. The location of this shock is shown in Fig. 2.10 as the heavy line marked by crosses.

The strength of the shock becomes gradually weaker with increasing distance from the center, because matter falls less deeply into the potential well of the forming star. In the outer parts of the disk, temperatures after shock passage are not high enough to vaporize the solids or to ionize the gas. Infalling matter can cross the shock without the destruction of the dust component. The metamorphosis of matter during passage of the accretion shock has been studied by Neufeld & Hollenbach (1994). They calculated for instance that the temperature of dust grains that first increases by frictional heating and ultraviolet radiation in the shock layer, later decreases again by radiative cooling. The minimum shock velocity required

Table 2.4 *Vaporization temperature* T_{vap}; *minimum shock velocity* v_{sh} *required for the onset of vaporization of solids, refractory and volatile organics, and water ice; and radius r inside of which the infall velocity component perpendicular to the disk surface exceeds* v_{sh}, *according to Neufeld & Hollenbach (1994)*

	Iron	Silicate	Troilite	Refractory organics	Volatile organics	Water ice
T_{vap} (K)	1395	1331	700	575	375	180
v_{sh} (km s^{-1})	>100	57	55	40	30	10
r (AU)	<0.015	0.046	0.049	0.092	0.16	1.5

[1] For a pre-shock density of $n_0 = 10^9\,\mathrm{cm}^{-3}$ and $M = 0.5\,M_\odot$

for vaporization, and the maximum radius out to which dust is vaporized during shock passage of infalling matter are shown in Table 2.4.

With respect to the gas-phase chemistry, Neufeld & Hollenbach (1994) found that for shock velocities $\gtrsim 20\,\mathrm{km\,s}^{-1}$ chemical reactions are activated that drive the composition of the gas phase to chemical equilibrium. In particular, after the passage of the accretion shock all C not bound in the organics is converted into CO and all O not bound into CO and silicates is bound in H_2O. The temperature of the organics stays lower than 150 K for shock velocities not exceeding $4\,\mathrm{km\,s}^{-1}$, which is satisfied for distances beyond $\approx 9\,\mathrm{AU}$. The isotopic anomalies in D and N carried by the organic material (see Section 2.2.5) will essentially be preserved during shock passage in that region because grains stay too cold for the activation of equilibration processes.

The dust composition entering the accretion disk, therefore, is identical to the composition of the dust in the molecular cloud from which the star forms, except for a region close to the star, where dust is destroyed during the passage through the accretion shock. Even ices survive in most parts of the disk. This means that the dust in the accretion disk is initially unmodified interstellar dust, coated possibly by ices formed in the molecular cloud. Because accretion of disk matter to the star by viscous transport processes liberates significant amounts of gravitational energy, which is converted into heat, the inner parts of the disk are hot; the dust is subject to a number of processes operating within the disk that change its structure and composition, or even destroy the dust, at much bigger distances as indicated in Table 2.4. These processes are the subject of several later chapters. Only the cool outer part of the disk may largely consist of interstellar dust. If so, the dust component of the cometary bodies should consist in large part of unmodified interstellar matter. This view, however, contrasts with results from recent laboratory studies of dust grains from comet Wild 2 returned by NASA's Stardust mission which demonstrated that this comet contains an un-equilibrated assortment of materials that have both presolar and Solar System origins (Brownlee *et al.* 2006; Wooden

Table 2.5 *Abundant dust species in the outer parts of protoplanetary accretion disks, the fraction of the abundant elements condensed in the dust species, the stoichiometric coefficient x for the silicates, and their respective melting temperatures (from Pollack et al. 1994)*

Name	Composition	Fractional abundance						Melting point (K)
		Mg	Fe	Si	S	C	x	
Olivine	$Mg_{2x}Fe_{2(1-x)}SiO_4$	0.83	0.42	0.63			0.7	2180
Pyroxene	$Mg_xFe_{1-x}SiO_3$	0.17	0.09	0.27			0.7	1850
Quartz	SiO_2			0.10				1996
Iron	Fe		0.10					1811
Troilite	FeS		0.39		0.75			1460
Kerogen	CHON					0.55		—

2008). Also, observations of crystalline silicates in the comets Halley (Swamy *et al.* 1988) and Tempel 1 (Harker *et al.* 2005) suggest that a large fraction of cometary dust was processed or formed in the hot inner regions of the protoplanetary disk and transported to the region where the comets formed.

Pollack et al. (1994) tried to estimate the likely dust mixture that entered the Solar System accretion disk. The model they proposed is shown in Table 2.5. It considers only the most abundant elements C, N, O, Mg, Si, Fe, and S. These are mainly condensed into dust that formed in the ISM; stardust is only a small component (Zhukovska *et al.* 2008). The less abundant refractory elements are neglected in the model, particularly the important elements Al, Ca, Ni, and Na, since nothing can presently be substantiated about their occurrence in dust grown in the ISM. They are depleted from the ISM gas phase, but it is not known which dust component bears these elements. The components in the model of Pollack *et al.* were selected on the basis of observations of cometary material (comet Halley), analysis of IDPs, and abundance constraints following from the known depletion of refractory elements. The material denoted as kerogen is a carbonaceous material containing a high fraction of H, O, and N (average composition $C_{100}H_{80}O_{20}N_4$). There are two components: a refractory with rather high vaporization temperature, and a volatile with lower vaporization temperature (see Table 2.4), but not as volatile as ices. From the volatile elements O, C and N which are involved in the formation of the abundant condensed phases, only some fraction is condensed in solids. The other part stays in the gas-phase N_2 or forms ices (H_2O, CO) at low temperature. Because of the rather uniform evolution of elemental abundances in the Milky Way, this dust mixture is likely to be representative for dust in all accretion disks around young stars. Substantially different mixtures can not be expected to exist anywhere.[5]

---

[5] Of course, accretion disks in binaries containing a very evolved object that transferred mass to the less evolved object, which then forms an accretion disk before being incorporated into the star, may be exceptions.

The model is certainly a very crude approximation of the initial composition of protoplanetary dust before it becomes subject to the numerous metamorphism processes that determine the composition and structure of the solids in small and big bodies of a planetary system, but no better model is presently known. Continued examination of IDPs and cometary material will probably improve the situation in the near future.

Acknowledgements This work was supported in part by Forschergruppe 759, sponsored by the Deutsche Forschungsgemeinschaft (DFG).

References

Akeson, R. L., Boden, A. F., Monnier, J. D., *et al.* 2003, *Astrophysical Journal*, **635**, 1173.

Alexander, C. M. O. D., Boss, A. P., Keller, L. P., Nuth, J. A., & Weinberger, A. 2007, in *Protostars and Planets V*, ed. B. Reipurth, D. Jewitt, & K. Keil, The University of Arizona Press, 801–813.

Amari, S., Anders, E., Virag, A., & Zinner, E. 1990, *Nature*, **345**, 238.

Amari, S., Nittler, L. R., Zinner, E., *et al.* 2001a, *Astrophysical Journal*, **546**, 248.

Amari, S., Nittler, L. R., Zinner, E., *et al.* 2001b, *Astrophysical Journal*, **551**, 1065.

Amari, S., Nittler, L. R., Zinner, E., Lodders, K., & Lewis, R. S. 2001c, *Astrophysical Journal*, **559**, 463.

Anders, E. & Zinner, E. 1989, *Meteoritics*, **28**, 490.

Beckwith, S. V. W., Sargent, A. J., Chini, I., & Güsten, R. 1990, *Astronomical Journal*, **99**, 924.

Bernatowicz, T., Amari, S., Zinner, E. K., & Lewis, R. S. 1991, *Astrophysical Journal*, **373**, L73.

Bernatowicz, T., Fraundorf, G., Ming, T., *et al.* 1987, *Nature*, **330**, 728.

Bianchi, S. & Schneider, R. 2007, *Monthly Notices of the Royal Astronomical Society*, **378**, 973.

Blitz, L., Fukui, Y., Kawamura, A., *et al.* 2007, in *Protostars and Planets V*, ed. B. Reipurt, D. Jewitt, & K. Keil, University of Arizona Press, 81–96.

Boothroyd, A. I. & Sackmann, I.-J. 1999, *Astrophysical Journal*, **510**, 232.

Boulanger, F. & Pérault, M. 1988, *Astrophysical Journal*, **330**, 964.

Bradley, J. 2003, in *Astromineralogy*, ed. T. Henning, Lecture Note in Physics, vol. 609, Springer 217–235.

Brownlee, D., Tsou, P., Aléon, J., *et al.* 2006, *Science*, **314**, 1711.

Busemann, H., Young, A. F., Alexander, C. M. O., *et al.* 2006, *Science*, **312**, 727.

Busso, M., Gallino, R., & Wasserburg, G. J. 1999, *Annual Review of Astronomy and Astrophysics*, **37**, 239.

Cameron, A. G. W. 1973, in *Interstellar Dust and Related Topics*, ed. J. M. Greenberg & H. C. van de Hulst, Reidel, 545–547.

Cassen, P. & Moosman, A. 1981, *Icarus*, **48**, 353.

Croat, T. K., Stadermann, F. J., & Bernatowicz, T. J. 2005, *Astrophysical Journal*, **631**, 976.

Crowther, P. A. 2007, *Annual Review of Astronomy and Astrophysics*, **45**, 177.

Demyk, K., Carrez, P., Leroux, H., *et al.* 2001, *Astronomy and Astrophysics*, **368**, L38.

Dickey, J. M. 1993, *Astronomical Society of the Pacific Conference Series*, **39**, 93.

Donn, B. 1968, *Astrophysical Journal*, **152**, L129.

Dorschner, J. 2003, in *Astromineralogy*, ed. T. Henning, Lecture Note in Physics, vol. 609, Springer, 1–54.

Draine, B. T. 1990, in *The Evolution of the Interstellar Medium and Intergalactic Medium.*, ed. L. Blitz ASP, 193–205.

Draine, B. T. 1995, *Astrophysics and Space Science*, **233**, 111.

Draine, B. T. 2003, *Annual Review of Astronomy and Astrophysics*, **41**, 241.

Draine, B. T. 2004, in *The Cold Universe*, ed. A. W. Blain, F. Combes, B. T. Draine, D. Pfenniger, & Y. Revaz, Saas-Fee Advanced Course 32, Springer, 213–304.

Draine, B. T. & Lee, H. M. 1984, *Astrophysical Journal*, **285**, 89.

Draine, B. T. & Salpeter, E. E. 1979, *Astrophysical Journal*, **231**, 438.

Dutrey, A., Guilloteau, S., & Simon, M. 2003, *Astronomy and Astrophysics*, **402**, 1003.

Ferrarotti, A. S. & Gail, H.-P. 2006, *Astronomy and Astrophysics*, **447**, 553.

Field, G. B. 1974, *Astrophysical Journal*, **187**, 453.

Fitzpatrick, E. L. & Massa, D. 2007, *Astrophysical Journal*, **663**, 320.

Friedemann, C. 1969, *Astronomische Nachrichten*, **291**, 177.

Gail, H.-P., Zhukovska, S. V., Hoppe, P., & Trieloff, M. 2009, *Astrophysical Journal*, **698**, 1136.

Gillett, F. C., Forrest, W. J., & Merrill, K. M. 1973, *Astrophysical Journal*, **183**, 87.

Gilman, R. C. 1969, *Astrophysical Journal*, **155**, L185.

Gounelle, M. & Meibom, A. 2008, *Astrophysical Journal*, **680**, 781.

Gyngard, F., Amari, S., Zinner, E., & Lewis, R. S. 2007, *Meteoritics and Planetary Science Supplement*, **42**, 5068.

Harker, D. E., Woodward, C. E., & Wooden, D. H. 2005, *Science*, **310**, 278.

Heck, P. R., Gyngard, F., Meier, M. M. M., et al. 2008, *Lunar and Planetary Institute Conference Abstracts*, **39**, 1239.

Hester, J. J. & Desch, S. J. 2005, *Astronomical Society of the Pacific Conference Series*, **341**, 107.

Higdon, J. C. & Lingenfelter, R. E. 2005, *Astrophysical Journal*, **628**, 738.

Hoppe, P., Amari, S., Zinner, E., Ireland, T., & Lewis, R. S. 1994, *Astrophysical Journal*, **430**, 870.

Hoppe, P., Annen, P., Strebel, R., et al. 1997, *Astrophysical Journal*, **487**, L101.

Hoppe, P., Strebel, R., Eberhardt, P., Amari, S., & Lewis, R. S. 2000, *Meteorititics and Planetary Science*, **35**, 1157.

Hoyle, F. & Wickramasinghe, N. C. 1962, *Monthly Notices of the Royal Astronomical Society*, **124**, 417.

Hoyle, F. & Wickramasinghe, N. C. 1969, *Nature*, **223**, 459.

Hoyle, F. & Wickramasinghe, N. C. 1970, *Nature*, **226**, 62.

Hutcheon, I. D., Huss, G. R., Fahey, A. J., & Wasserburg, G. J. 1994, *Astrophysical Journal*, **425**, L97.

Jadhav, M., Amari, S., Zinner, E., & Maruoka, T. 2006, *New Astronomy Reviews*, **50**, 591.

Jones, A. P., Tielens, A. G. G. M., & Hollenbach, D. J. 1996, *Astrophysical Journal*, **469**, 740.

Jones, A. P., Tielens, A. G. G. M., Hollenbach, D. J., & McKee, C. F. 1994, *Astrophysical Journal*, **433**, 797.

José, J., Hernanz, M., Amari, S., Lodders, K., & Zinner, E. 2004, *Astrophysical Journal*, **612**, 414.

Kemper, F., Vriend, W. J., & Tielens, A. G. G. M. 2005, *Astrophysical Journal*, **633**, 534.

Königl, A. & Pudritz, R. E. 2000, in *Protostars and Planets IV*, ed. V. Mannings, A. P. Boss, & S. S. Russell, University of Arizona Press, 759–787.

Königl, A. & Ruden, S. P. 1993, in *Protostars and Planets III*, ed. E. H. Levy & J. I. Lunine, University of Arizona Press, 641–687.

Lada, C. J. & Lada, E. A. 2003, *Annual Review of Astronomy and Astrophysics*, **41**, 57.

Larson, R. B. 1984, *Monthly Notices of the Royal Astronomical Society*, **206**, 197.

Léger, A. & Puget, J. L. 1984, *Astronomy and Astrophysics*, **137**, L5.

Lewis, R. S., Tang, M., Wacker, J. F., Anders, E., & Steel, E. 1987, *Nature*, **326**, 160.

Li, A. 2005, *Journal of Physics: Conference Series*, **6**, 229.

Lindblad, B. 1935, *Nature*, **135**, 133.

Lodders, K. & Amari, S. 2005, *Chemie der Erde*, **65**, 93.

Lugaro, M., Davis, A. M., Gallino, R., *et al.* 2003, *Astrophysical Journal*, **593**, 486.

Mathis, J. S., Rumpl, W., & Nordsieck, K. H. 1977, *Astrophysical Journal*, **217**, 425.

Matteucci, F. 2007, *The Chemical Evolution of the Galaxy*, Kluwer Academic Publishers.

McKee, C. F. & Ostriker, E. C. 2007, *Annual Review of Astronomy and Astrophysics*, **45**, 565.

McKeegan, K. D., Aléon, J., Bradley, J., *et al.* 2006, *Science*, **314**, 1724.

Merrill, P. W. 1952, *Astrophysical Journal*, **116**, 21.

Messenger, S. 2000, *Nature*, **404**, 968.

Messenger, S., Keller, L. P., & Lauretta, D. S. 2005, *Science*, **309**, 737.

Messenger, S., Keller, L. P., Stadermann, F., Walker, R. M., & Zinner, E. 2003, *Science*, **300**, 105.

Min, M., Waters, L. B. F. M., de Koter, A., *et al.* 2007, *Astronomy and Astrophysics*, **462**, 667; erratum: *Astronomy and Astrophysics*, **486**, 779 (2008).

Molster, F. J. & Waters, L. B. F. M. 2003, in *Astromineralogy*, ed. T. Henning, Lecture Note in Physics, vol. 609, Springer, 121–170.

Morton, D. C., Drake, J. F., Jenkins, E. B., *et al.* 1973, *Astrophysical Journal*, **181**, L103.

Muzerolle, J., Calvet, N., Briceno, C., Hartmann, L., & Hillenbrand, L. 2000, *Astrophysical Journal*, **535**, L47.

Nakamoto, T. & Nakagawa, Y. 1995, *Astrophysical Journal*, **445**, 330.

Nakamura-Messenger, K., Messenger, S., Keller, L. P., Clemett, S. J., & Zolensky, M. E. 2006, *Science*, **314**, 1439.

Neufeld, D. A. & Hollenbach, D. J. 1994, *Astrophysical Journal*, **428**, 170.

Nittler, L. R., Alexander, C. M. O., Gao, X., Walker, R. M., & Zinner, E. 1997, *Astrophysical Journal*, **483**, 475.

Nittler, L. R., Alexander, C. M. O. D., Gao, X., Walker, R. M., & Zinner, E. K. 1994, *Nature*, **370**, 443.

Nittler, L. R., Alexander, C. M. O. D., Gallino, R., *et al.* 2008, *Astrophysical Journal*, **682**, 1450.

Nittler, L. R., Hoppe, P., Alexander, C. M. O. D., *et al.* 1995, *Astrophysical Journal*, **453**, L25.

Nollett, K. M., Busso, M., & Wasserburg, G. J. 2003, *Astrophysical Journal*, **582**, 1036.

Oort, J. H. & van de Hulst, H. C. 1946, *Bulletin of the Astronomical Institutes of the Netherlands*, **10**, 187.

Pollack, J. B., Hollenbach, D., Beckwith, S., *et al.* 1994, *Astrophysical Journal*, **421**, 615.

Pont, F. & Eyer, L. 2004, *Monthly Notices of the Royal Astronomical Society*, **351**, 487.

Pudritz, R. E., Ouyed, R., Fendt, C., & Brandenburg, A. 2007, in *Protostars and Planets V*, ed. B. Reipurth, D. Jewitt, & K. Keil, University of Arizona Press, 277–294.

Qi, C., Kessler, J. E., Koerner, D. W., Sargent, A. I., & Blake, G. A. 2003, *Astrophysical Journal*, **597**, 986.

Rana, N. 1991, *Annual Review of Astronomy and Astrophysics*, **29**, 129.

Rauscher, T., Heger, A., Hoffman, R. D., & Woosley, S. E. 2002, *Astrophysical Journal*, **576**, 323.

Sargent, B., Forrest, W. J., D'Alessio, P., *et al.* 2006, *Astrophysical Journal*, **645**, 395.

Savage, B. D. & Sembach, K. R. 1996, *Annual Review of Astronomy and Astrophysics*, **34**, 280.

Savina, M. R., Davis, A. M., Tripa, C. E., *et al.* 2004, *Science*, **303**, 649.

Schalèn, C. A. W. 1934, in *Kungliga Svenska Vetenskapsakademier Handlingar*, Ser. III, **13**(2), 47.

Shu, F. 1977, *Astrophysical Journal*, **214**, 488.

Sloan, G. C., Kraemer, K. E., Matsuura, M., *et al.* 2006, *Astrophysical Journal*, **645**, 1118.

Snow, Jr., T. P. 1975, *Astrophysical Journal*, **202**, L87.

Stecher, T. P. & Donn, B. 1965, *Astrophysical Journal*, **142**, 1681.

Stern, S. A. 2003, *Nature*, **424**, 639.

Swamy, K. K. S., Sandford, S. A., Allamandola, L. J., Witteborn, F. C., & Bregman, J. D. 1988, *Icarus*, **75**, 351.

Tang, M. & Anders, E. 1988, *Geochimica et Cosmochimica Acta*, **52**, 1235.

Tielens, A. G. G. M. 1999, in *Formation and Evolution of Solids in Space*, ed. J. M. Greenberg & A. Li, Kluwer, 331–375.

Tielens, A. G. G. M. 2005, *The Physics and Chemistry of the Interstellar Medium*, Cambridge University Press.

Tielens, A. G. G. M., McKee, C. F., Seab, C. G., & Hollenbach, D. J. 1994, *Astrophysical Journal*, **431**, 321.

Travaglio, C., Gallino, R., Amari, S., *et al.* 1999, *Astrophysical Journal*, **510**, 325.

Treffers, R. & Cohen, M. 1974, *Astrophysical Journal*, **188**, 545.

Trümpler, R. J. 1930, *Publications of the Astronomical Society of the Pacific*, **42**, 214.

Tscharnuter, W. M. 1987, *Astronomy and Astrophysics*, **188**, 55.

Vollmer, C., Hoppe, P., & Brenker, F. 2008, *Astrophysical Journal*, **684**, 611.

Voors, R. H. M., Waters, L. B. F. M., de Koter, A., *et al.* 2000, *Astronomy and Astrophysics*, **356**, 501.

Weingartner, J. C. & Draine, B. T. 2001, *Astrophysical Journal*, **548**, 296.

Wickramasinghe, N. C. 1970, in *Ultraviolet Stellar Spectra and Related Ground-Based Observations*, ed. L. Houziaux & H. E. Butler, Reidel, 42–49.

Wickramasinghe, N. C. & Nandy, K. 1970, *Nature*, **227**, 51.

Wickramasinghe, N. C., Donn, B. D., & Stecher, T. P. 1966, *Astrophysical Journal*, **146**, 590.

Wilking, B. A., Lebovsky, M. J., Martin, P. G., Rieke, P. G., & Kemp, J. C. 1980, *Astrophysical Journal*, **235**, 905.

Witt, A. N. & Lillie, C. F. 1973, *Astronomy and Astrophysics*, **25**, 397.

Wooden, D. H. 2008, *Space Science Reviews*, **138**, 75.

Woolf, N. J. & Ney, E. P. 1969, *Astrophysical Journal*, **155**, L181.

Wuchterl, G. & Tscharnuter, W. M. 2003, *Astronomy and Astrophysics*, **398**, 1081.

Yorke, H., Bodenheimer, P., & Laughlin, G. 1993, *Astrophysical Journal*, **411**, 274.

Zhukovska, S. V., Gail, H.-P., & Trieloff, M. 2008, *Astronomy and Astrophysics*, **479**, 453.

Zijlstra, A. A., Matsuura, M., Wood, P. R., *et al.* 2007, *Monthly Notices of the Royal Astronomical Society*, **370**, 415.

Zinner, E. 2007, in *Meteorites, Comets and Planets*, ed. A. M. Davis, Treatise on Geochemistry, vol. 1, Elsevier, 1–33.

Zubko, V. G., Dwek, E., & Arendt, R. G. 2004, *Astrophysical Journal Supplement Series*, **152**, 211.

3

Evolution of protoplanetary disk structures

Fred J. Ciesla and Cornelis P. Dullemond

Abstract In this chapter, we review the general properties of protoplanetary disks and how the gaseous and solid components contained within evolve. We focus on the models that are currently used to describe them while highlighting the successes that these models have had in explaining the properties of disks and primitive materials in our Solar System. We close with a discussion of the open issues that must be addressed by future research in order to develop fully our understanding of protoplanetary disk structures.

Protoplanetary disks are natural consequences of star formation, being composed of the molecular cloud material that had too much angular momentum to fall directly onto their central stars. While protoplanetary disks are common around young stars, observations of these objects indicate that they exhibit a range of properties, and it is unclear which, if any, of the objects that have been observed are good analogs for the solar nebula – the protoplanetary disk from which our own planetary system formed. As an example of this caveat, it is important to note that none of the over 350 (and counting) exoplanetary systems provides a good dynamical or structural analog for our own. In this chapter, we take the approach that protoplanetary disk evolution is a universal process, which can be described by a common set of models and equations, and that the variety of structures and properties that have been observed reflect a range of different starting conditions from which, and environments within which, these objects evolved.

In order to understand fully the evolution of protoplanetary disks, it is necessary to develop an understanding of the dynamics and evolution of both the gas and solids that they contain. The detailed evolution of these components are often discussed separately, largely owing to our incomplete understanding of the processes involved. In introducing the different concepts in this chapter, we too will treat these components separately in order to present a clear picture of the fundamental processes that are believed to be at work. We will discuss the feedback that takes place between the solids and gas when such processes are thought to be well understood and stress the necessity that future models of disks consider the simultaneous evolution of both components.

3.1 Some properties of protoplanetary disks

3.1.1 Disk masses

As molecular cloud collapse gives way to a protostar and its disk, the density of the system is expected to be highest at the center and fall off with distance. The rate at which density falls off is determined mainly by the collapse process and the original angular momentum of the parent molecular cloud. As will be discussed further below, among the key factors in determining how a protoplanetary disk will evolve is the total mass of the disk and how it is distributed. The initial mass is difficult to quantify as there is no specific starting point at which the disk is termed "protoplanetary" – the boundary between protostellar and protoplanetary disks is somewhat arbitrary. As with nearly all models of protoplanetary disks, we simplify this issue by assuming that the central star has nearly reached its final mass and that the disk mass is $< 30\%$ of the mass of the star, otherwise it would be violently and globally unstable (Shu *et al.* 1990).

Among the first attempts to determine the structure of a protoplanetary disk were studies by Weidenschilling (1977b) and Hayashi (1981) who reconstructed the solar nebula based on the mass of solids that was left behind in the form of planets and asteroids. By taking the mass of each planet and distributing it across its "feeding zone" to form a smooth distribution, these authors then added the amount of hydrogen and helium needed at each location to produce a solar composition. This structure represents the least amount of material needed to form the Solar System, and thus was termed the minimum mass solar nebula (MMSN), which was described by:

$$\Sigma (r) = 1700 r_{AU}^{-3/2} \, \text{g cm}^{-2}, \tag{3.1}$$

$$T(r) = 280 r_{AU}^{-1/2} \, \text{K}, \tag{3.2}$$

where Σ represents the surface density of materials, r_{AU} is the distance from the Sun measured in AU, and T is the temperature. This given structure produced a disk with a mass between 0.01 and 0.07 $M_\odot$ out to 40 AU. The range of values reflects the uncertainty at the time in the compositions of the giant planets.

While the MMSN provides insight into a plausible structure of the solar nebula, the concept can be a bit misleading. The MMSN is assumed to be a purely passive disk, and issues such as disk evolution and solid transport, both of which are discussed at length in this chapter, were not considered in developing these models (in part due to the poor understanding of these processes at the time). As such, it is unclear how well the MMSN represents the true structure, at any time, of our solar nebula. Despite this, the MMSN concept remains a point of reference when describing the mass contained within protoplanetary disks.

Recently, estimates have been made of the masses of disks around other young stars, giving numbers that are consistent with the range estimated for the MMSN (Beckwith et al. 1990; Williams et al. 2005), and scaling roughly linearly with the mass of the central star (Klein et al. 2003). As the largely molecular hydrogen gas in these disks cannot be directly measured, these estimates are based on millimeter observations, which detect the presence of solid bodies up to ~0.1–1 cm. The gas mass in these disks is then inferred based on our current understanding of the opacity of such dust grains, their assumed size distribution, and, like the MMSN models, the assumption that the solids-to-gas mass ratio is constant throughout the disk (e.g. Andrews & Williams 2007). However, as bodies larger than a centimeter in size would be undetected, this is likely to underestimate of the total mass in the disk. Thus, while these estimates represent our best handle on disk masses, they, too, should be used cautiously as such estimates represent lower bounds on the values.

3.1.2 Disk sizes

The sizes of protoplanetary disks are generally more certain than their masses, and observations have found that disks generally extend to radial distances of hundreds of AU. Evidence for these large radii comes in the form of optical silhouettes of disks in the Orion Nebula (McCaughrean & O'dell 1996), mapping of disks in millimeter continuum or (sub)millimeter CO rotational lines (e.g. Dutrey et al. 2007), or through observations of scattered light from dust in the disk (e.g. Trilling et al. 2001). These observations are in direct contrast to the MMSN models which had our solar nebula ending at ~40 AU, where Pluto is located. While Kuiper Belt objects with larger semi-major axes have been identified further out, the amount of mass located beyond the orbit of Pluto is thought to be small. This then raises the question of whether or not our solar nebula also extended to such large sizes or was among the rare exceptions that were truncated at a small heliocentric distance.

Given the paucity of disks around solar-type stars that are truncated at relatively short radii, a number of theories have been offered to explain why the remaining mass in the Solar System is confined to relatively small radial distances despite the solar nebula beginning in a much more extended configuration. Among the possibilities are: (1) a passing star stripped off the outer material in the Solar System either before or after the gas in the solar nebula had dissipated (Pfalzner et al. 2005), (2) ultraviolet radiation from stars in the Sun's neighborhood heated the outer gas to temperatures that allowed the hydrogen molecules to reach the escape velocity in a process called photoevaporation (Hollenbach *et al.* 2000; see also Chapter 9 by Pascucci & Tachibana), and (3) solids decoupled from the gas and preferentially migrated inwards resulting in the increase in solid mass in the inner disk at the expense of the outer disk (Stepinski & Valageas 1996, 1997, see also the discussion in Section 3.3.2).

3.1.3 Disk clumpiness

In most cases, models of protoplanetary disks assume that the disk is axisymmetric and that the surface density falls off monotonically with the distance from the star. This is assumed because it is the general expectation for such disks based on collapse calculations, and, admittedly, for the sake of simplicity. However, some disks do show signs of radial or non-axisymmetric structure, indicating that such treatments are not always valid. Examples include AB Aurigae (Fukagawa *et al.* 2004; Piétu *et al.* 2005; see Fig. 3.1) or AA Tau with its rotating bump that periodically obscures the star (Bouvier *et al.* 2003).

In some cases disks may have structures that are unstable to perturbations driven by the gravitational force that the disk exerts on itself. This is quantified by the Toomre Q parameter, given by:

$$Q = \frac{\Omega_K c_s}{\pi G \Sigma}, \tag{3.3}$$

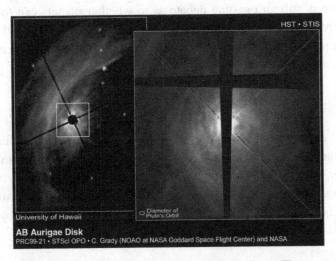

Figure 3.1 Hubble Space Telescope image of AB Aurigae. The central star is approximately 2.8 $M_\odot$ and the disk is approximately 400 AU in diameter (Corder *et al.* 2005). Spiral structure (density contrasts) can be seen in the outer regions of the disk where the disk is expected to be most susceptible to gravitational instabilities.

where Ω_K is the Keplerian orbital frequency ($\Omega_K^2 = GM_\odot/r^3$), G is the gravitational constant, and c_s is the local speed of sound. If $Q<1$ the disk is highly unstable to such perturbations, which would lead to rapid fragmentation of the disk. If $Q \gtrsim 2$ then the disk is expected to be stable to such perturbations, and capable of maintaining a relatively smooth structure. If $1<Q<2$, the disk is considered to be marginally unstable, allowing localized clumps to phase in and out of existence. Given the functional form of Q, its value may actually vary within a protoplanetary disk, with the cold, more slowly rotating gas in the outer disk being more likely to reach a gravitationally unstable state than the hot, rapidly rotating inner disk.

In some models of marginally unstable disks (e.g. Boss 2002; Boley *et al.* 2006), these clumps can grow massive enough to exert gravitational torques on other parts of the disk and drive large-scale motions and angular momentum transport within the disk. Also, such clumps can produce localized pressure maxima, whereas the pressure at the disk mid-plane falls off monotonically in smooth disks. These large-scale motions and localized pressure maxima could play a role in how solids are transported in disks, as will be discussed further below. Another important potential consequence of clumpy disks is the possible formation of giant planets through gravitational instability (also referred to as "disk fragmentation"). In these cases, a clump in a disk forms such that its self-gravity prevents it from being sheared apart by the differential rotation of the gas. The clump then contracts to give rise to a giant planet. The inefficiency of radiative cooling is, however, a major stumbling block for these models and because of the complexity of this radiation–hydrodynamic problem there is still an ongoing debate as to whether planets can form directly owing to gravitational fragmentation of the disk (e.g. Boss 2007) and whether or not the solar nebula itself went through such a marginally unstable phase at all. As such, this remains an active area of research.

3.2 Protoplanetary disk structure and evolution

The masses, sizes, and overall structure of protoplanetary disks are important to quantify as they set the total amount and the distribution of planet-forming materials. However, over time disks evolve and the dust contained within is transported, processed, and accreted into larger bodies. This evolution plays a critical role in determining both the physical and chemical properties of the dust, and by extension, of the planets that will eventually form.

3.2.1 Viscous evolution and mass accretion through the disk

Protoplanetary disks are not static objects. In addition to the orbital motion of the gas, there is radial motion as well. It is this radial motion that causes matter from the disk to move toward the star and eventually to land on the stellar surface. This contributes to the continuing growth of the star and releases a large amount of radiative energy in the process. From observations it is known that most stars younger than a few million years continue to accrete matter from their circumstellar disk, albeit at a low rate in the protoplanetary phase, i.e. once the main star-formation phase is over. The accretion rate can be estimated by observing the Hα equivalent width from hot gas near the stellar surface, or by measuring the excess ultraviolet flux above the expected stellar photospheric emission (Muzerolle *et al.* 2005). These accretion rates vary, but the typically observed range is roughly between $10^{-7} \, M_\odot \, yr^{-1}$ and $10^{-10} \, M_\odot \, yr^{-1}$ for stars and disks up to a few million years old. In rare cases, such as the so-called FU Orionis outburst events, the accretion rate can be orders of magnitude higher, but these phases are thought to last only a limited time (Bell *et al.* 1995).

Therefore, the question naturally arises: how does matter continue to accrete onto the star, even though the matter in the protoplanetary disk will have a specific angular momentum much larger than that of the surface of the star? The standard picture is that frictional forces work to redistribute the matter in the disk, allowing most of the matter to move inward toward the star, while pushing some outward to absorb the excess angular momentum (Lynden-Bell & Pringle 1974; Hartmann *et al.* 1998). The disk is therefore expected to grow in radius, but decrease in mass over time. Consequently the accretion rate from the inner edge of the disk onto the star also decreases with time, which appears to be consistent with the statistics of accretion rate of observed sources at different ages (e.g. Sicilia-Aguilar *et al.* 2004). Eventually, the disk will be destroyed by the process of photoevaporation (Hollenbach *et al.* 1994; Clarke *et al.* 2001; Alexander *et al.* 2006).

Due to the friction forces that cause the net inward motion, gravitational energy is converted into heat. This is usually called viscous heating or accretional heating of the disk. This heat is then radiated away at the disk's surface. The irradiation of the disk by the central star is also an important heating mechanism. While viscous heating tends to dominate in young disks ($\lesssim 1$ Myr) and at small distances from the star, irradiation dominates in older disks and at large radii ($\gtrsim 1$ AU). The temperature and density structure of the disk are thus continuously evolving. A detailed understanding of this evolution is critical in order to develop models for the chemical evolution of materials within a protoplanetary disk, where the pressure and temperature must be known as functions of time and location. While the specific details of this evolution remain the subject of ongoing research, models have been developed which have had success in matching or explaining various observations. In the next section, we describe the most common type of models to describe the evolution of protoplanetary disks.

3.2.2 A simple recipe for the radial structure

Modeling a disk by solving the full three-dimensional Navier–Stokes equations is a complicated task. Moreover, it is still not fully understood what is the cause of frictional forces in the disk. Molecular viscosity is by orders of magnitude too small to cause any appreciable accretion. Instead, the most widely accepted view is that instabilities within the disk drive turbulence that increases the effective viscosity of the gas (see Section 3.2.5). A powerful simplification of the problem is (a) to assume a parameterization of the viscosity, the so-called α-viscosity (Shakura & Syunyaev 1973) (β-viscosity in the case of shear instabilities, Richard & Zahn 1999) and (b) to split the disk into annuli, each of which constitutes an independent one-dimensional (1D) vertical disk structure problem. This then constitutes a 1+1D model: a series of 1D vertical models glued together in radial direction. Many models go even one step further in the simplification by considering only the vertically integrated or representative quantities such as the surface density $\Sigma(r) \equiv \int_{-\infty}^{+\infty} \rho(r, z)dz$ or mid-plane and surface temperatures (Shakura & Syunyaev 1973; Lynden-Bell & Pringle 1974; Lin & Papaloizou 1980). Here we outline a simple procedure for calculating a protoplanetary disk structure based on such a surface density description. Such a computational tool is useful for estimating the gas density and temperature throughout the disk.

The dynamical evolution of the disk as it undergoes such viscous evolution is described by the equation:

$$\frac{\partial \Sigma}{\partial t} = \frac{3}{r} \frac{\partial}{\partial r} \left[r^{\frac{1}{2}} \frac{\partial}{\partial r} \left(\Sigma \nu r^{\frac{1}{2}} \right) \right] \tag{3.4}$$

where ν is the turbulent viscosity of the gas. Typically the turbulence is assumed to be three-dimensional in nature, subsonic, and not to have eddies whose size is greater than the thickness of the disk. As a result, the turbulent viscosity is related to a characteristic velocity, usually taken as the local isothermal sound speed, $c_s = \sqrt{kT/\mu m_p}$; and to a characteristic length, usually taken as the local scale height of the disk, $H = c_s/\Omega$. As in Shakura & Syunyaev (1973), the viscosity is given by $\nu = \alpha c_s H$, where α is a dimensionless parameter ($\alpha < 1$).

Solving Eq. (3.4) time-dependently involves some numerical tinkering, but a steady-state solution, where $\dot{M}$ is constant with time and location in the disk, can be found more easily. As an input parameter we have the accretion rate $\dot{M}$ which is given by:

$$\dot{M} = -2\pi r \Sigma v_r \tag{3.5}$$

where v_r is the radial velocity of the gas. While in an evolving disk the large-scale flows associated with disk evolution would require v_r to be solved at every time interval in order to calculate the rate of mass transport, the velocity in a steady-state disk is given by $v_r = -(3\alpha c_s H/2r)(1 - \sqrt{r_{in}/r})^{-1}$ (Lynden-Bell & Pringle 1974). The steady-state solution is:

$$\Sigma v = \frac{1}{3\pi}\dot{M}\left(1 - \sqrt{\frac{r_{in}}{r}}\right) \tag{3.6}$$

where r_{in} is the inner radius of the disk. This equation explicitly assumes that there is no torque exerted on the inner boundary of the disk.

3.2.3 Heating/Cooling balance

To solve Eq. (3.4) or Eq. (3.6) we need to know the value of c_s, or in other words the temperature T_m near the mid-plane. This is determined by balancing the heating rates due to the viscous evolution of the disk and its irradiation by the central star with the cooling rate due to the thermal emission from the disk. The vertically integrated heating rate due to viscosity (i.e. accretional heating) is:

$$Q_{visc} = \frac{9}{4}\Sigma v\Omega^2 = \frac{3}{4\pi}\dot{M}\Omega^2\left(1 - \sqrt{\frac{r_{in}}{r}}\right) \tag{3.7}$$

where the second equality only holds for steady-state disks ($\dot{M}$= constant everywhere). The heating due to irradiation is more difficult to determine, because this depends on the value of the grazing incidence angle by which the stellar radiation enters the disk. This, in turn, depends on the precise geometry of the disk. Let us, for simplicity, assume that the disk has a flaring geometry, that is one in which the vertical structure of the disk is concave (upwards, H/r increases with r), such that the angle between the stellar radiation and the disk surface is $\phi = 0.05$ (for discussion of "flaring" vs. "flat" disks see Section 3.4.1). Let us also assume that the star can be regarded as a point source at the center. Then, the heating rate of the disk is:

$$Q_{irr} = 2\phi\frac{L_*}{4\pi r^2} \tag{3.8}$$

The factor of 2 is because the disk has two sides. The cooling rate, assuming blackbody cooling, is:

$$Q_{cool} = 2\sigma T_e^4 \tag{3.9}$$

where σ is the Stefan–Boltzmann constant and T_e is the effective surface temperature of the disk. For a disk that is dominated by irradiation heating we therefore get an effective surface temperature of $T_{e,irr} = (\phi L_*/4\pi\sigma)^{1/4}r^{-1/2}$. For a steady disk that is dominated by accretional heating we obtain $T_{e,visc} = (3GM_*\dot{M}(1 - \sqrt{r_{in}/r})8\pi\sigma)^{1/4}r^{-3/4}$. Since $T_{e,visc}$ drops faster with radius than $T_{e,irr}$ it follows

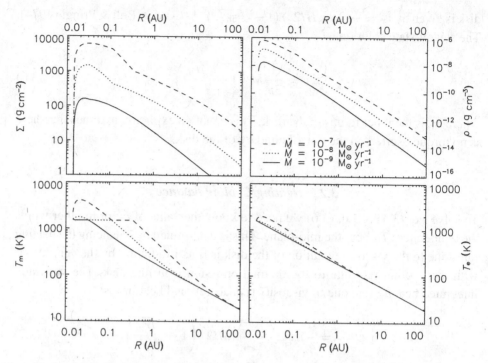

Figure 3.2 Radial runs of various disk variables according to the simple steady-state toy disk model described in the main text for three different global accretion rates. The central star is Sun-like. We have assumed a gray opacity of $\kappa = 1$ in regions where $T_m < 1500\,\mathrm{K}$. In regions where $T_m > 1500\,\mathrm{K}$ we switched to $\kappa = 0.01$ to mimic the effect of dust evaporation. Since dust evaporation reduces the mid-plane temperature there will be a region where dust is only partly evaporated to keep T_m at $1500\,\mathrm{K}$. Dust evaporation acts as a thermostat here.

that the surface temperature is likely to be dominated by irradiation at large radii and/or low accretion rates. Figure 3.2 shows how the above formulae work out for a stationary disk around a solar-type star for three different accretion rates.

Using $T_e(r)$ we can derive some observational quantities for these disks. For instance, the infrared flux from the entire disk at some wavelength λ is

$$F_\lambda = 2\pi \frac{\cos i}{d^2} \int_{r_{in}}^{r_{out}} B_\lambda(T_e(r))\, r\, dr \qquad (3.10)$$

where $B_\lambda(T)$ is the Planck function, λ the wavelength of observation, d is the distance to the observer and i is the inclination. If we forget about the $(1 - \sqrt{r_{in}/r})$ factor in the temperature we get $T_e \propto r^{-q}$ with $q = 1/2$ for the irradiated case and $q = 3/4$ for the accretion-dominated case, then we will find that the slope of the infrared part of the spectrum is $\lambda F_\lambda \propto \lambda^{(2-4q)/q}$, i.e. $\lambda F_\lambda = \mathrm{const}$ for the irradiated case and $\lambda F_\lambda \propto \lambda^{-4/3}$ for the accretion-dominated case. The latter slope is also found for irradiation-dominated disks with a flat geometry, in which the angle $\phi \propto 1/r$ so that $T_e \propto r^{-3/4}$ is in the accretion-dominated case. The difference in observed infrared slopes between the near-infrared and the mid- or far-infrared is therefore also often interpreted as a difference in geometry of the disk (Meeus *et al.* 2001). This is particularly clearly seen in the spectral energy distributions (SEDs) of disks around Herbig Ae stars (Meeus *et al.* 2001), but different slopes are also seen in the SEDs of T Tauri stars (Furlan *et al.* 2006), although they are overall steeper (more toward flatter disks) than for the Herbig stars. Since the disk geometry is set by hydrostatic equilibrium (see below), which is only weakly dependent on temperature, the strong differences in geometry may point to the sedimentation of dust grains that flattens the disk as seen in dust continuum (e.g. Miyake & Nakagawa 1995; Chiang *et al.* 2001; Dullemond & Dominik 2004).

For theories of planet formation and disk chemistry, the mid-plane temperature, T_m, is more important than the surface temperature. Also, to solve Eq. (3.4) or Eq. (3.6) we need to know ν, which depends on T_m, not on T_e. For irradiation-dominated disks T_m is roughly equal to the surface effective temperature T_e. For disks in which accretional heating is important this is not the case. A rough estimate for the mid-plane temperature valid for any case is: $T_m^4 = (3/8)T_{e,\text{visc}}^4 \tau_R + T_{e,\text{irr}}^4$, where $T_{e,\text{visc}}$ and $T_{e,\text{irr}}$ are given above and $\tau_R = \Sigma\kappa_R$ with κ_R the Rosseland mean opacity of the gas–dust mixture at the temperature T_m. The expression for T_m is therefore not explicit, as it is also used for κ_R. For realistic dust opacities it requires a numerical iterative solution procedure. For a gray opacity, or for piecewise power-law opacities, such as those in Bell *et al.* (1997), it can be solved analytically.

Now that we have a procedure for finding T_m for given Σ and α we can determine Σ time-dependently from Eq. (3.4) or stationary for given $\dot{M}$ from Eq. (3.6). The latter is a local algebraic equation at each radius r, solved analytically if the opacities are power-law formulae, or numerically otherwise.

3.2.4 Vertical structure

The above treatments provide a way of determining the surface density of a proto-planetary disk, which is a convenient value in terms of astronomical observations as it easily translates into quantities such as the optical depth. However, models of chemistry within protoplanetary disks require knowledege of the pressure of the gas and how it varies with time and location. Such a quantity is found by solving the equation of hydrostatic equilibrium.

In hydrostatic equilibrium, the force on a parcel of gas due to the weight of the material above it is balanced by the internal pressure of that parcel. Mathematically, this is written:

$$\frac{dP}{dz} = -\rho g_z = \rho \Omega_K^2 z \qquad (3.11)$$

where P is the gas pressure, ρ is the gas density, and g_z is the vertical component of the gravitational pull from the central star and we assume $z \ll r$. If we assume that the disk is vertically isothermal, solving the equation gives:

$$P(z) = P_0 e^{-z^2/2H^2} \qquad (3.12)$$

where P_0 is the mid-plane pressure, with $H = (kT_m r^3/\mu m_p GM_\odot)^{1/2}$ or C_s/Ω_K. As discussed above, the temperature does vary with z, but in general such an approximation is used to simplify matters. The density of the disk will be described by a similar equation with $\rho_0 = \Sigma/\sqrt{2\pi} H$. Using the ideal gas law then allows the pressure and temperature to be calculated everywhere within the disk, given the output of dynamical disk models. The top-right panel of Figure 3.2 shows how the mid-plane gas density varies in the steady-state models derived here.

3.2.5 Sources of viscosity

While the viscous model for the evolution of protoplanetary disks has had some success in matching some of the general properties of protoplanetary disks, such as the observed mass accretion rates and effective temperatures, the exact source of the viscosity remains the subject of ongoing studies. Currently, the most popular candidates for driving the mass transport in protoplanetary disks are the magneto-rotational instability (MRI) and gravitational instability. A third candidate, shear instability, has also been proposed based on laboratory experiments of rotating fluids (Richard & Zahn 1999), but questions remain as to whether these results can be extended to the scale of protoplanetary disks.

The MRI arises in a disk when free charges (ions and electrons) are present to couple to a magnetic field. If two parcels of ionized gas in close proximity, but on different orbits, were to couple to a given magnetic field line, the differential rotation of the disk would cause the separation between them to grow with time. Because the charges "drag" the magnetic field with them, this results in a tensional force between the parcels of gas (as if the field line were a spring). This force causes the gas on the slightly smaller orbit to lose angular momentum (move inward), while the gas on the larger orbit gains angular momentum (moves outward). Along with this angular-momentum transport, energy dissipation leads to the generation of turbulence within the disk. This is a very simplified picture of the mechanism, but a fully rigorous linear stability analysis indeed points to such an instability driving angular momentum transport in a way that is consistent with our understanding of disk evolution (Balbus & Hawley 1991).

A key requirement for the MRI is thus that ions must be present in order for the angular-momentum exchange to operate. Even at a low abundance (at an ion fraction of $f_i \gtrsim 10^{-14}$), the ions could affect the neutral molecules through collisions as they themselves are affected by the MRI. However, protoplanetary disks are largely expected to be cold, neutral objects. Sufficient ionization can develop in regions of the disk which are hot enough ($>1,300\,$K) to result in collisional ionization; however, this is expected to represent only a small region of the disk. High-energy photons (X-rays, ultraviolet rays) and cosmic rays can also ionize the gas in protoplanetary disks, though ions are only produced in the disk regions penetrated by photons, which again may only be a relatively small portion of the disk. X-rays are currently thought to be the best source of ionization in the cool regions of protoplanetary disks due to high X-ray activity of young stars and the large penetration depth of hard X-rays (Glassgold *et al.* 1997).

While X-rays can thus provide a source of ionization in the cool outer regions of the disk, they cannot penetrate to other regions of the disk, thus leaving it too neutral to be affected by the MRI. This region is termed the "Dead Zone" (Gammie 1996). The size of the Dead Zone is unknown, as the ionization fraction is determined by the balance of the rate of ionization (dependent on the X-ray flux) and the recombination rates, which are uncertain. Further, the recombination rates will depend strongly on the available surface areas on dust grains, meaning that the role of the MRI in a protoplanetary disk is intimately linked to the dust distribution (Sano *et al.* 2000).

An alternative for driving disk evolution are the gravitational torques produced within a marginally gravitationally unstable disk (Boss 2002; Boley *et al.* 2006). When massive clumps form within such disks, gas is "pulled" in various directions, leading to large-scale motions. As the clumps are believed to phase in and out of existence, the resulting motions are likely to be more episodic and variable (particularly for massive disks) than those predicted for a disk evolving under the influence of MRI-like processes.

Understanding what produces the viscosity in a protoplanetary disk is key to determining the appropriate value of α to use in evolutionary models. Currently, models which look at the localized effects of the MRI predict that values of α would range from ~ 0.001 to 0.01 (Fleming & Stone 2003). Surprisingly, if a Dead Zone is present in the disk, it is possible that stresses will be "stirred up" within it by the surrounding MRI activity, and thus, rather than being completely stagnant, may transport material at values of $\alpha < 10^{-4}$ (Fleming & Stone 2003). Under the influence of gravitational torques, Boley *et al.* (2006) estimate that the effective α would vary with time and location, but would be in the range 0.0001–0.01. The ensuing release of energy in such a disk may power some turbulent convection and diffusivity even if the associated viscous evolution is small. Further, viscous evolution models are able to match the observed properties of protoplanetary disks when values of α between 0.0001 and 0.01 are used (Hartmann *et al.* 1998; Hueso & Guillot 2005).

In short, there is still much to learn about the specific processes that drive disk evolution. Currently, the two best candidates, the MRI and gravitational torques, likely produce values of α that vary with time and location in a protoplanetary disk. In order to account properly for these variations more detailed models than those discussed in this chapter are required, and due to their complexity and computational rigor, the amount of model time that could be investigated is limited to 10^3–10^4 yr. While adopting a constant or "effective" value of α overlooks the details of these variations, it greatly reduces the numerical complexity of disk models, allowing the evolution of disks for times $> 10^6$ yr to be calculated. This provides a way for the timescales that are needed to study meteoritic materials or dust around other stars to be modeled.

3.2.6 Disk evolution and photoevaporation

The time-dependent evolution of a viscous disk is described by Eq. (3.4), but so far we have mainly focused on stationary solutions. In practice a disk will evolve in the sense that it loses mass through accretion onto the star, while at the same time spreading out to larger radii to globally conserve angular momentum. According to the analytic solutions of Lynden-Bell & Pringle (1974), as discussed by Hartmann *et al.* (1998), the accretion rate drops with time as $\dot{M} \propto t^{-(\xi+1)/(\xi+1/2)}$ if we assume that the mid-plane temperature goes as $T_m \propto r^{-\xi}$. Note that for irradiation-dominated disks we have approximately $\xi = q$ because the mid-plane temperature T_m and surface effective temperature T_e are nearly the same for that case. Assuming a constant incidence angle gives $\xi = q = -1/2$, meaning $\dot{M} \propto t^{-3/2}$. In the left panel of Fig. 3.3 it is shown that such a decline of accretion rate as a function of time is at least consistent with the measured accretion rates of stars of varying age. This distribution is, however, very broad, and therefore it is hard to say whether this can be interpreted as evidence for the viscous accretion model.

One could go one step back and include the formation process of the disk through infalling cloud material into the model. In that case the continuum equation for the disk acquires a source term for the matter falling onto the disk from the infalling envelope (Hueso & Guillot 2005). The precise way in which this matter falls onto the disk can be, to keep things simple, assumed to follow a rotational ballistic infall model (Ulrich 1976). The matter from the envelope falls onto the disk within a certain radius, and while part of this accretes onto the star, another part is pushed toward larger radii in order globally to conserve angular momentum. In this case the matter in the cold parts of the protoplanetary disk may originally have been

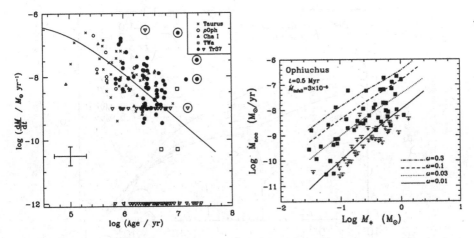

Figure 3.3 Left: measured accretion rate as a function of age (Sicilia-Aguilar *et al.* 2006). The line is the prediction of a single Lynden–Bell & Pringle model as a function of time. Right: measured accretion rates as a function of stellar mass for a cluster of a given age, in this case in Ophiucus. The lines are model predictions for a given initial dimensionless rotation rate of the parent molecular cloud core from which these stars were formed. From Dullemond *et al.* (2006b).

much closer to the star, and thereby thermally processed (Dullemond *et al.* 2006a). It is, however, not clear if the disk can spread appreciably against the ram pressure from the infalling matter. This is something that remains to be investigated and is not treated in simplified models. Whether true or not, one can make predictions for the measured accretion rate as a function of the stellar mass of the star, which can be compared against observations (Dullemond *et al.* 2006b). This is shown in the right panel of Fig. 3.3. Also here the general trend is reproduced, and while encouraging, the spread is so large that no definitive conclusion can be drawn.

While the $\dot{M} \propto t^{-3/2}$ is a relatively fast decline of the accretion rate, it is still not consistent with the observed fact that very old clusters with ages over 10 Myr hardly show any accreting sources, nor sources with infrared excess emission indicating the presence of disks. Something must destroy the disk faster than the normal viscous accretion process. It is believed that this is the process of photoevaporation. The extreme ultraviolet (EUV) radiation of the central star ionizes a very thin layer of gas at the surface of the disk. This gas has a temperature of about 10^4 K, meaning that beyond about 1 AU this hot gas can flow off the disk and away from the system through a Parker-wind-like flow (Hollenbach *et al.* 1994; Liffman 2003). Since the illumination of the disk by EUV is diluted further out in the disk by scattering, the photoevaporation process is maximally efficient just at about 1 AU, drilling a hole in the disk (Alexander *et al.* 2006).

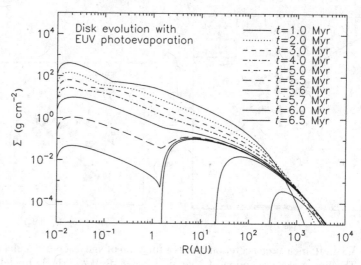

Figure 3.4   Viscous evolution of a protoplanetary disk around a solar-type star. The initial disk radius is 30 AU and initial disk mass is 0.01 $M_\odot$. The EUV flux from the central star is 10^{41} photons per second.

Figure 3.4 shows the evolution of a disk due to viscous spreading and destruction through EUV photoevaporation. The net effect is to reduce the surface density of the disk, preferentially by removing the gaseous component. The overall effects of this gas removal again are the focus of ongoing work. One potential outcome of this photoevaporation is that the inner regions of a disk may be cleared after the viscous evolution has slowed below some critical value, allowing the so-called "transition disks" – disks with a lack of near-infrared flux due to the lack of material inward of some radius – to form (Forrest *et al.* 2004). More detailed discussion of disk dispersal mechanisms are given in Chapter 9.

3.3 Particle dynamics

Solid particles are suspended within the gas of a protoplanetary disk and experience motions as a result of their interactions with the gas. These interactions depend on such factors as the particle size and local properties of the gas (pressure, density, and their respective gradients). An important measure of how a solid particle is affected by the gas is its stopping time, which is given by:

$$t_s = \frac{\rho_s a}{\rho c_s} \tag{3.13}$$

where ρ_s is the material density of the solid particle and a is its radius. Strictly speaking, this expression holds true only for particles whose radius is less than the mean free path of the gas (which is greater than tens of centimeters under most disk conditions). As most primitive materials that we concern ourselves with in meteorites or in astronomical observations are in this size range, we will use the above form of the stopping time for the rest of this chapter.

The dynamical evolution of solids in protoplanetary disks is controlled by the large-scale flows that develop due to disk evolution, diffusion associated with the turbulence that is related to disk evolution or shear instabilities, gas-drag-induced motions due to different orbital velocities of the gas and solids, and settling towards the mid-plane due to the vertical component of the central star's gravity. As the large-scale flows have already been discussed, we now discuss the other sources of particle motions.

3.3.1 Turbulent diffusion

The resulting transport of gas in an evolving protoplanetary disk is unlikely to have been perfectly ordered. Instead, the release of large amounts of gravitational energy associated with such evolution would likely manifest itself in small-scale fluctuations and random motions within the gas. These random motions represent the turbulence that is often discussed in terms of the evolution of protoplanetary disks, and they give rise to a diffusivity within the gas.

The exact relation between the evolution of the disk and the turbulent motions that develop is the subject of ongoing research. The ratio of the viscosity to the diffusivity is generally taken to be unity. Such a treatment allows us to define the diffusivity of a tracer species to be $\mathcal{D} = \alpha c_s H$. Moreover, as discussed above, the dynamics of solid particles are determined by how strongly coupled they are to the gas. In turbulent disks, a measure of this coupling is given by the dimensionless parameter called the Stokes number, defined as $St = t_s/t_e$, where t_e is the turnover time of the largest, local turbulent eddy, which is expected to be comparable to the orbital period ($t_e = \Omega^{-1}$). The effective diffusivity of solids due to the turbulent motions is given by $\mathcal{D}_a = \mathcal{D}/(1 + St)$ (Cuzzi & Weidenschilling 2006). (Note Youdin & Lithwick (2007) have argued that the dependence is $\mathcal{D}_a = \mathcal{D}/(1 + St^2)$. In the case of $St < 1$, which is the focus here, this results in little difference for the diffusivity.) Small particles ($St \ll 1$) will be more strongly affected by the random motions in the gas, and therefore their dynamics will be most similar to a tracer species within the disk. Large particles ($St \gg 1$) would be minimally affected by the small-scale fluctuations in the disk and are stirred much less. It is generally assumed that the turbulence in a disk is isotropic, meaning that the random motions in the gas are, on average, the same, regardless of the direction of motion, though suggestions have been made that the radial diffusion coefficient may be greater than that for vertical diffusion (Johansen *et al.* 2006).

Diffusion tends to smooth out heterogeneities within the disk by allowing materials to be transported down concentration gradients. As will be discussed below, such diffusion appears to be critical to balance the effects of inward motions and gravitational settling in order to explain the mineralogy of meteorites and comets, as well as observed properties of protoplanetary disks. Indeed, for even moderate levels of turbulence ($\alpha \geq 10^{-4}$), there will be an equilibrium between sedimentation and vertical mixing established on a timescale much shorter than disk evolution times (Dubrulle *et al.* 1995). Therefore, dust can be regarded mostly as being in such an equilibrium state.

On a more local scale, the effects of turbulence are to impart random motions to particles that are then added to the motions that those particles experience owing to other effects. The details of the resulting motions are beyond the scope of this chapter (comprehensive discussion and equations are provided in Cuzzi & Weidenschilling 2006; Ormel & Cuzzi 2007). A characteristic velocity that describes these motions is the root-mean-square turbulent velocity, $\sqrt{\alpha}c_s$ which is the overturn velocity of the largest eddies, and an estimate of the maximum velocity two particles (both of $St = 1$) would develop with respect to one another (Cuzzi & Hogan 2003). Under nominal conditions, such velocities can exceed ~ 100 m s^{-1} (for $\alpha = 0.01$). These velocities are important to consider when developing models for dust coagulation and planetesimal formation (see Section 3.4.1 and Chapters 7 and 10).

3.3.2 Gas-drag-induced motions

Solids in a gas-free orbit around a star will follow Keplerian orbits, meaning that their orbital velocity is found by balancing the centrifugal force of their motions with the central force of gravity from the star, or:

$$\frac{V_K^2}{r} = \frac{GM_*}{r^2} \tag{3.14}$$

Gas, however, feels a slightly reduced central force due to the outward pressure gradient that arises within the disk (hot, dense gas near the star and cool, sparse gas further out). The gas velocity, V_g, thus obeys:

$$\frac{V_g^2}{r} = \frac{GM_*}{r^2} + \frac{1}{\rho}\frac{\partial P}{\partial r} \tag{3.15}$$

Because the radial pressure gradient is expected to be generally negative in a protoplanetary disk, the result is that the gas will orbit the star slower than the solids (Adachi *et al.* 1976; Weidenschilling 1977a). Large solids, therefore, will feel a headwind as they attempt to follow Keplerian orbits. This drag force will cause them to lose energy and angular momentum to the gas, causing the particles to drift inwards with time. Smaller particles, whose stopping times are much less than an orbital period, instead orbit at approximately the same velocity as the gas. Because they do not feel the pressure gradient that the gas does, this results in an imbalance between their centrifugal force and the central force of gravity from the star. These solids also drift inwards as a result. It should be noted that if the pressure gradient locally switches sign, as a result of a clump forming in a marginally unstable disk, the gas in part of that region may orbit more rapidly than Keplerian, producing a tailwind on the dust particles and causing them to migrate outwards for a period of time (Haghighipour & Boss 2003). Again, this would only be a local effect, and would likely be accompanied by increases in the magnitude of the negative pressure gradient on the opposite side of the clump, which would result in more rapid inward drift of solids.

Figure 3.5 shows the radial drift velocities for spherical particles of different sizes for a region of the disk in which the differential velocity between the gas and the solids is 70 m s^{-1}. The dependence of the drift velocity on particle size is due to the drag force being proportional to the effective cross-section ($\sim a^2$) of the particle, while the motions in response depend on the mass ($\sim a^3$). Typically, it is the bodies with sizes between tens of centimeters to approximately a meter that experience the largest inward drift velocities (see Weidenschilling 1977a; Cuzzi & Weidenschilling 2006; and Chapter 10).

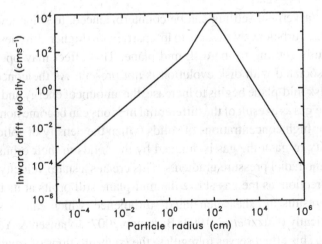

Figure 3.5 Plotted are the inward drift velocities of particles of different sizes in a disk where the velocity differential between the gas and a Keplerian orbit is 70 m s^{-1}. The kink at ~ 10 cm is due to the change in the gas-drag law as the particles exceed the mean free path of the gas.

3.3.3 Vertical motions

In addition to the radial motions described thus far, solids will settle towards the disk mid-plane under the influence of gravity. Ignoring the mass of the disk, a particle will feel a component of gravity from the central star pulling it towards the mid-plane given by:

$$F_z = -m\frac{GM_\odot}{r^3}z = -m\Omega_K^2 z \tag{3.16}$$

As the particle begins to move through the gas at a rate, v, it feels a resistive force given by:

$$F_d = \frac{4}{3}\rho_g \sigma v c_s \tag{3.17}$$

where σ is the effective cross-section of the dust grain (Dullemond & Dominik 2004). (Again, note this particular form of the drag force is only applicable to small dust particles.) Balancing these two forces gives the settling velocity of a dust particle:

$$v_{sett} = \frac{3\Omega_K^2 z}{4\rho_g c}\frac{m}{\sigma} \tag{3.18}$$

With this, we can define a timescale for dust settling:

$$t_{sett} = \frac{z}{v_{sett}} = \frac{4}{3}\frac{\sigma}{m}\frac{\rho_g c}{\Omega_K^2} \tag{3.19}$$

For typical disk parameters and for micron-sized particles at 1 AU the settling timescales are $\sim 10^6$ years .

The effects of vertical settling can be counterbalanced, to some level, if turbulence is present. Turbulence will serve to loft particles to higher altitudes, preventing grains from fully settling to the disk mid-plane. This effect may operate even if turbulence associated with disk evolution is not present. As the concentration of solids at the disk mid-plane begins to increase, the amount of energy and momentum imparted to the gas as a result of the differential motions can become non-negligible. As a result, for high concentrations of solids (when the density of solids is greater than the density of gas), the gas is dragged by the solids in their orbits, offsetting the effects of the radial pressure gradients. This creates a sharp velocity gradient in the vertical direction, as the gas above the mid-plane still orbits at its unperturbed velocity; this results in "shear-generated" turbulence, which causes solids to be dispersed vertically (Cuzzi *et al.* 1993; Barranco 2007; Johansen & Youdin 2007; Chiang 2007). This effect serves to regulate the concentrations of small solids that may be produced at the disk mid-plane. It had long been been recognized that this turbulence would inhibit the formation of planetesimals through gravitational instabilities (as reviewed in Cuzzi & Weidenschilling 2006), though recent simulations show that growth of Ceres-sized bodies in turbulent environments may be possible as a result of localized pressure maxima that develop and the resulting dynamics of the particles (Johansen *et al.* 2007) or preferential concentration of equal-sized particles in between turbulent eddies (Cuzzi *et al.* 2008).

While turbulence can counterbalance the effects of vertical settling, this effect is strongly size-dependent, and as particles grow in the surface layers of a disk they will rapidly settle to lower layers around the mid-plane. While small particles can then diffuse up to higher altitudes, the larger particles remain locked at lower altitudes. Thus, settling serves to deplete the upper layers of the disk in dust, leaving only a population of fine-grained materials. The effects of this settling are necessary to account for millimeter-wave emission, SEDs, scattered light images, and silicate emission features of protoplanetary disks (D'Alessio *et al.* 2006). In many cases, the upper layers of the disk have <10% of the canonical dust-to-gas ratio, and what dust is present generally must not exceed 1 μm, otherwise the observed emission features would not be present.

3.4 Protoplanetary disk dynamics and dust evolution

Thus far we have introduced and provided brief overviews of the dynamical effects believed to be responsible for shaping the environments in which dust would be processed within protoplanetary disks. In the remaining part of this chapter we discuss how these processes may have left their fingerprints on the properties of dust, as revealed from telescopic observations and from laboratory analysis of primitive samples. This should not be taken as a definitive discussion, but rather as an examination of how current models of protoplanetary disk evolution can be used to make sense of the properties of meteorites, comets, and the dust that is present around young stars.

3.4.1 Dust distribution

Observations of protoplanetary disks indicate that these objects remain optically thick for timescales of millions of years, meaning that a population of dust is sustained for that period of time (see Chapter 9 for a detailed discussion of disk lifetimes and dispersal mechanisms). As will be discussed in Chapter 10, the timescale for dust growth and incorporation into planetesimals is less than this time period. Additionally, the timescale for dust settling is much less than the age of these disks. However, the apparent contradictions between these timescales and the observations can be explained within the context of the processes described thus far.

In terms of aggregating into larger bodies, dust grains are expected to grow through collisions and sticking to other particles. In a laminar disk, the rate at which growth occurs is determined by the differential motions of the particles due to the effects of Browninan motion, gas drag, and vertical settling. When such calculations have been performed, it is found that planetesimals form, and dust is depleted, on timescales of 10^4–10^5 years (Weidenschilling 1980, 1997; Dullemond & Dominik 2005; Tanaka *et al.* 2005). While turbulence would serve as an additional source of relative velocity, and therefore lead to more frequent collisions between solids, the resulting collisions would be more energetic and, therefore, be more likely to lead to disruption than growth. As a result, turbulence serves to constantly replenish dust by inducing energetic collisions between larger bodies. This also effectively slows the rate at which planetesimals form (Weidenschilling 1984; Dullemond & Dominik 2005; Ciesla 2007a). Thus, the fact that a dust population is sustained on timescales of millions of years is taken as circumstantial evidence for the presence of turbulence in protoplanetary disks.

Even in the absence of growth, however, dust in protoplanetary disks is expected to settle out of the upper layers of the disk on timescales of $<10^6$ years. Yet many disks exhibit evidence of dust grains being present at high altitudes. Such disks are termed "flared," as opposed to "flat," where the grains had settled significantly towards the mid-plane. The distinction between the two geometries is made by looking at the temperature of the dust. Dust at high altitudes can intercept a larger fraction of radiation from the central star (have a greater flaring angle as discussed in Section 3.2.4), achieving higher temperatures at a given distance from the star than if the disk were flat. The presence of dust at high altitudes at these later stages of disk evolution requires some process to counteract the settling towards the disk mid-plane. Vertical diffusion by turbulence is the best candidate.

3.4.2 Chondritic meteorites

Chondritic meteorites have long been thought to represent relatively unaltered products of our solar nebula as the relative abundances of the elements they contain (apart from the most volatile ones) mirror that which is observed in the Sun. Because of their primitive nature, it is of interest that the chondritic meteorites are divided into 13 distinct types, with the distinguishing features being their bulk chemistry, the ratios of the oxygen isotopes they contain, and the relative abundances and sizes of their individual components (see Section 1.1). These individual components – refractory inclusions, chondrules, and matrix – each record different formation environments and processes. Understanding how the different chondrites formed and acquired their distinctive properties thus requires understanding how their different components formed and came together to be accreted into common meteorite parent bodies.

The most common type of refractory inclusions are the calcium–aluminum-rich inclusions (CAIs) whose mineralogy is consistent with the minerals that are predicted to condense first from a gas of solar composition (Grossman 1972; Yoneda & Grossman 1995). This suggests that these objects formed in environments which exceeded ~1700 K. The CAIs were also subjected to episodes of vaporization and condensation that required temperatures in excess of ~1300 K (see Chapter 8 and e.g. Richter *et al.* 2007). The duration of each individual heating episode is uncertain, though it appears that the total time during which CAIs were subjected to such heating is at most ~3×10^5 years (Young *et al.* 2005). The CAIs also record the most ^{16}O-rich environments in our Solar System.

Chondrules, like CAIs, also record very high temperatures in the solar nebula. However, chondrules record such temperatures as being the result of transient heating processes (see Chapter 8), where low-temperature ferro-magnesium silicate precursors (<650 K as inferred from the presence of FeS) were rapidly heated to temperatures ~2000 K, and were allowed to cool over periods of hours or days (Hewins 1997; Ciesla 2005). As a result of this high-temperature processing, chondrules are igneous spherules, and despite being processed at high temperatures, they contain non-negligible amounts of volatile material (those which would condense at temperatures below the melting point of these objects). Chondrules exhibit a range of oxygen isotope ratios, but are significantly enriched in the heavy isotopes relative to CAIs, clustering around the same oxygen isotopic mixture as found in Earth's oceans (see also Section 4.4).

The matrix of chondritic meteorites (see also Chapter 7) is less understood than the refractory inclusions or chondrules that it surrounds. This is in part due to the fact that the matrix is composed of small (<10 μm) grains which are difficult to separate and study as individual objects. As a result, only the bulk properties of the matrix have been characterized in detail. It has been found that the matrix has a complementary relationship with the chondrules that it surrounds, in that while neither the chondrules or matrix by themselves have solar compositions, the sum of their compositions is nearly solar (Wood 1985; Bland *et al.* 2005). This has led to the suggestion that matrix was formed or processed in the same events in which chondrules were formed (Wood 1985; Klerner & Palme 2000; Bland *et al.* 2005; Scott & Krot 2005). This would allow any elements that were lost from the chondrules during heating to recondense on the matrix grains. A contemporaneous origin for the chondrules and matrix is further supported by the fact that the two components have similar oxygen isotopic compositions in a given meteorite. The idea of matrix and chondrules being perfectly co-eval does run into problems though, in explaining the presence of presolar grains in the matrix of chondrites. Such grains are thought to be unprocessed materials that were inherited from the molecular cloud from which the Sun formed, and due to their fragile nature, must have escaped processing in the thermal events in which chondrules formed (Huss 2004). In short, matrix itself represents a collection of materials that were processed at high temperatures and materials that escaped processing altogether, even at moderate temperatures.

One of the challenges that chondritic meteorites have posed for models of Solar System formation is identifying how materials that record such different formation conditions could have been accreted into a common parent body. However, our current picture of protoplanetary disk evolution allows us to begin to outline a manner by which this happened. For example, in order to sustain the temperatures required for CAI formation within the solar nebula, an energy source in addition to irradiation is required. Mass transport through the disk is an appealing source of this energy, as high mass accretion rates, which would produce the highest temperatures in the disk, occur at the earliest stages of disk evolution. This would be consistent with the refractory CAIs being the oldest objects in our Solar System. Even if these high temperatures were limited to being close to the Sun (<1 AU), a consequence of this evolution would be large-scale mixing, possibly allowing for CAIs to be transported outward to large distances by radial diffusion and delivered to the region where chondritic meteorite parent bodies formed (2–4 AU). Presolar grains could diffuse inward from the outer disk where the lower temperatures and more gentle infall onto the nebula allowed them to retain their interstellar signatures. Chondrules and matrix would then be composed of the materials present in the chondrite-formation region when a chondrule-forming event occurred. The distinctive chemical and isotopic properties of the chondrite classes may be imprinted in these events and reflect changing chemical abundances in the nebula, possibly related to evaporation fronts (see below). As the chondrules and matrix in a chondrite appear to be unique to that particular chondrite class, this suggests that the formation of parent bodies occurred on timescales that were short compared to the mixing timescales for this region of the solar nebula. (These issues are discussed at greater length in the review by Cuzzi *et al.* (2005).)

3.4.3 Cometary grains

With the success of the Stardust mission, tests of our models for solar nebula, and thus protoplanetary disk evolution, are no longer limited to asteroidal bodies (meteorites), but now can be applied to cometary bodies as well. Stardust returned dust grains that were ejected from the surface of comet Wild 2, a Jupiter-family comet that is thought to have formed at distances of >20 AU from the Sun (Brownlee *et al.* 2006). Thus, we now have samples of materials from the outer solar nebula that can be studied in detail.

Among the goals for the Stardust mission was identifying the origin of the crystalline silicates in comets, whose presence in comets had been known from observations of comets Halley and Hale–Bopp (as reviewed in Bockelée-Morvan *et al.* 2002). That crystalline silicates are so abundant in comets has been puzzling as the original inventory of silicates in the molecular cloud from which the Solar System formed is largely expected to have been amorphous (<1% crystalline; see erratum to Kemper *et al.* 2004). Conversion of these amorphous precursors to crystalline grains requires temperatures in excess of ~1000 K (see Chapters 5 and 8 for details on thermal annealing and also Hallenbeck *et al.* 2000), temperatures which are far above those expected in the regions where comets formed (<150 K in order for water ice to be present). Interestingly, crystalline grains have been identified at high abundances in the cool outer regions of other protoplanetary disks (van Boekel *et al.* 2004; Apai *et al.* 2005), suggesting that the presence of such grains at increased abundances is a fundamental consequence of protoplanetary disk evolution.

These crystalline grains were thought to be the result of either transient heating events in the outer solar nebula, which may be related to those that formed chondrules (Harker & Desch 2002), or large-scale outward transport from the warm, inner nebula (Nuth *et al.* 2000; Gail 2001; Bockelée-Morvan *et al.* 2002). The Stardust results point to transport as being responsible due to the great similarities between the cometary materials and those found in chondrites, including a ~10 μm refractory grain dubbed "Inti." Inti is similar to CAIs in both its mineralogy and its oxygen isotopes (McKeegan *et al.* 2006; Zolensky *et al.* 2006), meaning it formed in the same environment as the chondritic refractory inclusions and was subsequently transported outward to be accreted by Comet Wild 2. Other high-temperature materials, like crystalline silicates, would have accompanied Inti to the outer Solar System. In doing so, the cometary materials would remain in contact with the nebular gas, allowing any volatiles lost by the grains during processing to recondense prior to the formation of the comet.

This scenario differs from that envisioned by the X-wind models (Shu *et al.* 1996, 1997) in which it was proposed that grains near the inner edge of the disk would be launched in jets above the disk and heated to high temperatures as a result of direct illumination from the young star. In fact, Shu *et al.* (1996) specifically predicted that comets would contain CAIs. However, the jets in these models carry the gas from the inner edge of the disk out into space while the solids decouple from the flow and rain back down to be incorporated into asteroids and comets. Because of the high temperatures in these jets, volatiles would not recondense onto the grains and therefore the comets and asteroids which accreted them would be depleted in their volatile abundance. Preliminary investigations indicate that Comet Wild 2 exhibits no such depletions (Flynn *et al.* 2006).

While transport in the nebula appears to be a viable explanation of how these grains were brought to the comet-formation region, it is difficult to understand how such a large quantity of grains was transported in this manner based on the basic models introduced in this chapter. In a basic steady-state disk, grains would diffuse outwards in a time, t, a distance of $(\alpha c_s H t)^{1/2}$, but be pushed inwards by the large-scale flows $3\alpha c_s H t/2r$. Consequently, diffusion only wins out for short time periods – eventually the advective flows win and limit how far a grain moves outwards (though the results depend sensitively on the ratio of the disk viscosity and radial diffusion: Clarke & Pringle 1988; Pavlyuchenkov & Dullemond 2007). Thus, until recently, the only way that outward transport could explain the presence of crystalline grains in the outer regions of protoplanetary disks at tens of percent by volume level was to have high temperatures exist far out in the disk. This would reduce the total distance grains needed to be transported (Bockelée-Morvan et al. 2002). Alternatively, transient heating events could have operated out beyond 10 AU, though at the low gas densities expected there, it is unknown what such events may be (Nuth & Johnson 2006).

More recently, it has been suggested that the advective flows may not be as great an impediment to outward diffusion as once believed. Upon its formation and before it can be approximated by a steady-state model, the distribution of mass of a protoplanetary disk will reflect the initial angular momentum of the molecular cloud from which it forms. Slowly rotating clouds will give rise to disks whose mass is concentrated near the star and falls off rapidly with increasing distance. Viscous evolution of such disks leads to much greater outward transport of materials than predicted for steady-state disks, allowing for grains that were present in the hot inner disk to be carried outwards to the icy regions where they could be incorporated in comets (Dullemond et al. 2006a).

In addition, two-dimensional studies of viscous protoplanetary disks indicate that the large-scale flows associated with disk evolution are not uniform with height above the disk mid-plane. Specifically, the viscous stresses within a disk rapidly drive mass inward near the surface of the disk, but the rate of inflow decreases closer to the mid-plane. Immediately around the mid-plane, the flows are directed outwards, to compensate for the conservation of angular momentum. This provides a window through which grains can diffuse outwards without being pushed inwards again by advection (Ciesla 2007b). As above, the size of this outward flow region, and thus the level of outward transport, depends on the structure of the protoplanetary disk, meaning that different levels of outward transport are expected, which depend on the formation conditions of the disk. Thus, both of these studies suggest that the abundance of crystalline silicates provides information on the initial conditions of disk formation and evolution.

3.4.4 Evaporation fronts

Thus far, the discussion has focused on the dynamics of the dust particles in protoplanetary disks, as this is the material that remained behind in our Solar System and is the most easily observable component of disks around other stars. However, many of the dynamical processes described in this chapter apply to the vapor component as well. Further, the dynamical evolution of the vapor in a disk is ultimately linked to the dynamical evolution of the solids. This evolution must be understood when considering the chemical evolution of the dust in a disk.

The innermost radius in a protoplanetary disk where temperatures are low enough for a particular chemical species to exist as a solid has traditionally been termed the "condensation front" of that species. The most commonly used example of such a front is the "snowline," which represents the point inside of which water exists as a vapor and outside of which it exists as ice. This term largely was used in static disk models where dynamical processes were ignored.

More recently, work has focused on understanding the effects of the largely inward motion of solids across these boundaries where they vaporize, giving rise to the term "evaporation front" (e.g. Cuzzi & Zahnle 2004). Specifically, as dust grains coagulate into larger bodies that are strongly affected by gas drag, they drift inward with a much greater mass flux than if they had remained as small particles. As these bodies vaporize after crossing the evaporation front, they locally introduce vapor at a rate greater than it would be if such growth and transport were ignored. This vapor is then redistributed by diffusion and by the large-scale flows associated with disk evolution, both of which generally operate on longer timescales than the inward drift of solids from the outer solar nebula. As a result, there is a "pile-up" of vapor inside the evaporation front that then gets redistributed throughout the inner solar nebula, allowing the inner nebula to be enhanced in the concentration of this species. As the inward flux of drifting bodies decreases, either due to further growth of bodies in the outer disk which are only weakly affected by gas drag, or due to the depletion of solids by the inward migration, the rate at which the vapor is redistributed by diffusion and advection exceeds the rate at which it is resupplied. Thus, the inner nebula begins to be depleted in this species, and it is possible that the abundance will drop below the "canonical" (nebula-wide average) value if the transport timescales are less than the disk lifetime (e.g. Stevenson & Lunine 1988).

In the case of water, the above evolution could have significant implications for the chemical evolution of primitive materials. As water is a major oxidizing agent, its increased abundance in the inner nebula may help explain the higher-than-expected fayalite content of minerals in chondritic meteorites (Krot *et al.* 2000; Fedkin & Grossman 2006). As water may have been the carrier for heavy oxygen in the solar nebula (Yurimoto & Kuramoto 2004; Lyons & Young 2005), these enhancements may have also played a role in the shift in oxygen isotopes seen from primitive CAIs to chondrules and matrix. The subsequent depletion of water in the inner nebula may then be responsible for the very reduced minerals in enstatite chondrites (Cyr *et al.* 1999; Hutson & Ruzicka 2000). The timescales for such fluctuations are a few million years, consistent with the ages of chondritic meteorites (Ciesla & Cuzzi 2006). In addition, observed abundances of water in the inner protoplanetary disk range from levels that are lower than predicted in equilibrium studies (Carr *et al.* 2004) to greater than expected under canonical conditions (Carr & Najita 2008).

This latter study is particularly noteworthy. Carr & Najita (2008) used the Spitzer Space Telescope to identify spectral emission features of water and simple organic molecules in the gas of the protoplanetary disk around AA Tau out to a distance of 3 AU. As these emission features likely only originate from the very upper atmosphere of the disk, the inferred abundances are greater than those expected for a stagnant disk. Carr & Najita (2008) argue that these molecules are stirred up by turbulence from the mid-plane, where the molecules are being resupplied by volatile-rich boulders that are drifting inward from the outer disk as predicted by Cuzzi & Zahnle (2004) and Ciesla & Cuzzi (2006).

The study by Carr & Najita (2008) suggests we are moving into a new era where the gaseous chemistry of protoplanetary disks can be studied in detail. Similar observations of other disks as were carried out for AA Tau will allow correlations between the chemical composition, physical structure, and dust distribution to be made. These correlations will serve as detailed tests for our models of protoplanetary disk evolution and material transport. For example, Cuzzi & Zahnle (2004) and Ciesla & Cuzzi (2006) predict that after an initial influx of water ice enhances the concentration of water in the inner regions of a protoplanetary disk, there should be systematic decreases in the abundance of water with time. This suggests that there should be systematic variations in the water abundance with stellar age. Further, Ciesla & Cuzzi (2006) predict that larger disks will contain more water ice beyond the snowline, providing more mass to drift inwards and allowing larger water-vapor enhancements to be achieved than for more truncated disks. These predictions should be tested in coming years by future observations.

Besides water, the above evolution could occur for silicates (Cuzzi *et al.* 2003), organics, or more volatile ices in the outer solar nebula. The effects of these changes in abundances on the composition of primitive materials remains to be studied.

3.5 Summary

Observations show that protoplanetary disks are dynamic, evolving objects, whose density and temperature structures change with time due to the transport of mass and angular momentum. Chondritic meteorites and comets record a dynamic history of our own solar nebula as they contain materials that formed in a wide variety of chemical and physical environments. It remains unclear how the dynamical evolution of a protoplanetary disk would give rise to such an array of birth environments for primitive dust. We have outlined the basic ideas behind the models for how protoplanetary disks evolved and highlighted ways in which these models are consistent with the general properties of meteorites, comets, and dust around other stars. In the near future, more data will be available as further analysis of the Stardust samples are carried out and new instruments and techniques are applied to studying primitive materials in laboratories. In addition, new telescopic facilities, in particular the Atacama Large Millimeter/submillimeter Array (ALMA), will allow for greater resolution studies of protoplanetary disks and star-formation regions than have ever been realized. This increase in data volume will certainly allow for detailed and comprehensive testing of our models for protoplanetary disks, and possibly require their complete revision.

Until then, models must focus on developing a better understanding of the sources and consequences of global disk evolution and how different regions of the disk are affected. In addition, the dynamical processes described in this chapter depend sensitively on the sizes of solids under consideration, and the details of particle growth must be further developed. Finally, as described above, the dynamical transport of solids will impact the dynamical evolution of the gas through changes in optical depth and heating or through the exchange of angular momentum and energy. In the future, models must address these feedbacks in order to better identify how the physical evolution of a disk will affect the chemical environments it contains.

Acknowledgments We appreciate the comments provided by Jeff Cuzzi, Frank Hersant, and an anonymous reviewer on an initial draft of this chapter.

References

Adachi, I., Hayashi, C., & Nakazawa, K. 1976, *Progress of Theoretical Physics*, **56**, 1756.

Alexander, R. D., Clarke, C. J., & Pringle, J. E. 2006, *Monthly Notices of the Royal Astronomical Society*, **369**, 229.

Andrews, S. M. & Williams, J. P. 2007, *Astrophysical Journal*, **659**, 705.

Apai, D., Pascucci, I., Bouwman, J., *et al.* 2005, *Science*, **310**, 834.

Balbus, S. A. & Hawley, J. F. 1991, *Astrophysical Journal*, **376**, 214.

Barranco, J. A. 2007, ArXiv e-prints, 711.

Beckwith, S. V. W., Sargent, A. I., Chini, R. S., & Guesten, R. 1990, *Astronomical Journal*, **99**, 924.

Bell, K. R., Cassen, P. M., Klahr, H. H., & Henning, T. 1997, *Astrophysical Journal*, **486**, 372.

Bell, K. R., Lin, D. N. C., Hartmann, L. W., & Kenyon, S. J. 1995, *Astrophysical Journal*, **444**, 376.

Bland, P. A., Alard, O., Benedix, G. K., *et al.* 2005, *Proceedings of the National Academy of Sciences*, **102**, 13755.

Bockelée-Morvan, D., Gautier, D., Hersant, F., Huré, J.-M., & Robert, F. 2002, *Astronomy and Astrophysics*, **384**, 1107.

Boley, A. C., Mejía, A. C., Durisen, R. H., *et al.* 2006, *Astrophysical Journal*, **651**, 517.

Boss, A. P. 2002, *Astrophysical Journal*, **576**, 462.

Boss, A. P. 2007, *Astrophysical Journal*, **661**, L73.

Bouvier, J., Grankin, K. N., Alencar, S. H. P., *et al.* 2003, *Astronomy and Astrophysics*, **409**, 169.

Brownlee, D., Tsou, P., Aléon, J., *et al.* 2006, *Science*, **314**, 1711.

Carr, J. S. & Najita, J. R. 2008, *Science*, **319**, 1504.

Carr, J. S., Tokunaga, A. T., & Najita, J. 2004, *Astrophysical Journal*, **603**, 213.

Chiang, E. 2007, *Astrophysical Journal*, **675**, 1549.

Chiang, E. I., Joung, M. K., Creech-Eakman, M. J., *et al.* 2001, *Astrophysical Journal*, **547**, 1077.

Ciesla, F. J. 2005, *Astronomical Society of the Pacific Conference Series*, **341**, 811.

Ciesla, F. J. 2007a, *Astrophysical Journal*, **654**, L159.

Ciesla, F. J. 2007b, *Science*, **318**, 613.

Ciesla, F. J. & Cuzzi, J. N. 2006, *Icarus*, **181**, 178.

Clarke, C. J., Gendrin, A., & Sotomayor, M. 2001, *Monthly Notices of the Royal Astronomical Society*, **328**, 485.

Clarke, C. J. & Pringle, J. E. 1988, *Monthly Notices of the Royal Astronomical Society*, **235**, 365.

Corder, S., Eisner, J., & Sargent, A. 2005, *Astrophysical Journal*, **622**, L133.

Cuzzi, J. N., Ciesla, F. J., Petaev, M. I., *et al.* 2005, *Astronomical Society of the Pacific Conference Series*, **341**, 732.

Cuzzi, J. N., Davis, S. S., & Dobrovolskis, A. R. 2003, *Icarus*, **166**, 385.

Cuzzi, J. N., Dobrovolskis, A. R., & Champney, J. M. 1993, *Icarus*, **106**, 102.

Cuzzi, J. N. & Hogan, R. C. 2003, *Icarus*, **164**, 127.

Cuzzi, J. N., Hogan, R. C., & Shariff, K. 2008, *Astrophysical Journal*, **687**, 1432.

Cuzzi, J. N. & Weidenschilling, S. J. 2006, in *Meteorites and the Early Solar System II*, ed. D. S. Lauretta & H. Y. McSween, University of Arizona Press, 353–381.

Cuzzi, J. N. & Zahnle, K. J. 2004, *Astrophysical Journal*, **614**, 490.

Cyr, K. E., Sharp, C. M., & Lunine, J. I. 1999, *Journal of Geophysics Research*, **104**, 19003.

D'Alessio, P., Calvet, N., Hartmann, L., Franco-Hernández, R., & Servín, H. 2006, *Astrophysical Journal*, **638**, 314.

Dubrulle, B., Morfill, G., & Sterzik, M. 1995, *Icarus*, **114**, 237.

Dullemond, C. P., Apai, D., & Walch, S. 2006a, *Astrophysical Journal*, **640**, L67.

Dullemond, C. P. & Dominik, C. 2004, *Astronomy and Astrophysics*, **421**, 1075.

Dullemond, C. P. & Dominik, C. 2005, *Astronomy and Astrophysics*, **434**, 971.

Dullemond, C. P., Natta, A., & Testi, L. 2006b, *Astrophysical Journal*, **645**, L69.

Dutrey, A., Guilloteau, S., & Ho, P. 2007, in *Protostars and Planets V*, ed. B. Reipurth, D. Jewitt, & K. Keil, University of Arizona Press, 495–506.

Fedkin, A. V. & Grossman, L. 2006, in *Meteorites and the Early Solar System II*, ed. D. S. Lauretta & H. Y. McSween, University of Arizona Press, 279–294.

Fleming, T. & Stone, J. M. 2003, *Astrophysical Journal*, **585**, 908.

Flynn, G. J., Bleuet, P., Borg, J., *et al.* 2006, *Science*, **314**, 1731.

Forrest, W. J., Sargent, B., Furlan, E., *et al.* 2004, *Astrophysical Journal Supplement*, **154**, 443.

Fukagawa, M., Hayashi, M., Tamura, M., *et al.* 2004, *Astrophysical Journal*, **605**, L53.

Furlan, E., Hartmann, L., Calvet, N., *et al.* 2006, *Astrophysical Journal Supplement*, **165**, 568.

Gail, H.-P. 2001, *Astronomy and Astrophysics*, **378**, 192.

Gammie, C. F. 1996, *Astrophysical Journal*, **457**, 355.

Glassgold, A. E., Najita, J., & Igea, J. 1997, *Astrophysical Journal*, **480**, 344.

Grossman, L. 1972, *Geochimica et Cosmochimica Acta*, **36**, 597.

Haghighipour, N. & Boss, A. P. 2003, *Astrophysical Journal*, **598**, 1301.

Hallenbeck, S. L., Nuth, III, J. A., & Nelson, R. N. 2000, *Astrophysical Journal*, **535**, 247.

Harker, D. E. & Desch, S. J. 2002, *Astrophysical Journal*, **565**, L109.

Hartmann, L., Calvet, N., Gullbring, E., & D'Alessio, P. 1998, *Astrophysical Journal*, **495**, 385.

Hayashi, C. 1981, *Progress of Theoretical Physics Supplement*, **70**, 35.

Hewins, R. H. 1997, *Annual Review of Earth and Planetary Sciences*, **25**, 61.

Hollenbach, D., Johnstone, D., Lizano, S., & Shu, F. 1994, *Astrophysical Journal*, **428**, 654.

Hollenbach, D. J., Yorke, H. W., & Johnstone, D. 2000, *Protostars and Planets IV*, ed. V. Mannings, A. P. Boss, & S. S. Russell, University of Arizona Press, 401.

Hueso, R. & Guillot, T. 2005, *Astronomy and Astrophysics*, **442**, 703.

Huss, G. R. 2004, *Antarctic Meteorite Research*, **17**, 132.

Hutson, M. & Ruzicka, A. 2000, *Meteoritics and Planetary Science*, **35**, 601

Johansen, A., Klahr, H., & Mee, A. J. 2005, *Monthly Notices of the Royal Astronomical Society*, **370**, L71.

Johansen, A., Oishi, J. S., Low, M.-M. M., *et al.* 2007, *Nature*, **448**, 1022.

Johansen, A. & Youdin, A. 2007, *Astrophysical Journal*, **662**, 627.

Kemper, F., Vriend, W. J., & Tielens, A. G. G. M. 2004, *Astrophysical Journal*, **609**, 826; erratum, 2005, **633**, 534.

Klein, R., Apai, D., Pascucci, I., Henning, T., & Waters, L. B. F. M. 2003, *Astrophysical Journal*, **593**, L57.

Klerner, S. & Palme, H. 2000, *Meteoritics and Planetary Science Supplement*, **35**, 89.

Krot, A. N., Fegley, Jr., B., Lodders, K., & Palme, H. 2000, *Protostars and Planets IV*, ed. V. Mannings, A. P. Boss, & S. S. Russell, University of Arizona Press, 1019–1054.

Liffman, K. 2003, *Publications of the Astronomical Society of Australia*, **20**, 337.

Lin, D. N. C. & Papaloizou, J. 1980, *Monthly Notices of the Royal Astronomical Society*, **191**, 37.

Lynden-Bell, D. & Pringle, J. E. 1974, *Monthly Notices of the Royal Astronomical Society*, **168**, 603.

Lyons, J. R. & Young, E. D. 2005, *Nature*, **435**, 317.

McCaughrean, M. J. & O'dell, C. R. 1996, *Astronomical Journal*, **111**, 1977.

McKeegan, K. D., Aléon, J., Bradley, J., *et al.* 2006, *Science*, **314**, 1724.

Meeus, G., Waters, L. B. F. M., Bouwman, J., *et al.* 2001, *Astronomy and Astrophysics*, **365**, 476.

Miyake, K. & Nakagawa, Y. 1995, *Astrophysical Journal*, **441**, 361.

Muzerolle, J., Luhman, K. L., Briceño, C., Hartmann, L., & Calvet, N. 2005, *Astrophysical Journal*, **625**, 906.

Nuth, J. A., Hill, H. G. M., & Kletetschka, G. 2000, *Nature*, **406**, 275.

Nuth, J. A. & Johnson, N. M. 2006, *Icarus*, **180**, 243.

Ormel, C. W. & Cuzzi, J. N. 2007, *Astronomy and Astrophysical*, **466**, 413.

Pavlyuchenkov, Y. & Dullemond, C. P. 2007, *Astronomy and Astrophysics*, **471**, 833.

Pfalzner, S., Umbreit, S., & Henning, T. 2005, *Astrophysical Journal*, **629**, 526.

Piétu, V., Guilloteau, S., & Dutrey, A. 2005, *Astronomy and Astrophysics*, **443**, 945.

Richard, D. & Zahn, J.-P. 1999, *Astronomy and Astrophysics*, **347**, 734.
Richter, F. M., Janney, P. E., Mendybaev, R. A., Davis, A. M., & Wadhwa, M. 2007, *Geochimica et Cosmochimica Acta*, **71**, 5544.
Sano, T., Miyama, S. M., Umebayashi, T., & Nakano, T. 2000, *Astrophysical Journal*, **543**, 486.
Scott, E. R. D. & Krot, A. N. 2005, *Astrophysical Journal*, **623**, 571.
Shakura, N. I. & Syunyaev, R. A. 1973, *Astronomy and Astrophysics*, **24**, 337.
Shu, F. H., Shang, H., Glassgold, A. E., & Lee, T. 1997, *Science*, **277**, 1475.
Shu, F. H., Shang, H., & Lee, T. 1996, *Science*, **271**, 1545.
Shu, F. H., Tremaine, S., Adams, F. C., & Ruden, S. P. 1990, *Astrophysical Journal*, **358**, 495.
Sicilia-Aguilar, A., Hartmann, L. W., Briceño, C., Muzerolle, J., & Calvet, N. 2004, *Astronomical Journal*, **128**, 805.
Sicilia-Aguilar, A., Hartmann, L. W., Fure'sz, G., *et al.* 2006, *Astrophysical Journal*, **132**, 2135.
Stepinski, T. F. & Valageas, P. 1996, *Astronomy and Astrophysics*, **309**, 301.
Stepinski, T. F. & Valageas, P. 1997, *Astronomy and Astrophysics*, **319**, 1007.
Stevenson, D. J. & Lunine, J. I. 1988, *Icarus*, **75**, 146.
Tanaka, H., Himeno, Y., & Ida, S. 2005, *Astrophysical Journal*, **625**, 414.
Trilling, D. E., Koerner, D. W., Barnes, J. W., Ftaclas, C., & Brown, R. H. 2001, *Astrophysical Journal*, **552**, L151.
Ulrich, R. K. 1976, *Astrophysical Journal*, **210**, 377.
van Boekel, R., Min, M., Leinert, C., *et al.* 2004, *Nature*, **432**, 479.
Weidenschilling, S. J. 1977a, *Monthly Notices of the Royal Astronomical Society*, **180**, 57.
Weidenschilling, S. J. 1977b, *Astrophysics and Space Science*, **51**, 153.
Weidenschilling, S. J. 1980, *Icarus*, **44**, 172.
Weidenschilling, S. J. 1984, *Icarus*, **60**, 553.
Weidenschilling, S. J. 1997, *Icarus*, **127**, 290.
Williams, J. P., Andrews, S. M., & Wilner, D. J. 2005, *Astrophysical Journal*, **634**, 495.
Wood, J. A. 1985, in *Protostars and Planets II*, ed. D. C. Black & M. S. Matthews, University of Arizona Press, 687–702.
Yoneda, S. & Grossman, L. 1995, *Geochimica et Cosmochimica Acta*, **59**, 3413.
Youdin, A. N. & Lithwick, Y. 2007, *Icarus*, **192**, 588.
Young, E. D., Simon, J. I., Galy, A., *et al.* 2005, *Science*, **308**, 223.
Yurimoto, H. & Kuramoto, K. 2004, *Science*, **305**, 1763.
Zolensky, M. E., Zega, T. J., Yano, H., *et al.* 2006, *Science*, **314**, 1735.

4

Chemical and isotopic evolution of the solar nebula and protoplanetary disks

Dmitry Semenov, Subrata Chakraborty, and Mark Thiemens

Abstract In this chapter we review recent advances in our understanding of the chemical and isotopic evolution of protoplanetary disks and the solar nebula. Current observational and meteoritic constraints on physical conditions and chemical composition of gas and dust in these systems are presented. A variety of chemical and photochemical processes that occur in planet-forming zones and beyond, both in the gas phase and on grain surfaces, are overviewed. The discussion is based upon radio-interferometric, meteoritic, space-borne, and laboratory-based observations, measurements and theories. Linkage between cosmochemical and astrochemical data are presented, and interesting research puzzles are discussed.

Circumstellar disks surrounding young low- and intermediate-mass stars and brown dwarfs possess sufficient material to accrete planetary systems (e.g. Lin & Papaloizou 1980; Lissauer 1987; Payne & Lodato 2007). The chemical composition of the constituent gas and the inherent physical properties of dust regulate the mechanism and timescale for planetary formation. Molecules and dust are significant heating and cooling agents of the gas connected through collisions (see Chapter 3), while dust grains also determine the opacity of the medium. Ionization that allows the coupling of the gas to magnetic fields is regulated by photochemical and molecular reactions and magneto-hydrodynamical processes. To develop suitable evolutionary models of the solar nebula, a variety of relevant physical and chemical processes must be identified and resolved.

Over the past decade significant progress has been achieved in detailed understanding of disk chemistry, both theoretically and observationally. Since the dominant molecular component of the gas (H_2) is not directly observable apart from high-temperature regions ($\gtrsim 500$–$1000\,K$, e.g. Bitner *et al.* 2007; Martin-Zaïdi *et al.* 2007), other trace species are used to probe chemical composition as well

97

Table 4.1 *Molecular species commonly utilized to study disks*

Tracer	Quantity	Zone of ices	Zone of molecules	Zone of radicals	Inner zone
^{12}CO, ^{13}CO	Temperature	mm*	mm	mm	IR
H_2	—	—	—	—	IR
NH_3	—	cm	cm	—	—
CS, H_2CO	Density	—	mm	—	IR
CCH, HCN, CN	Photochemistry	—	mm	—	IR
HCO^+	Ionization	mm	mm	—	—
N_2H^+, H_2D^+	—	—	mm	—	—
C^+	—	—	mm	IR	IR
Metal ions	—	—	—	—	IR
Complex organics	Surface processes	IR**	IR, mm	—	IR, mm
DCO^+, DCN, H_2D^+	Deuterium fractionation	mm	mm	—	—

* – "mm/cm" and "IR" mean radio-interferometric and infrared observations, respectively.
** – Complex molecules frozen onto dust surfaces could be detected through absorption lines in infrared, while the gas-phase counterparts emit at (sub)millimeter frequencies.

as thermal and density structure of disks utilizing (sub)millimeter interferometry and infrared observations (Table 4.1). Molecular lines possess inherent information regarding disk kinematics that permit derivation of masses of young stars (e.g. Simon *et al.* 2000). Along with emission lines, multi-wavelength observations of thermal dust emission are employed to constrain dust properties and infer disk masses (Beckwith & Sargent 1993; Bergin *et al.* 2007; Dutrey *et al.* 2007a; Natta *et al.* 2007). Due to the limited sensitivity and spatial resolution of modern interferometers such studies are presently scarce and restricted to the outer-disk regions ($\gtrsim 30$ AU). Recently, direct observation of the inner, planet-forming disk zone with the Spitzer Space Telescope in (ro-)vibrational emission/absorption lines has become possible. Solid-state molecular emission bands of ices and a variety of silicates and polyaromatic hydrocarbons (PAHs) have also been detected in the infrared (Pontoppidan *et al.* 2005; Lahuis *et al.* 2006; Pascucci *et al.* 2007).

Detailed information may be derived from interferometric data by modeling the disk structure and the radiation transport via molecular lines and continuum, also including time-dependent chemistry (van Zadelhoff *et al.* 2003; Semenov *et al.* 2005; Isella *et al.* 2007). To date, primarily static chemical models have been used to investigate disk chemical evolution, though several dynamical studies have been reported (e.g. Aikawa *et al.* 1999; Ilgner *et al.* 2004; Willacy *et al.* 2006). According to the models the evolving disk may be parsed into four chemically distinct zones (see Fig. 4.1), controlled by temperature (from equatorial plane to the surface):

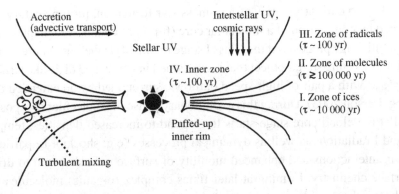

Figure 4.1 Physical and chemical structure of a protoplanetary disk. Timescales are order-of-magnitude estimates and given for a radius of about 100 AU.

(1) the cold mid-plane, (2) a warm molecular layer, (3) a hot dilute atmosphere, and (4) an inner accretion-heated region where planets may form.

The astrophysical models of protoplanetary disks based on optical observations and laboratory experiments and meteoritic measurements provide the basis for theories of nebular evolution. The best and most precise relevant measurements are from meteoritic analysis. Meteorites from the Asteroid Belt of our Solar System are the best record of the evolution of the solar nebula from a gas–dust mixture to an organized planetary system. The addition of cometary and solar-wind sample analysis complement these data. Combination of fundamental laboratory-based experiments and modeling efforts has led to a highly resolved understanding of the chemical conditions and processes in the primordial solar nebula (see Chapter 6). In this chapter an overview of recent advances in our understanding of the chemical and isotopic evolution of the early Solar System and protoplanetary disks is presented.

4.1 Protoplanetary disks

4.1.1 Global evolutionary picture of disk chemical structure

The current paradigm of the chemical structure of a disk based on advanced theoretical models and observational results is summarized in Fig 4.1 (see e.g. Aikawa & Herbst 1999; Willacy & Langer 2000; Bergin *et al.* 2003; van Zadelhoff *et al.* 2003; Semenov *et al.* 2004). The disk can be divided into four distinct chemical regions, primarily determined by temperature. The "inner zone" corresponds to radii < 20 AU and is currently barely observable by radio-interferometry, while the other three regions, "zone of ices," "zone of molecules," and "zone of radicals," represent the outer disk regions ($r > 20$–50 AU).

According to current theories of low-mass star formation, protoplanetary disks are formed by contraction of a molecular core (for more details see Chapters 1–3). This is multi-stage process. At the Class I stage (~ 0.5 Myr) the initial core is transformed into a young stellar object that is embedded in a flattened disk-like structure ($\sim 0.1 M_{Sun}$) with a pair of high-speed outflows, surrounded by an extended envelope (e.g. Larson 2003). During this evolutionary phase the molecular composition reached at the cloud core stage starts to respond to increased densities, temperatures, and irradiation, as well as dynamical processes (e.g. shocks). In particular, gas–grain interactions and enhanced mobility of surface radicals begin to drive a rich surface chemistry, forming at later times complex (organic) molecules such as alcohols, ethers, and long carbon chains and cyanopolyynes. Often the Class I phase is not implicitly considered in disk chemical models and pure atomic or molecular cloud core abundances are adopted as the initial conditions for disk chemistry.

The next several million years are essential to the formation of planets that occur in a low-mass dense rotating disk ($\sim 0.01 \, M_\odot$) around a freshly born pre-main-sequence star (Class II source; Lada 1999). During this stage dust grains grow substantially, and active gas-phase and gas–grain chemistry almost completely reprocesses material from previous stages, forming a prominent chemically layered structure and a large amount of various ices. Given that this phase is relatively long compared to the disk viscous evolution timescale, transport processes become a crucial factor for chemical evolution. Turbulent mixing and large-scale advective flows tend to smear out abundance gradients caused by variations of physical conditions (e.g. Ilgner *et al.* 2004; Semenov *et al.* 2006; Willacy *et al.* 2006; Tscharnuter & Gail 2007). This is especially true for the dense planet-forming zone where initial isotopic variations are likely to be smeared out in a sequence of condensation–evaporation events.

During the ~ 3–7 Myr of disk evolution, the dense mid-plane is totally opaque to incoming radiation, apart from the highest energy cosmic-ray particles, and remains cold and essentially neutral (10–20 K; "zone of ices"). Chemical complexity in this region is initially achieved due to ion–molecular reactions in the gas, followed by accretion onto grains and potentially surface reactions with hydrogen and other light atoms. Once formed, a molecule is seldom re-emitted into the gas phase. Typical chemical timescales for this region are determined by freeze-out and surface reactions, and are approximately 10^3–10^4 yr in the outer disk ($\gtrsim 10$–50 AU). In the innermost part the mid-plane is hot due to viscous accretion heating (near the puffed-up rim; see Fig. 4.1, bottom). In this "inner zone" ($\lesssim 10$ AU) temperatures are above 100 K and gas–grain interactions are unimportant, so the chemical timescale is due to gas-phase, primarily endothermic reactions (~ 100 yr). The inner zone can be considered to be in chemical equilibrium, which allows the investigation of the formation and destruction of solids by using simple condensation–evaporation thermodynamical models.

Adjacent to the mid-plane is a less dense, warmer layer ($T \sim 30$–$70\,$K), which is only partly shielded from stellar and interstellar ultraviolet (UV)/X-ray radiation. Young stars emit intense non-thermal UV and thermal and variable X-ray radiation fields that strongly influence disk chemistry at intermediate heights (e.g. Bergin *et al.* 2007). Ultraviolet excess in T Tauri stars is related to their chromospheric activity, which is due to channeling and heating of the accreting disk material in the magneto-spheric funnels (e.g. Bouvier *et al.* 2007). Soft X-ray radiation has likely the same origin, while harder X-rays are produced in magnetic reconnection loops generated by a dynamo operating in the fully convective stellar interior or due to jets (e.g. Güdel *et al.* 2007).

These energetic photons dissociate and ionize gas and photodesorb surface species, thus enriching the gas composition and driving a rich chemistry. Abundances of most molecules reach high concentrations in this zone, and numerous molecular lines are excited and observable emission produced ("zone of molecules"). Chemistry does not reach steady state in this region and a typical timescale, as determined by surface chemistry, exceeds 10^5 yr.

Above the intermediate layer a hot, rarefied, and heavily irradiated disk atmosphere exists ($T \gtrsim 100\,$K). This is a molecule-deficient region (apart from H_2) where only simple light hydrocarbons, their ions, and other radicals, such as CCH and CN, are able to survive ("zone of radicals"). Chemical timescales are short ($\sim 100\,$yr) and defined by ionization and irradiation.

After $\sim 10\,$Myr small dust grains are transformed into larger bodies while gas is blown out or collected by giant planets, leaving only a trace amount of collisionally generated dust visible ("debris disk"; Class III object). These disks are optically thin and their gas and dust temperatures decouple, leading to restricted, steady-state ion chemistry typical of photon-dominated regions (e.g. dominated by C^+, H^+, and light radicals).

For the purposes of this chapter, we will focus on the Class II stage because it is the best-studied phase of the evolution of the solar nebula.

4.1.2 Observational constraints on composition of gas and ices

To date ~ 150 species have been detected in the interstellar space.[1] Among these species only a handful of molecules have been spatially resolved in disks: CO (and the isotopomers ^{13}CO and C^{18}O), CN, HCN, DCN, HNC, H_2CO, C_2H, CS, HCO^+, $H^{13}CO^+$, DCO^+, and N_2H^+ (Koerner & Sargent 1995; Dutrey *et al.* 1997;

---

[1] http://astrochemistry.net/

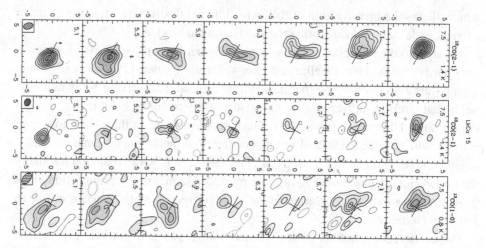

Figure 4.2 The molecular disk around LkCa 15 observed in lines of the CO isotopomers with the Plateau de Bure interferometer by Piétu *et al.* (2007).

Kastner *et al.* 1997; Aikawa *et al.* 2003; Dutrey *et al.* 2007a; Qi *et al.* 2008). Observational facilities, such as the Plateau de Bure interferometer (France), Owens Valley Radio Observatory (USA), Submillimeter Array (USA), and Nobeyama Millimeter Array (Japan), have permitted measurements of nearby systems such as DM Tau, LkCa 15, AB Aur, and TW Hya, see e.g. Fig. 4.2.

Typically, studies of disk physics are initiated with observations of strong low- and high-lying CO lines. These lines are thermally excited at densities $\sim 10^3 - 10^4 \, \text{cm}^{-3}$. The ^{12}CO lines are optically thick and their intensities measure kinetic temperature in the upper disk layer (e.g. Dutrey *et al.* 1997). The lines of less abundant ^{13}CO and C^{18}O are typically optically thin or partially optically thick and are sensitive to both temperature and corresponding column densities throughout the entire disk. Strong CO lines are most suitable for accurate determination of disk kinematics, as well as orientation and geometry. Their measured widths indicate that turbulence in disks is subsonic, with typical velocities of about $0.05 - 0.2 \, \text{km s}^{-1}$.

It has been found that disks appear progressively larger from observations of dust continuum, onwards to C^{18}O, ^{13}CO, and ^{12}CO, respectively, with typical values of 300–1000 AU. This is likely a manifestation of selective isotopic photodissociation that is discussed later. Moreover, most disks exhibit a vertical temperature gradient, ranging from $\sim 10 \, \text{K}$ at the mid-plane to ~ 50–$100 \, \text{K}$ in the atmosphere region, as determined by physical models (e.g. Dartois *et al.* 2003; Qi *et al.* 2006; Isella *et al.* 2007; Piétu *et al.* 2007). By comparing the intensities of 6–5 to 2–1 CO transitions it was found by Qi *et al.* (2006) that the TW Hya disk has a surface region that is superheated, so that an additional heating mechanism is required, possibly stellar

X-ray radiation (Glassgold *et al.* 1997). Dartois *et al.* (2003) and Piétu *et al.* (2007) have reported a large reservoir of cold CO and HCO^+ gases at 10 K in DM Tau – temperatures at which these molecules should fully deplete onto dust surfaces, but still puzzlingly exist in the gas.

The second most readily observed molecular species in disks is HCO^+. The low-lying transitions of this ion are thermalized at densities of about $10^5 \, cm^{-3}$. This is one of the most abundant charged molecules in disks, the other being C^+ (not observable at millimeter wavelengths). Another, less abundant, detected ion is N_2H^+, with the 1–0 transition that possesses a number of hyperfine components. Using chemical models and these ions, it was found that the ionization degree is $\sim 10^{-9}$ inside the warm molecular disk layer, in agreement with derived cosmic-ray ionization rates (Qi *et al.* 2003; Dutrey *et al.* 2007b). The ionization degree is $\sim 10^{-4}$ in the disk atmosphere and can be as low as $\sim 10^{-12}$ in the disk mid-plane.

Less strong lines of C_2H, CN, and HCN are sensitive to the intensity and shape of the incident UV spectrum and are excellent tracers of photochemistry (e.g. Bergin *et al.* 2003). Similarly to molecular clouds, deuterium fractionation is effective in disks, leading to a significantly higher D/H ratio than the canonical cosmic value of $\sim 0.001\%$. Recently detected DCO^+ and DCN have abundances that are about 1–10% of the HCO^+ and HCN densities (Qi *et al.* 2008). It remains to be verified whether such a large degree of deuteration is a heritage of cloud chemistry or produced *in situ*.

Rotational lines of CS and H_2CO are difficult to observe and resolve, even in bright nearby disks. The CS lines, excited at $n \sim 10^5 - 10^6 \, cm^{-3}$, are sensitive to density, while the simplest organic molecule, H_2CO, has a slightly asymmetric top structure and thus serves as both a densitometer and thermometer (Mangum & Wootten 1993).

Observed molecular lines from almost all disks exhibit a signature of perfect Keplerian rotation, $V(r) \propto r^{-0.5}$. This indicates that over ~ 1 Myr, viscous evolution smears out any large-scale clumps in the disk structure that may remain from early, unstable evolutionary stages. A major observational result is that molecular abundances are "depleted" by factors 5–100 compared to the values in the Taurus Molecular Cloud (Dutrey *et al.* 2007a). Since disks have higher densities up to $10^7 - 10^{10} \, cm^{-3}$ and, enshrouded in stronger ionizing radiation fields, this "depletion" can be attributed to a combined effect of photodissociation and freeze-out.

The composition of dust constituents and ices in disks are studied by ground-based and space-borne infrared spectroscopy, and by observing various vibrational bands in absorption (when a disk is seen edge-on) or emission (when a disk is seen face-on). The results from the Infrared Space Observatory and Spitzer Space

Telescope have proven the existence of a significant amount of frozen material and various types of silicates and PAHs in disks (van den Ancker *et al.* 2000; van Dishoeck 2004; Bouwman *et al.* 2008). The PAH features at $\sim$3–12 μm probe the incident radiation field and density distribution of the upper disk (e.g. Habart *et al.* 2004).

Ices are present in the cold, outer disk region beyond the "snowline," at 100 K. The major component is water ice, with an abundance of about $\sim 10^{-4}$ (relative to the total hydrogen density). Usually water ice is intermixed with other, more volatile ices of e.g. CO, CO_2, NH_3, CH_4, H_2CO, and HCOOH (Zasowski *et al.* 2007). Typical abundances of these minor constituents are about 0.5–10% of that of water. Trapping of the volatile ices depends on the crystallinity of the water ice and the history of thermal processing (Sonnentrucker *et al.* 2008).

The recent detection of the [NeII] line emission at 12.81 μm from several disks by the Spitzer Space Telescope (e.g. Pascucci *et al.* 2007) has confirmed theoretical predictions that the disk atmosphere is heavily ionized and superheated, either by X-rays (Glassgold *et al.* 2007) or by extreme UV irradiation (Pascucci *et al.* 2007). However, X-rays and cosmic-ray particles (CRPs) may not be able to penetrate further toward the mid-plane of the planet-forming disk zone ($r \sim 3$–20 AU), which makes the mid-plane essentially neutral and thus stable against accretion ("Dead Zone"; Gammie 1996; Dolginov & Stepinski 1994).

Detected rotational H_2 and ro-vibrational CO, CO_2, C_2H_2, HCN, as well as H_2O and OH lines trace hot gas in the inner, planet-forming disk zone with $T \gtrsim 300$ K (Brittain *et al.* 2003; Lahuis *et al.* 2006; Salyk *et al.* 2008), see Fig. 4.3. These lines are a good measure of temperature and high-energy radiation fields, and presumably sensitive to disk accretion, which could be a stress-test for advanced chemo-dynamical models. In Table 4.1 the various molecules used to study protoplanetary disks are overviewed.

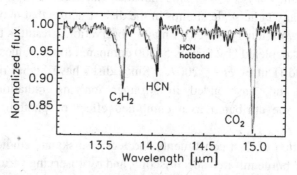

Figure 4.3 The IRS 46 spectrum obtained with Spitzer that shows the C_2H_2, HCN, and CO_2 ro-vibrational absorption bands (Lahuis *et al.* 2006).

4.1.3 Nebular photoprocesses

The higher the efficiency of photoionization or dissociation of a molecule the closer to the disk mid-plane it is located. Stellar and interstellar UV photons are able to penetrate deep into flaring disks by scattering (van Zadelhoff *et al.* 2003). Young T Tauri stars ($T_{eff} \simeq 4000\,K$) emit intense non-thermal UV radiation that has a spectrum different from the interstellar UV field (Bergin *et al.* 2003), while hot ($T_{eff} \gtrsim 10000\,K$) Herbig Ae/Be stars produce a lot of thermal UV emission. The overall intensity of the stellar UV radiation at 100 AU from the star can be as high as 500 and 10^5 times the intensity of the interstellar UV field (Habing 1968) for a T Tauri and a Herbig Ae star, respectively.

Many molecules, such as CO, H_2, and CN, are dissociated by radiation at short wavelengths ($\lambda \lesssim 1100\,\text{Å}$), while photodissociation of other species, such as HCN, occurs at longer wavelengths (for an introduction see van Dishoeck & Black 1988; van Dishoeck *et al.* 2006). Some molecules (e.g., H_2 and CO) dissociate via absorption of UV photons at discrete lines to the excited (Rydberg) states, whereas other molecules are dissociated either by the continuum (e.g. CH_4) or by the continuum and lines (e.g. C_2) (van Dishoeck 1988; van Dishoeck *et al.* 2006). The high ratio of CN to HCN abundances, observed in disks (Dutrey *et al.* 1997), can be explained by more intense photodissociation of HCN if part of the UV flux comes as Lyα photons (1216 Å; Nuth & Glicker 1982; Bergin *et al.* 2003; Pascucci *et al.* 2009).

Since the dissociation of the abundant H_2 and CO molecules result from photoabsorption at discrete wavelengths, isotopically selective photodissociation based on self-shielding is possible. Two conditions are required for this: (1) dissociation via line absorption for each isotopically substituted molecule, and (2) differential photolysis that depends upon the isotopic abundances. Self-shielding occurs when the spectral lines leading to dissociation of the major isotopic species optically saturate, while the other residual lines relevant for dissociation of the minor isotopes remain transparent. As a consequence, such photolysis depends on nucleic abundance rather than the mass of a molecule; see Fig. 4.4 (Langer 1977; Thiemens & Heidenreich 1983).

The absorption cross-sections for the isotopically substituted species are the same and thus for a given photon flux and path length, the photon absorption scales with the abundance of the isotopolog. Under conditions of the interstellar medium and solar nebula, isotope-selective chemistry may occur for abundant CO molecules (e.g. Dalgarno & Stephens 1970; Thiemens & Heidenreich 1983; Draine & Bertoldi 1996; Lee *et al.* 1996; van Dishoeck & Black 1988).

There are other isotopic effects that may occur from photolysis; these will be discussed in the context of the interpretation of meteoritic anomalies. It is apparent

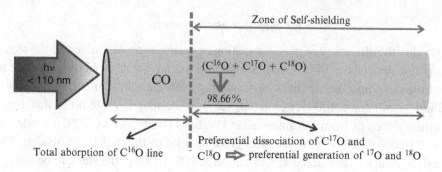

Figure 4.4 A sketch of self-shielding in the isotopologs of CO.

from the experimental results of Chakraborty *et al.* (2008) that these processes dominate self-shielding effects.

4.1.4 Cold gas-phase chemistry

Models of chemical evolution of protoplanetary disks employ a set of chemical reactions of relevance to astrophysical conditions. Modern astrochemical databases include up to 500 species involved in 5000 gas-phase, gas–grain, and surface reactions. Only 10–20% of the reaction rates have been studied in the laboratory or calculated theoretically and thus models are prone to inherent uncertainties (see, e.g. Herbst 1980; Wakelam *et al.* 2006; Vasyunin *et al.* 2008). Three ratefiles are widely applied in astrochemistry: the University of Manchester (UMIST,[2] Millar *et al.* 1997; Le Teuff *et al.* 2000; Woodall *et al.* 2007), the Ohio State University (OSU,[3] Smith *et al.* 2004), and a network incorporated in the "Photo-Dissociation Region" code from Meudon[4] (Le Petit *et al.* 2006). Several data sets of surface reactions have also been compiled (e.g. Allen & Robinson 1977; Tielens & Hagen 1982; Hasegawa *et al.* 1992; Hasegawa & Herbst 1993; Garrod & Herbst 2006). All these reactions can be divided into four distinct groups that are dominant in different disk regions, see Table 4.2 (after van Dishoeck & Black 1988).

Under local thermodynamical equilibrium the reaction rates can be written in the Arrhenius form:

$$\alpha_{AB} = \alpha_0 \left(\frac{T}{300 \text{ K}} \right)^{\beta} \exp \left(-\frac{\gamma}{T[\text{K}]} \right), \tag{4.1}$$

where T is the gas temperature, α_0 is the reaction rate constant at room temperature, the parameter β is the temperature dependence of the rate, and γ is the activation barrier (in Kelvin). For exothermic reactions the latter parameter is zero.

---

[2] http://www.udfa.net
[3] http://www.physics.ohio-state.edu/~eric/research.html
[4] http://aristote.obspm.fr/MIS/pdr/charge.html

Table 4.2 *Chemical reactions active in disks*

Reaction type	Example	Zone of ices	Zone of molecules	Zone of radicals	Inner zone
Bond formation					
Radiative association	$A + B \rightarrow AB + h\nu$	✓	✓	✓	✓
Associative detachment	$A^- + B \rightarrow AB + e^-$	✓	✓	—	✓
Surface formation	$A + B \| gr \rightarrow AB + gr$	✓	✓	—	—
Three-body	$A + B + M \rightarrow AB + M$	—	—	—	✓
Bond destruction					
Photodissociation	$AB + h\nu \rightarrow A + B$	—	✓	✓	✓
Dissociation by CRP	$AB + CRP \rightarrow A + B$	✓	✓	—	—
Dissociation by X-rays	—	—	—	✓	✓
Collisional dissociation	$AB + M \rightarrow A + B + M$	—	—	—	✓
Dissociative recombination	$AB^+ + e^- \rightarrow A + B$	✓	✓	✓	✓
Bond restructuring					
Neutral–neutral	$A + BC \rightarrow AB + C$	✓	✓	✓	✓
Ion–molecule	$A^+ + BC \rightarrow AB^+ + C$	✓	✓	✓	✓
Charge transfer	$A^+ + BC \rightarrow A + BC^+$	✓	✓	✓	✓
Unchanged bond					
Photoionization	$AB + h\nu \rightarrow AB^+ + e^-$	—	✓	✓	✓
Ionization by CRP	$AB + CRP \rightarrow AB^+ + e^-$	✓	✓	—	—
Ionization by X-rays	—	—	—	✓	✓

Apart from the very dense inner zone, all reactions in disks are two-body processes. Three-body reactions become competitive only at $\lesssim 10\,\mathrm{AU}$, where $n \gtrsim 10^{10}\,\mathrm{cm}^{-3}$ (Aikawa *et al.* 1999). The processes leading to formation of molecular bonds are radiative association, associative detachment, and surface reactions. Reactions of associative detachment are not efficient despite their high reaction rates ($\alpha_0 \sim 10^{-9}\,\mathrm{cm}^{-3}\,\mathrm{s}^{-1}$), mainly due to low abundances of negative ions (but see also Herbst 1981; Millar *et al.* 2000; McCarthy *et al.* 2006).

The most important process for the formation of the initial polyatomic species is radiative association. Upon collision, a large unstable molecule in an excited state ("collisional complex") may form, which is stabilized with a low probability by the emission of a photon (e.g. Bates 1951; Williams 1972; Herbst & Klemperer 1973). For example, the formation of light hydrocarbons starts with radiative association of C^+ and H_2, leading to excited CH_2^+ ($\alpha_0 = 4.0 \cdot 10^{-16}\,\mathrm{cm}^{-3}\,\mathrm{s}^{-1}$ and $\beta = -0.2$, Herbst 1985). Despite their importance, rates of many radiative association processes remain unknown because they are difficult to measure in the laboratory or predict theoretically (Herbst & Dunbar 1991).

Ion–neutral reactions dominate disk chemistry and constitute the largest fraction of astrochemical models. These reactions are exothermic, with high rate coefficients $\sim 10^{-9}$ cm^{-3} s^{-1}, which often increase toward low temperatures ($\beta < 0$, e.g. Dalgarno & Black 1976). This inverse temperature effect is due to the long-distance Coulomb attraction of an ion and a molecule with a high dipole moment. Ion–neutral reactions result in bond restructuring of the reactants. One of the most important reactions of this variety in disks are protonation reactions, such as: $H_3^+ + O \rightarrow OH^+ + H_2$, $OH^+ + H_2 \rightarrow H_2O^+ + H$, $H_2O^+ + H_2 \rightarrow H_3O^+ + H$, which leads to formation of water.

Molecular ions are efficiently destroyed in reactions of dissociative recombination with electrons and negative ions. These processes are especially fast at low temperatures, with typical rates of about 10^{-7} cm^{-3} s^{-1} at 10 K and negative temperature dependence, $\beta \approx -0.5$ (Woodall *et al.* 2007). For nearly all observed species, dissociative recombination is an important formation pathway (e.g. water and hydrocarbons). Often, at later evolutionary times, $\gtrsim 10^5$ years, dissociative recombination is balanced by protonation reactions, e.g. $CO + H_3^+ \rightarrow HCO^+ + H_2$ followed by $HCO^+ + e^- \rightarrow CO + H$. Dissociative recombination rates are not difficult to derive but branching ratios and products are not easily predictable (e.g. Bates & Herbst 1988; Spanel & Smith 1994).

In protoplanetary disks, a number of neutral–neutral reactions can also be fast, and not restricted to the warm inner region. In particular, neutral–neutral reactions between radicals and radicals, radicals and open-shell atoms, and radicals and unsaturated molecules are quite effective (van Dishoeck 1998). The typical rate coefficient for these reactions is $\sim 10^{-11}$–10^{-10} cm^{-3} s^{-1}, i.e. only about an order of magnitude lower than for the ion–molecule processes (e.g. Clary *et al.* 1994; Smith *et al.* 2004). One of the most interesting reactions of this type is the formation of HCO^+ upon collision between O and CH ($\alpha_0 = 2.0 \cdot 10^{-11}$ cm^{-3} s^{-1} and $\beta = 0.44$; Woodall *et al.* 2007). Another vital neutral–neutral reaction is the formation of formaldehyde: $CH_3 + O \rightarrow H_2CO + H$ ($\alpha_0 = 1.3 \cdot 10^{-10}$ cm^{-3} s^{-1} and $\beta = 0$; Woodall *et al.* 2007).

4.1.5 Freeze-out of gas-phase molecules

Due to large geometrical cross-sections and high densities, interactions of gaseous species with grains are efficient. In cold disk regions grains serve as a passive sink for heavy molecules and electrons, while providing free electrons to dissociate positive ions. In the darkest disk zone dust grains become the dominant charged species (Semenov *et al.* 2004). At low temperatures of $\lesssim 20$–50 K, many molecules adhere to a grain with a nearly 100% sticking probability because their kinetic energy is much smaller than the binding energy (see, e.g. d'Hendecourt *et al.* 1985; Buch & Zhang 1991). This sticking probability also depends on the grain surface properties, such as porosity and distribution of surficial chemi- and physisorption sites

(Mattera 1978; Leitch-Devlin & Williams 1985; Buch & Zhang 1991). Chemisorption requires the formation of a chemical bond between a surface species and a grain, and such a species does not evaporate easily (Cazaux *et al.* 2008). Physisorption is weak binding by van der Waals forces, and such a species can be returned to gas (e.g. Williams 1993).

4.1.6 Desorption of grain mantles

The most effective desorption processes for disk chemistry are thermal evaporation, cosmic-ray-induced desorption, and photodesorption. Thermal evaporation occurs if a molecule has energy that exceeds its binding energy. Typical binding energies of physisorbed species are about 1000 K for light molecules like CO and N_2 (Bisschop *et al.* 2006), and much larger for heavier cyanopolyynes and carbon chains. Chemisorbed species do not desorb until very high temperatures of 100–1000 K are reached.

In dark disk regions cosmic-ray particles provide an effective desorptive energy. A relativistic iron nucleus may eventually collide with a grain and impulsively heat it to about 70 K, releasing a portion of the volatile component (e.g. Watson & Salpeter 1972; Leger *et al.* 1985; Hartquist & Williams 1990).

In less opaque disk regions, penetrating UV photons lead to the photoevaporation of surface species. The probability of evaporation per one UV photon has been measured in the laboratory for some simple molecules, such as CO, H_2O, CH_4, and NH_3, and is about 10^{-6}–10^{-2} (e.g. Greenberg 1973; Bourdon *et al.* 1982; Westley *et al.* 1995; Öberg *et al.* 2007). The dilute UV radiation field produced by cosmic rays in the disk mid-plane (Prasad & Tarafdar 1983) can also be important for high values of the photodesorption rates. There were other non-thermal desorption mechanisms proposed, such as turbulent diffusion, as means to reconcile the inconsistency between heavy freeze-out in static chemical models and large concentrations of very cold CO and HCO^+ observed in disks (Dartois *et al.* 2003; Semenov *et al.* 2006). However, current observations of molecules in disks can be understood relatively well assuming that only canonical desorption mechanisms are at work there, and that the UV photodesorption yields are high, $\sim 10^{-3}$.

4.1.7 Surface formation of complex species

The surfaces of dust grains serve as catalysts for many reactions that do not proceed efficiently in the gas phase, in particular those involving intermediate excited-state complexes (e.g. radiative association). The most notable example is the formation of molecular hydrogen, which occurs almost entirely on dust surfaces (e.g. Hollenbach & Salpeter 1971; Watson & Salpeter 1972). It is assumed that a typical small $0.1 \mu m$ amorphous silicate grain has $\sim 10^6$ surface sites available for accretion. An accreted atom or light radical, if it is not chemisorbed, may migrate over the surface from site to site by thermal hopping, when its energy exceeds the barriers for particle motions, and react with other species. It was proposed that for the lightest species (e.g. H, H_2) this process can become less efficient than direct quantum-mechanical tunneling through the potential barriers (e.g. Williams 1993). This leads to enhanced conversion of surface molecules, such as CO, O, and N, to their saturated analogs like NH_3, H_2O, CH_4 (e.g. Tielens & Hagen 1982). However, detailed analysis of the the recent laboratory experiments has shown that the tunneling of hydrogen is unlikely to occur, at least on amorphous surfaces (Katz *et al.* 1999).

There is an increasing body of evidence from laboratory measurements that complex molecules like methanol cannot be produced in the gas phase via radiative association forming a large protonated precursor followed by its dissociative recombination (Geppert *et al.* 2005). Thus, surface reactions due to thermal hopping remain the only viable formation pathway for production of complex (organic) molecules in protoplanetary disks, found also in meteoritic samples. Further discussion of surface reaction mechanisms relevant for the oxygen isotopic anomalies in meteorites is presented in Section 4.4.3. In the following sections the focus shifts from the astronomical observations to isotopic analysis of meteorites to further explore the physico-chemical processes taking place in the solar nebula.

4.2 Chemical constraints from early Solar System materials

Several types of the early Solar System materials are available for laboratory analysis (see Chapter 1 and Table 1.1 and Fig. 1.1). Each material has unique characteristics and provides specific constraints on the chemistry of the solar nebula. Major components of this sample are meteorites, fragments of asteroids, that serve as an excellent archive of the early Solar System conditions. Primitive chondritic meteorites contain glassy spherical inclusions termed chondrules, some of the oldest solids in the Solar System. Most chondrites were modified by aqueous alteration or metamorphic processes in parent bodies but there are some chondrites that are minimally altered (un-equilibrated chondrites, UCs). They have yielded a wealth of information on the chemistry, physics, and evolution of the young Solar System.

Calcium–aluminum-rich refractory inclusions (CAIs) in chondrites are the oldest Solar System solids (2–3 Myr older than chondrules, see Chapters 1 and 9). The mineralogy, petrographic, chemical, and isotopic characteristics of these primitive solids constrain the physical and chemical processes through which these materials have been processed (Zhu *et al.* 2001; Becker & Walker 2003; Bizzarro *et al.* 2004; Chaussidon *et al.* 2008). The primordial presolar grains were mostly altered by the thermal processes in the early solar nebula, while a small fraction have surprisingly survived in primitive chondrites, interplanetary dust particles, and comet samples (Bradley 2003; Huss & Draine 2007). Presolar grains exhibit a range of chemical and thermal resistance, so their relative abundance can be used to probe the conditions of the early nebular environment (e.g. temperature, radiation field, and mixing). As an example, heating in a highly oxidizing environment would affect carbonaceous components more than oxides, whereas a reducing environment would increase the survival of carbonaceous components. Thus, relative abundances of different kinds of presolar grains could be related to the thermal processing in the solar nebula that produced the bulk compositions of different classes of meteorite from primordial materials (Huss *et al.* 2003; Huss & Draine 2007). In the following sections we outline the isotopic anomalies found in different primitive meteoritic materials and provide a detailed discussion of the oxygen isotopic anomaly, which has been at the center of debate for the last three decades in the cosmochemistry community (Thiemens 2006; Clayton 2007).

4.3 Isotopic anomalies and condensation sequence

The observed overabundance of deuterated species in molecular clouds and outer disks compared to the measured interstellar D/H ratio of $\sim 10^{-5}$ is well established. A classical isotopic deuterium fractionation is possible at low temperatures of ~ 10–20 K owing to disbalance between forward and reversed reaction efficiencies: $H_3^+ + HD \leftrightarrows H_2D^+ + H_2 + 232$ K (e.g. Millar *et al.* 1989; Gerlich *et al.* 2002). The temperature dependency in an isotope exchange reaction is a consequence of the zero-point vibrational energy difference for the isotopically substituted molecules (Bigeleisen & Mayer 1947; Urey 1947). This leads to an elevated ratio of H_2D^+/H_3^+ compared to HD/H_2, which is quickly transferred into other molecules by ion–molecule reactions (see e.g. Roberts & Millar 2000; Roberts *et al.* 2003). For example, the dominant reaction pathway to produce DCO^+ is via ion–molecule reactions of CO with H_2D^+. In disks it results in a DCO^+ to HCO^+ ratio that increases with radius owing to the outward decrease of temperature (Aikawa & Herbst 2001; Willacy 2007; Qi *et al.* 2008).

Other important fractionation reactions effective at higher temperatures (up to 70 K) are: $CH_3^+ + HD \leftrightarrows CH_2D^+ + H_2 + 390$ K (Asvany *et al.* 2004) and $C_2H_2^+ + HD \leftrightarrows C_2HD^+ + H_2 + 550$ K (Herbst *et al.* 1987). Both reactions lead to DCN by the ion–molecule reaction: $N + CH_2D^+ \rightarrow DCN^+$, followed by reaction with H_2 and dissociative recombination (Roueff *et al.* 2007). An alternative route is surface addition of D to CN (Hiraoka *et al.* 2006). The measured abundance ratio of DCN to HCN is about 1% in the disk around TW Hya (Qi *et al.* 2008). Effective deuterium fractionation on dust grains in cold disk regions is also possible owing to the effective catalytic exchange reactions on a surface. Subsequent surface-facilitated addition of D to CO-bearing radicals, and substitution of H by D in intermediate products, are all viable processes (Hiraoka *et al.* 2005).

There is evidence from chondrites that the solar nebula was well mixed between 0.1 and 10 AU during its first several million years of the evolution, as shown by the homogeneity in concentrations of many isotopes of refractory elements (Boss 2004; Chapter 9). This is likely caused by the evaporation and recondensation of solids in the very hot inner nebula, followed by outward transport due to turbulent diffusion and angular momentum removal. Materials out of which terrestrial planets and asteroids are built have been heated to temperatures above 1300 K and are thus depleted in volatile elements. The inner solar nebula, with some exceptions, does not retain memories of the pristine interstellar medium (ISM) chemical composition (Palme 2001; Trieloff & Palme 2006).

This observation is also supported by the fact that silicates in comets and other protoplanetary disks are partly crystalline and thus require high-temperature processing, which is not possible in the cool outer region of the solar nebula (Chapters 5 and 9, e.g. van Boekel *et al.* 2004; Wooden *et al.* 2007). However, isotopic anomalies in certain elements, such as short-lived radio nuclides ^{26}Al and ^{41}Ca, suggest temporal and spatial heterogeneity in the nebula (Boss 2004). The source of this anomaly is not certain and is attributed to injection from nearby supernovae in a short time span (Goswami & Vanhala 2000) or by *in situ* production (Gounelle *et al.* 2001). Certain groups of meteorites also show minor variations in their chemical composition from the bulk solar nebula, reflecting the local conditions of their formation. An excellent and comprehensive review of the subject has been provided by Grossman (2008).

All these compositional variations and fractionation effects can be better understood if one invokes the condensation sequence of the elements in the early Solar System (see Chapters 2 and 8). Assuming thermodynamical equilibrium between solids and the gas – true for the inner, hot and dense solar nebula but not for outer cold regions – it can be derived which minerals condense out first and at what temperature (Grossman 1972). The C/O ratio in the local environment played a key role in the condensation of particular types of minerals, depending on oxygen fugacity (Grossman 2008), which is largely determined by the relative abundances of carbon, oxygen, and hydrogen in nebular gases.

According to Palme (2001), four major condensation components can be isolated, see Table 4.3. While highly refractory elements have been locked in CAIs already at $T \sim 1800$ K, the bulk of meteoritic samples are made of less refractory materials, like silicates and metals ($T < 1400$ K). Iron condenses almost entirely in metallic form, and silicates are mostly as Mg-rich forsterite and enstatite. At about 700 K

Table 4.3 *Condensation sequence for the early solar nebula*

Component	Elements	Condensation temperature*
Refractory	Al-, Ca-, Ti-oxides, trace (Zr, V, Wo, etc.)	1850–1400 K
Main	Mg-rich silicates, metallic Fe, Ni, Co, Li, Cr	1350–1250 K
Volatile	Sulfides, silicates, metals (Mn, P, Na, Rb, K, F, Zn, Au, Cu, Ag, etc.)	1250–640 K
Highly volatile	C, H, O, N, Cl, Ne, Ar, Xe, Kr, etc.	< 640 K

* Condensation temperatures are given for the pressure of 10^{-4} bar.

sulfur is condensed in the form of troilite (FeS). At even lower temperatures other elements (C, H, O, N) locked in various molecules "condense" (or accrete) from the gas phase, e.g. water ($T \sim 100\,K$), ammonia ($T \sim 70\,K$), CO ($T \sim 20\,K$), etc. Note, that volatile and highly volatile elements are depleted in almost all materials of the inner Solar System (see Chapter 10), so these differ chemically from the bulk composition retained in the outer disk regions, where comets have formed.

4.4 Oxygen isotopes

Oxygen is the third most abundant element in the Solar System. Oxygen possesses three stable isotopes, $^{16,17,18}O$, with concentrations that are different between distinct families of meteoritic minerals. The simultaneous presence in both gaseous and solid phases renders oxygen isotopes as important tracers of physical and chemical processes in the solar nebula (Clayton 2007). Isotopic partitioning depends on the zero-point energy difference and involves chemical kinetics and thermodynamic equilibrium, properties which vary widely from molecule to molecule. These changes, by convention, are measured through the so-called δ-notation (in permil):

$$\delta^{17,18}O = \left[\left(^{17,18}O/^{16}O \right)_{sample} / \left(^{17,18}O/^{16}O \right)_{std} - 1 \right] \times 1000 \quad (4.2)$$

where subscripts "sample" and "std" represent sample and reference material, respectively. For oxygen isotopes, the reference material is the Vienna Standard Mean Ocean Water (VSMOW). As has been shown for equilibrium as well as for

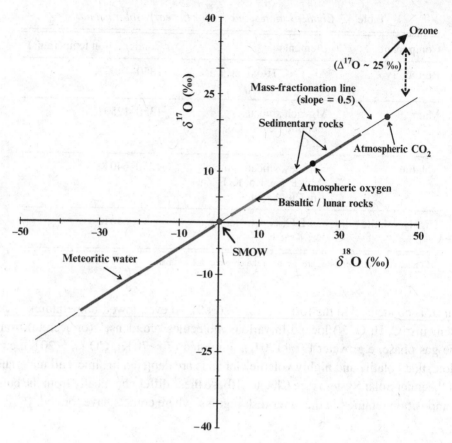

Figure 4.5 The oxygen isotopic composition of various terrestrial and lunar reservoirs.

kinetic processes (Bigeleisen & Mayer 1947; Urey 1947; Young *et al.* 2002), the variation between $^{18}O/^{16}O$ and $^{17}O/^{16}O$ differs by a factor of ≈ 0.5:

$$\delta^{17}O \simeq 0.5 \times \delta^{18}O$$

The oxygen isotopes of terrestrial materials mostly fall along with what is termed a mass-fractionation line (MFL). The MFL is defined by a slope of 0.5 passing through VSMOW as shown in Fig. 4.5. The oxygen isotopic composition is unique for different types of meteorites, which are broadly classified as chondrites and achondrites.

In Fig. 4.6 the oxygen isotopic composition of these two meteoritic types are compared. The bulk oxygen isotopic composition of various groups of achondrites

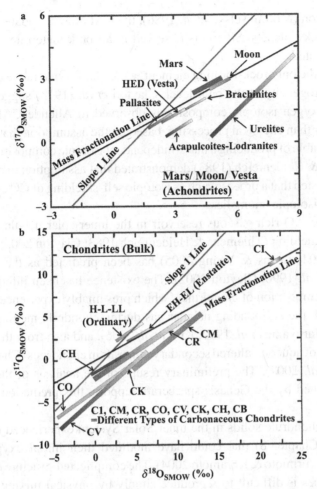

Figure 4.6 Bulk oxygen isotopic compositions of (a) achondrites and meteorites from Mars, the Moon, and Vesta; (b) chondrites (after Yurimoto *et al.* 2006).

(Fig. 4.6a) and chondrites (Fig. 4.6b) occupy distinct positions in the three-isotope space, which reflects the influence of environmental conditions.

The oxygen isotopic composition of CAIs is unique and more depleted in heavy oxygen isotopes $^{17,18}O$ compared to chondrules. It plots along the line of slope ~ 1 and exhibits a large range of $\Delta^{17}O$ ($= \delta^{17}O - 0.5\delta^{18}O$) values, ranging from < -20 to $+5‰$ (Clayton 1993; Aléon *et al.* 2002; Itoh *et al.* 2004). Some igneous texture CAIs and chondrules are $^{17,18}O$-enriched. Although matrix minerals possess variations in oxygen isotopic abundances (Kunihiro *et al.* 2005), the bulk O-isotopic composition of CAIs is similar to that of chondrules (Yurimoto *et al.* 2006). The heterogeneity in oxygen isotopes is not reflected in the isotopes of its companion element, silicon (Molini-Velsko *et al.* 1986), the next most abundant element after oxygen in rocky planets. Silicon is observed in all bulk meteorites to be strictly mass fractionated.

Oxygen is the only rock-forming element, which has a wide range of anomalous isotopic heterogeneity at the bulk level. Clayton *et al.* (1973) suggested that the anomalous oxygen isotopic compositions observed in Allende CAIs are due to nuclear rather than chemical processing, based on the assumption that no chemical process is capable of producing a mass-independent isotopic composition. The work of Thiemens & Heidenreich (1983) demonstrated that assumption to be wrong and instead suggested that a process such as isotopic self-shielding of CO could produce the observed isotopic anomalies.

An extreme ^{16}O-rich gaseous reservoir in the inner solar nebula arising from the CO self-shielding (Thiemens & Heidenreich 1983; Clayton 2002; Yurimoto & Kuramoto 2004; Lyons & Young 2005) has been predicted as the precursor of CAIs (Shu *et al.* 1996; Clayton 2002). The existence has been inferred from the ^{17,18}O-rich composition of chondrules, which presumably experienced O-isotopic exchange with the surrounding nebular gas during chondrule melting events (Yu *et al.* 1995; Maruyama *et al.* 1999; Krot *et al.* 2006), and also from the ^{17,18}O-rich compositions of aqueous-altered secondary minerals in chondrites (Choi *et al.* 1998; Sakamoto *et al.* 2007). The preliminary results of the analysis of the solar wind samples returned by the Genesis spacecraft supports this prediction (McKeegan *et al.* 2008).

Most non-chondrule solids in the inner Solar System experienced thermal processing (see Chapter 8) that could have modified their initial oxygen isotopic composition (Yurimoto & Kuramoto 2004). The complicated structure of meteoritic oxygen isotopes is difficult to reproduce simply by physical mixing of different reservoirs. Apart from thermal processing (e.g. melting, vaporization, condensation), a large mass-independent chemical process is required. The exact mechanism for this likely photochemical process is yet unknown, but the available constraints leave only a few pathways open.

4.4.1 CO self-shielding

Three CO isotopologs, ^{12}C^{16}O, ^{12}C^{17}O and ^{12}C^{18}O, have relative abundances of 98.65, 0.0375, and 0.202%, respectively. They dissociate by photon absorption from more than 40 discrete lines within a 92 to 108 nm spectral window (van Dishoeck & Black 1988; Eidelsberg *et al.* 1991, 1992). The concept of self-shielding in the solar nebula of O$_2$ and CO was first suggested by Thiemens & Heidenreich (1983) based on their experiments and on molecular cloud observations (Langer 1977; Bally & Langer 1982). The CO self-shielding has been invoked to explain the enhanced

^{13}C and ^{18}O atomic signatures observed by radio astronomers (Glassgold *et al.* 1985; van Dishoeck & Black 1988; Langer & Penzias 1990; Sheffer *et al.* 2002; Dartois *et al.* 2003; Piétu *et al.* 2007) and anomalous meteoritic δ^{15}N composition (Thiemens & Heidenreich 1983; Clayton 2002).

Recently, self-shielding of CO has been revived to account for the source of isotopically anomalous oxygen in the solar reservoir and as a mechanism for production of meteoritic oxygen isotopic compositions falling along the CCAM (carbonaceous chondrite anhydrous mineral) line (Clayton 2002; Yurimoto & Kuramoto 2004; Lyons & Young 2005). Selective photodissociation of CO theoretically results in a heavy O-atom reservoir with a characteristic composition of $\delta^{17}O = \delta^{18}O$. In the inner zone of the nebula it occurs along a modest path length of a few AU. All of the CO absorption lines may not be equally effective in dissociation owing to screening by strongly absorbing hydrogen lines in the UV range. The CO absorption line at 105.17 nm is one of the most effective lines for the $C^{18}O$ dissociation, responsible for 58% of its destruction (van Dishoeck & Black 1988).

Recently, experiments to measure the isotopic fractionation associated with the CO photodissociation have been performed at the Advanced Light Source (ALS, Chakraborty *et al.* 2008). Photodissociation of CO by UV synchrotron photons (10^{16} photons per second) at four different wavelengths ($\lambda = 107.61$, 105.17, 97.62, 94.12 nm) leads to C and O atoms, which subsequently recombine to form CO_2. In the experiments, optical depths of relevant $C^{16}O$ lines were several orders of magnitude thicker compared to the optical depth of $C^{18}O$ and $C^{17}O$, which were optically thin.

The oxygen isotopic composition of the product CO_2 was heavily enriched (of the order of a few thousands of permil) compared to the initial CO (Fig. 4.6). This extent of isotopic fractionation is one of the largest reported for any process. The slope values in a three-isotope oxygen plot ($\delta^{18}O$ vs. $\delta^{17}O$) is wavelength-dependent for CO_2. It was anticipated previously that a self-shielding should be present at 105.17 and 94.12 nm (van Dishoeck & Black 1988), whereas there should be no self-shielding for the 107.62 and 97.03 nm lines (Warin *et al.* 1996). A slope value of unity is expected on the three-isotope oxygen plot from the CO self-shielding.

To the contrary, the experimental results show a unique slope value of 1.38 for 105.17 and 107.61 nm (Fig. 4.7). Therefore, the experimental results *do not show the effect of self-shielding* but instead reveal an extremely large wavelength-dependent isotopic fractionation. Isotope-selective effects during CO dissociation at various wavelengths are caused by distinct dissociation dynamics at higher electronic states upon absorption of a photon.

These experimental results for the first time demonstrate that a heavy-isotope-enriched and anomalously fractionated O-reservoir within the solar nebula (or

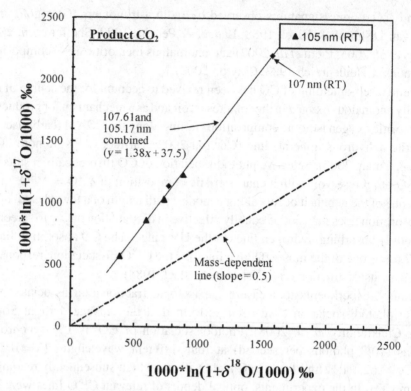

Figure 4.7 Oxygen isotopic composition of product CO_2 generated through UV (105.17 and 107.61 nm) photolysis of CO. The results are plotted in three-oxygen isotope space on a logarithmic scale to accommodate the non-linearity in δ scale for large δ values.

within the precursor molecular cloud) is possible through wavelength-dependent photodissociation of CO. The fate of this isotopically anomalous O-atom may be the production of water through chain reactions with H_2 and H_3^+ (Yurimoto & Kuramoto 2004; Lyons & Young 2005). As a consequence, the formation of an anomalously fractionated water ice reservoir is possible in the outer disk mid-plane far from the proto-Sun owing to the CO photodissociation and dynamical transport.

However, the mechanism through which this anomaly is sequestered inside the solids is not clear. It is postulated that, as the nebular disk evolved, this isotopically enriched ice migrates inward (and mixes with the isotopically normal water) and volatilizes in the inner nebula, driving gas-phase oxidation processes to form solids, during which the isotopic anomaly is passed to condensates (Yurimoto & Kuramoto 2004). A major issue is that the silicate materials do not show the required level of heavy-oxygen fractionation, so another mass-independent mechanism is required.

4.4.2 Oxygen isotopic fractionation in silicates

A chemically based, mass-independent fractionation process was first observed during ozone formation through the gas-phase recombination reaction (Thiemens & Heidenreich 1983): $O + O_2 + M \rightarrow O_3 + M$. The product ozone possesses equally enriched heavy-oxygen isotopes $^{17,18}O$, by approximately 100‰ with respect to the initial oxygen, with a slope value of unity in a three-isotope oxygen plot. This discovery led to the conclusion that a symmetry-dependent reaction can produce meteoritic isotopic anomalies (Thiemens 1999, 2006). Recently, theoretical calculations of Gao & Marcus (2001) established the major role of symmetry in isotopolog-specific stabilization of vibrationally excited ozone molecules that give rise to the mass-independent compositions.

The formation of a stable product is a two-step process: $O + O_2 \rightarrow O_3^*; O_3^* + M \rightarrow O_3 + M$. After the formation of the vibrationally excited molecule, the subsequent redistribution of the energy among its vibrational–rotational modes proceeds at some finite rate and may be incomplete during the typical lifetime of the molecule. This non-statistical concept was used to model the laboratory data with a mathematically introduced non-statistical factor "η." A value of 1.18 for "η" matches the observations well.

Ozone is not the only molecule that shows a symmetry-based mass-independent effect; the reaction: $CO + O \rightarrow CO_2$ (Bhattacharya & Thiemens 1989; Pandey & Bhattacharya 2006) also shows similar behavior and emphasizes the role of symmetry as being of general nature in isotopic fractionation processes. This is particularly significant for the nebula as the oxidative reactions leading to solid symmetric silicates undoubtedly occurred there, e.g. $SiO + O \rightarrow OSiO$.

Kimura *et al.* (2007) reported the first reproducible production of mass-independent oxygen in iron and silicate smokes from gas-phase reactions (Fig. 4.8). It is possible to misinterpret these data as being associated with ozone, which is known to have a large mass-independent composition. Comparing kinetics of the rates of the two reaction channels competing for reaction with O-atoms: $O + O_2 + M \rightarrow O_3 + M$ and $O + H_2 \rightarrow OH + H$, one finds a factor of twenty higher probability for a H_2 sink compared to ozone formation.

In fact, the observed values of the silicate oxygen are depleted with respect to the initial oxygen (with a slope of 0.65) and do not fall over the mixing line of ozone and silicate (100 and 10‰ as two end-members, respectively, see Fig. 4.9). This implies that the relevant oxidation step may involve OH (or HO_2) when H_2 is present as a reactant and generate the mass-independent oxygen composition by: $SiO + OH \rightarrow SiO_2 + H; SiO + HO_2 \rightarrow SiO_2 + OH$.

The support for this OH-based oxidation as a source of mass-independent effect comes from experimentally studied symmetry-based oxidations of CO with OH (Roeckmann *et al.* 1998; Thiemens 2006). Therefore, OH oxidation may be an

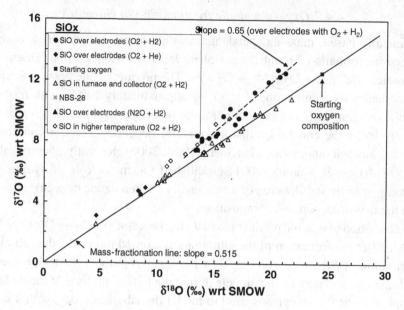

Figure 4.8 Oxygen isotopic composition of silicates formed in the smoke chamber of the Goddard Space Flight Center in a three-isotope plot.

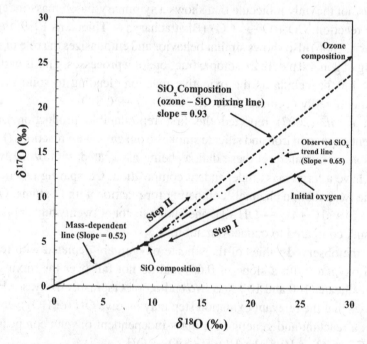

Figure 4.9 Schematic illustration displaying the observed oxygen isotopic composition of silicate and that expected via reaction with ozone in a three-isotope plot. Oxidation was postulated as a two-step fractionation process (Kimura *et al.* 2007) as shown in the diagram.

alternative way to form mass-independently fractionated solids in the solar nebula as the concentration of OH in the gas phase in the inner nebula is significant (Lyons *et al.* 2007).

4.4.3 Isotopic anomalies due to oxidation on grain surfaces

Marcus (2004) hypothesized that CAI formation through surface oxidation can lead to ^{16}O enhancements in solid products. This silicate-forming oxidation process bypasses the drawback of very low densities (and thus long timescales) for gas-phase recombination reactions under nebular conditions. The reaction scheme is displayed in Fig. 4.10.

During growth of CAI grains, equilibrium is postulated between adsorbed reactants: $XO_{(ads)} + O_{(ads)} \rightarrow XO^*_{2(ads)}$, where $XO^*_{2(ads)}$ is a vibrationally excited adsorbed dioxide molecule and X may be Si, Al, Ti, or any other metal. The surface of a growing grain serves as a catalytic site, enhancing the effective reactive concentrations by several orders of magnitude relative to the gas phase (entropic factor). A surface η effect yields $XO^*_{2(ads)}$ that is mass-independently enriched in ^{17}O and

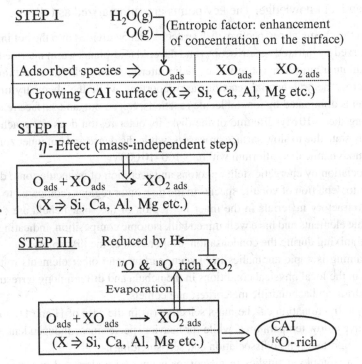

Figure 4.10 Schematic of a surface-assisted ^{16}O-rich CAI formation scheme of Marcus (2004).

^{18}O, analogous to that observed for ozone. This yields $XO_{(ads)}$ and $O_{(ads)}$, which are mass-independently depleted in ^{17}O and ^{18}O as the $XO_{2(ads)}^*$ is deactivated by vibrational energy loss to the grain. Subsequently, evaporation occurs, which may lead to an additional (mass-dependent) fractionation. The other species, $O_{(ads)}$ and $XO_{(ads)}$, are ^{17}O- and ^{18}O-poor (^{16}O-rich) and react with adsorbed metal atoms or metallic monoxides to form CAIs, become mass-independently rich in ^{16}O, and lie along the slope-one line. The CAIs directly condensed from a hot nebular gas may be formed by this chemical mechanism, but not igneous CAIs.

Overall, the origin of the oxygen isotopic anomalies in the Solar System is still a matter of debate (Thiemens 2006). After the failure of the supernova hypothesis (Clayton *et al.* 1973), these anomalies most likely originated through chemical or physical processes within the nebula.

4.5 Summary

In this chapter we have reviewed our current understanding of the chemical evolution of the solar nebula and protoplanetary disks using astrophysical and cosmochemical knowledge. The key points are summarized as follows.

(i) From the chemical perspective, disk structure can be divided into the hot inner region (observed by infrared spectroscopy) and the cold mid-plane, warm intermediate layer, and dilute atmosphere (observed by radio-interferometry). In the dense inner region numerous neutral–neutral reactions are important, while the chemistry of the outer region is dominated by ion–molecule reactions, freeze-out, and surface reactions.

(ii) During the $\lesssim 10$ Myr lifetime of the disk, its outer region does not reach chemical steady state due to slow surface processes, while the dense and hot inner zone reaches thermodynamical equilibrium within $\lesssim 100$–$10\,000$ yr.

(iii) Dissociation by energetic stellar photons and accretion of molecules onto dust grains leads to depletion of volatile species in the disk outer regions (compared to the ISM).

(iv) The refractory materials in the inner hot region of the solar nebula are depleted in volatile elements and have well-mixed bulk isotopic composition, indicating effective radial mixing during the condensation–evaporation of the first solids.

(v) Remaining isotopic anomalies in oxygen, carbon, and other elements reflect variations in the local physical conditions in the nebula and differ among terrestrial rocks, chondritic and achondritic meteorites, and comets.

(vi) Isotopic fractionation can be mass selective (as in the case of H and D) and thus be effective at low temperatures, while for heavier species mass-independent effects are more important (photodissociation and self-shielding).

(vii) Oxygen isotopic anomalies in meteorites cannot be understood in terms of simple mixing of different O-bearing reservoirs and require (photo)chemical processing.

References

Aikawa, Y. & Herbst, E. 1999, *Astronomy and Astrophysics*, **351**, 233.

Aikawa, Y. & Herbst, E. 2001, *Astronomy and Astrophysics*, **371**, 1107.

Aikawa, Y., Momose, M., Thi, W.-F., *et al.* 2003, *Publications of the Astronomical Society of the Pacific*, **55**, 11.

Aikawa, Y., Umebayashi, T., Nakano, T., & Miyama, S. M. 1999, *Astrophysical Journal*, **519**, 705.

Aléon, J., Krot, A. N., & McKeegan, K. D. 2002, *Meteoritics and Planetary Science*, **37**, 1729.

Allen, M. & Robinson, G. W. 1977, *Astrophysical Journal*, **212**, 396.

Asvany, O., Schlemmer, S., & Gerlich, D. 2004, *Astrophysical Journal*, **617**, 685.

Bally, J. & Langer, W. D. 1982, *Astrophysical Journal*, **261**, 747.

Bates, D. R. 1951, *Monthly Notices of the Royal Astronomical Society*, **111**, 303.

Bates, D. R. & Herbst, E. 1988, in *Rate Coefficients in Astrochemistry*, ed. T. J. Millar & D. A. Williams, Kluwer, 41–48.

Becker, H. & Walker, R. J. 2003, *Nature*, **425**, 152.

Beckwith, S. V. W. & Sargent, A. I. 1993, in *Protostars and Planets III*, ed. E. H. Levy & J. I. Lunine, University of Arizona Press, 521–541.

Bergin, E., Calvet, N., D'Alessio, P., & Herczeg, G. J. 2003, *Astrophysical Journal*, **591**, L159.

Bergin, E. A., Aikawa, Y., Blake, G. A., & van Dishoeck, E. F. 2007, in *Protostars and Planets V*, ed. B. Reipurth, D. Jewitt, & K. Keil, University of Arizona Press, 751–766.

Bhattacharya, S. K. & Thiemens, M. H. 1989, *Zeitschrift Naturforschung A: Physical Sciences* **44**, 435.

Bigeleisen, J. & Mayer, M. 1947, *Journal of Chemical Physics*, **15**, 261.

Bisschop, S. E., Fraser, H. J., Öberg, K. I., van Dishoeck, E. F., & Schlemmer, S. 2006, *Astronomy and Astrophysics*, **449**, 1297.

Bitner, M. A., Richter, M. J., Lacy, J. H., *et al.* 2007, *Astrophysical Journal*, **661**, L69.

Bizzarro, M., Baker, J. A., & Haack, H. 2004, *Nature*, **431**, 275.

Boss, A. P. 2004, *Astrophysical Journal*, **616**, 1265.

Bourdon, E. B., Prince, R. H., & Duley, W. W. 1982, *Astrophysical Journal*, **260**, 909.

Bouvier, J., Alencar, S. H. P., Harries, T. J., Johns-Krull, C. M., & Romanova, M. M. 2007, in *Protostars and Planets V*, ed. B. Reipurth, D. Jewitt, & K. Keil, University of Arizona Press, 479–494.

Bouwman, J., Henning, T., Hillenbrand, L. A., *et al.* 2008, *Astrophysical Journal*, **683**, 479.

Bradley, J. P. 2003, *Treatise on Geochemistry*, **1**, 689.

Brittain, S. D., Rettig, T. W., Simon, T., *et al.* 2003, *Astrophysical Journal*, **588**, 535.

Buch, V. & Zhang, Q. 1991, *Astrophysical Journal*, **379**, 647.

Cazaux, S., Caselli, P., Cobut, V., & Le Bourlot, J. 2008, *Astronomy and Astrophysics*, **483**, 435.

Chakraborty, S., Ahmed, M., Jackson, T. L., & Thiemens, M. H. 2008, *Science*, **321**, 1328.

Chaussidon, M., Libourel, G., & Krot, A. N. 2008, *Geochimica et Cosmochimica Acta*, **72**, 1924.

Choi, B.-G., McKeegan, K. D., Krot, A. N., & Wasson, J. T. 1998, *Nature*, **392**, 577.

Clary, D. C., Haider, N., Husain, D., & Kabir, M. 1994, *Astrophysical Journal*, **422**, 416.

Clayton, R. N. 1993, *Annual Review of Earth and Planetary Sciences*, **21**, 115.

Clayton, R. N. 2002, *Nature*, **415**, 860.

Clayton, R. N. 2007, *Annual Review of Earth and Planetary Sciences*, **35**, 1.

Clayton, R. N., Grossman, L., & Mayeda, T. K. 1973, *Science*, **182**, 485.
Dalgarno, A. & Black, J. H. 1976, *Reports of Progress in Physics*, **39**, 573.
Dalgarno, A. & Stephens, T. L. 1970, *Astrophysical Journal*, **160**, L107.
Dartois, E., Dutrey, A., & Guilloteau, S. 2003, *Astronomy and Astrophysics*, **399**, 773.
d'Hendecourt, L. B., Allamandola, L. J., & Greenberg, J. M. 1985, *Astronomy and Astrophysics*, **152**, 130.
Dolginov, A. Z. & Stepinski, T. F. 1994, *Astrophysical Journal*, **427**, 377.
Draine, B. T. & Bertoldi, F. 1996, *Astrophysical Journal*, **468**, 269.
Dutrey, A., Guilloteau, S., & Guelin, M. 1997, *Astronomy and Astrophysics*, **317**, L55.
Dutrey, A., Guilloteau, S., & Ho, P. 2007a, in *Protostars and Planets V*, ed. B. Reipurth, D. Jewitt, & K. Keil, University of Arizona Press, 495–506.
Dutrey, A., Henning, T., Guilloteau, S., *et al.* 2007b, *Astronomy and Astrophysics*, **464**, 615.
Eidelsberg, M., Benayoun, J. J., Viala, Y., *et al.* 1992, *Astronomy and Astrophysics*, **265**, 839.
Eidelsberg, M., Viala, Y., Rostas, F., & Benayoun, J. J. 1991, *Astronomy and Astrophysics Supplement*, **90**, 231.
Gammie, C. F. 1996, *Astrophysical Journal*, **457**, 355.
Gao, Y. Q. & Marcus, R. A. 2001, *Science*, **293**, 259.
Garrod, R. T. & Herbst, E. 2006, *Astronomy and Astrophysics*, **457**, 927.
Geppert, W. D., Hellberg, F., Österdahl, F., *et al.* 2005, in *Astrochemistry: Recent Successes and Current Challenges*, ed. D. C. Lis, G. A. Blake, & E. Herbst, IAU Symposia, no. 231, 117–124.
Gerlich, D., Herbst, E., & Roueff, E. 2002, *Planetary and Space Science*, **50**, 1275.
Glassgold, A. E., Huggins, P. J., & Langer, W. D. 1985, *Astrophysical Journal*, **290**, 615.
Glassgold, A. E., Najita, J., & Igea, J. 1997, *Astrophysical Journal*, **480**, 344.
Glassgold, A. E., Najita, J. R., & Igea, J. 2007, *Astrophysical Journal*, **656**, 515.
Goswami, J. N. & Vanhala, H. A. T. 2000, in *Prostostars and Planets IV*, ed. V. Mannings, A. P. Boss, & S. S. Russell, University of Arizona Press, 963–990.
Gounelle, M., Shu, F. H., Shang, H., *et al.* 2001, *Astrophysical Journal*, **548**, 1051.
Greenberg, J. M. 1973, in *Molecules in the Galactic Environment*, ed. M. Gordon & L. Snyder, John Wiley and Sons, 93–142.
Grossman, L. 1972, *Geochimica et Cosmochimica Acta*, **36**, 597.
Grossman, L. 2008, in *Oxygen in the Solar System*, ed. G. J. MacPherson, The Mineralogical Society of America, 93–140.
Güdel, M., Skinner, S. L., Mel'Nikov, S. Y., *et al.* 2007, *Astronomy and Astrophysics*, **468**, 529.
Habart, E., Natta, A., & Krügel, E. 2004, *Astronomy and Astrophysics*, **427**, 179.
Habing, H. J. 1968, *Bulletin of the Astronical Institute of the Netherlands*, **19**, 421.
Hartquist, T. W. & Williams, D. A. 1990, *Monthly Notices of the Royal Astronomical Society*, **247**, 343.
Hasegawa, T. I. & Herbst, E. 1993, *Monthly Notices of the Royal Astronomical Society*, **263**, 589.
Hasegawa, T. I., Herbst, E., & Leung, C. M. 1992, *Astrophysics Journal Supplement Series*, **82**, 167.
Herbst, E. 1980, *Astrophysical Journal*, **241**, 197.
Herbst, E. 1981, *Nature*, **289**, 656.
Herbst, E. 1985, *Astrophysical Journal*, **291**, 226.
Herbst, E., Adams, N. G., Smith, D., & Defrees, D. J. 1987, *Astrophysical Journal*, **312**, 351.

Herbst, E. & Dunbar, R. C. 1991, *Monthly Notices of the Royal Astronomical Society*, **253**, 341.

Herbst, E. & Klemperer, W. 1973, *Astrophysical Journal*, **185**, 505.

Hiraoka, K., Ushiama, S., Enoura, T., *et al.* 2006, *Astrophysical Journal*, **643**, 917.

Hiraoka, K., Wada, A., Kitagawa, H., *et al.* 2005, *Astrophysical Journal*, **620**, 542.

Hollenbach, D. & Salpeter, E. E. 1971, *Astrophysical Journal*, **163**, 155.

Huss, G. R. & Draine, B. T. 2007, *Highlights of Astronomy*, **14**, 353.

Huss, G. R., Meshik, A. P., Smith, J. B., & Hohenberg, C. M. 2003, *Geochimica et Cosmochimica Acta*, **67**, 4823.

Ilgner, M., Henning, T., Markwick, A. J., & Millar, T. J. 2004, *Astronomy and Astrophysics*, **415**, 643.

Isella, A., Testi, L., Natta, A., *et al.* 2007, *Astronomy and Astrophysics*, **469**, 213.

Itoh, S., Kojima, H., & Yurimoto, H. 2004, *Geochimica et Cosmochimica Acta*, **68**, 183.

Kastner, J. H., Zuckerman, B., Weintraub, D. A., & Forveille, T. 1997, *Science*, **277**, 67.

Katz, N., Furman, I., Biham, O., Pirronello, V., & Vidali, G. 1999, *Astrophysical Journal*, **522**, 305.

Kimura, Y., Joseph, N. A.and Chakraborty, S., & Thiemens, M. H. 2007, *Meteoritics and Planetary Science*, **42**, 1429.

Koerner, D. W. & Sargent, A. I. 1995, *Astronomical Journal*, **109**, 2138.

Krot, A. N., Libourel, G., & Chaussidon, M. 2006, *Geochimica et Cosmochimica Acta*, **70**, 767.

Kunihiro, T., Nagashima, K., & Yurimoto, H. 2005, *Geochimica et Cosmochimica Acta*, **69**, 763.

Lada, C. J. 1999, in *The Origin of Stars and Planetary Systems*, ed. C. J. Lada & N. D. Kylafis, Kluwer, NATO ASIC Proceedings, 540, 143–192.

Lahuis, F., van Dishoeck, E. F., Boogert, A. C. A., *et al.* 2006, *Astrophysical Journal*, **636**, L145.

Langer, W. D. 1977, *Astrophysical Journal*, **212**, L39.

Langer, W. D. & Penzias, A. A. 1990, *Astrophysical Journal*, **357**, 477.

Larson, R. B. 2003, *Reports of Progress in Physics*, **66**, 1651.

Le Petit, F., Nehmé, C., Le Bourlot, J., & Roueff, E. 2006, *Astrophysics Journal Supplement Series*, **164**, 506.

Le Teuff, Y. H., Millar, T. J., & Markwick, A. J. 2000, *Astronomy and Astrophysics Supplement*, **146**, 157.

Lee, H.-H., Herbst, E., Pineau des Forets, G., Roueff, E., & Le Bourlot, J. 1996, *Astronomy and Astrophysics*, **311**, 690.

Leger, A., Jura, M., & Omont, A. 1985, *Astronomy and Astrophysics*, **144**, 147.

Leitch-Devlin, M. A. & Williams, D. A. 1985, *Monthly Notices of the Royal Astronomical Society*, **213**, 295.

Lin, D. N. C. & Papaloizou, J. 1980, *Monthly Notices of the Royal Astronomical Society*, **191**, 37.

Lissauer, J. J. 1987, *Icarus*, **69**, 249.

Lyons, J. R., Boney, E., & Marcus, R. A. 2007, *Lunar and Planetary Institute Conference Abstracts*, **38**, 2382.

Lyons, J. R. & Young, E. D. 2005, *Nature*, **435**, 317.

Mangum, J. G. & Wootten, A. 1993, *Astrophysics Journal Supplement Series*, **89**, 123.

Marcus, R. A. 2004, *Journal of Chemical Physics*, **121**, 8201.

Martin-Zaïdi, C., Lagage, P.-O., Pantin, E., & Habart, E. 2007, *Astrophysical Journal*, **666**, L117.

Maruyama, S., Yurimoto, H., & Sueno, S. 1999, *Earth and Planetary Science Letters*, **169**, 165.

Mattera, L. 1978, Ph.D. thesis, University of Waterloo, Canada.

McCarthy, M. C., Gottlieb, C. A., Gupta, H., & Thaddeus, P. 2006, *Astrophysical Journal*, **652**, L141.

McKeegan, K. D., Jarzebinski, G. J., Kallio, A. P., *et al.* 2008, *Lunar and Planetary Institute Conference Abstracts*, **39**, 2020.

Millar, T.-J., Bennett, A., & Herbst, E. 1989, *Astrophysical Journal*, **340**, 906.

Millar, T. J., Farquhar, P. R. A., & Willacy, K. 1997, *Astronomy and Astrophysics Supplement*, **121**, 139.

Millar, T. J., Herbst, E., & Bettens, R. P. A. 2000, *Monthly Notices of the Royal Astronomical Society*, **316**, 195.

Molini-Velsko, C., Mayeda, T. K., & Clayton, R. N. 1986, *Geochimica et Cosmochimica Acta*, **50**, 2719.

Natta, A., Testi, L., Calvet, N., *et al.* 2007, in *Protostars and Planets V*, ed. B. Reipurth, D. Jewitt, & K. Keil, University of Arizona Press, 767–781.

Nuth, J. A. & Glicker, S. 1982, *Journal of Quantitative Spectroscopy and Radiative Transfer*, **28**, 223.

Öberg, K. I., Fuchs, G. W., Awad, Z., *et al.* 2007, *Astrophysical Journal*, **662**, L23.

Palme, H. 2001, *Royal Society of London Philosophical Transactions Series A*, **359**, 2061.

Pandey, A. & Bhattacharya, S. K. 2006, *Journal of Chemical Physics*, **124**, 234–301.

Pascucci, I., Apai, D., Luhman, K., *et al.* 2009, *Astrophysical Journal*, **696**, 143.

Pascucci, I., Hollenbach, D., Najita, J., *et al.* 2007, *Astrophysical Journal*, **663**, 383.

Payne, M. J. & Lodato, G. 2007, *Monthly Notices of the Royal Astronomical Society*, **381**, 1597.

Piétu, V., Dutrey, A., & Guilloteau, S. 2007, *Astronomy and Astrophysics*, **467**, 163.

Pontoppidan, K. M., Dullemond, C. P., van Dishoeck, E. F., *et al.* 2005, *Astrophysical Journal*, **622**, 463.

Prasad, S. S. & Tarafdar, S. P. 1983, *Astrophysical Journal*, **267**, 603.

Qi, C., Kessler, J. E., Koerner, D. W., Sargent, A. I., & Blake, G. A. 2003, *Astrophysical Journal*, **597**, 986.

Qi, C., Wilner, D. J., Aikawa, Y., Blake, G. A., & Hogerheijde, M. R. 2008, *Astrophysical Journal*, **681**, 1396.

Qi, C., Wilner, D. J., Calvet, N., *et al.* 2006, *Astrophysical Journal*, **636**, L157.

Roberts, H., Herbst, E., & Millar, T. J. 2003, *Astrophysical Journal*, **591**, L41.

Roberts, H. & Millar, T. J. 2000, *Astronomy and Astrophysics*, **361**, 388.

Roeckmann, T., Brenninkmeijer, C. A. M., Saueressig, G., *et al.* 1998, *Science*, **281**, 544.

Roueff, E., Parise, B., & Herbst, E. 2007, *Astronomy and Astrophysics*, **464**, 245.

Sakamoto, N., Setu, Y., Itoh, S., *et al.* 2007, *Science*, **317**, 231.

Salyk, C., Pontoppidan, K. M., Blake, G. A., *et al.* 2008, *Astrophysical Journal*, **676**, L49.

Semenov, D., Pavlyuchenkov, Y., Schreyer, K., *et al.* 2005, *Astrophysical Journal*, **621**, 853.

Semenov, D., Wiebe, D., & Henning, T. 2004, *Astronomy and Astrophysics*, **417**, 93.

Semenov, D., Wiebe, D., & Henning, T. 2006, *Astrophysical Journal*, **647**, L57.

Sheffer, Y., Lambert, D. L., & Federman, S. R. 2002, *Astrophysical Journal*, **574**, L171.

Shu, F. H., Shang, H., & Lee, T. 1996, *Science*, **271**, 1545.

Simon, M., Dutrey, A., & Guilloteau, S. 2000, *Astrophysical Journal*, **545**, 1034.

Smith, I. W. M., Herbst, E., & Chang, Q. 2004, *Monthly Notices of the Royal Astronomical Society*, **350**, 323.

Sonnentrucker, P., Neufeld, D. A., Gerakines, P. A., *et al.* 2008, *Astrophysical Journal*, **672**, 361.

Spanel, P. & Smith, D. 1994, *Chemical Physics Letters*, **229**, 262.

Thiemens, M. H. 1999, *Science*, **283**, 341.

Thiemens, M. H. 2006, *Annual Review of Earth and Planetary Sciences*, **34**, 217.

Thiemens, M. H. & Heidenreich, III, J. E. 1983, *Science*, **219**, 1073.

Tielens, A. G. G. M. & Hagen, W. 1982, *Astronomy and Astrophysics*, **114**, 245.

Trieloff, M. & Palme, H. 2006, *The Origin of Solids in the Early Solar System Planet Formation*, Cambridge University Press.

Tscharnuter, W. M. & Gail, H.-P. 2007, *Astronomy and Astrophysics*, **463**, 369.

Urey, H. C. 1947, *Journal of the Chemical Society*, **99**, 2116.

van Boekel, R., Min, M., Leinert, C., *et al.* 2004, *Nature*, **432**, 479.

van den Ancker, M. E., Bouwman, J., Wesselius, P. R., *et al.* 2000, *Astronomy and Astrophysics*, **357**, 325.

van Dishoeck, E. F. 1988, in *Rate Coefficients in Astrochemistry*, ed. T. Millar & D. Williams, Kluwer, 49–72.

van Dishoeck, E. F. 1998, in *The Molecular Astrophysics of Stars and Galaxies*, ed. T. W. Hartquist & D. A. Williams, Clarendon Press, 53–100.

van Dishoeck, E. F. 2004, *Annual Review of Astronomy and Astrophysics*, **42**, 119.

van Dishoeck, E. F. & Black, J. H. 1988, *Astrophysical Journal*, **334**, 771.

van Dishoeck, E. F., Jonkheid, B., & van Hemert, M. C. 2006, *Faraday Discussions*, **133**, 231.

van Zadelhoff, G.-J., Aikawa, Y., Hogerheijde, M. R., & van Dishoeck, E. F. 2003, *Astronomy and Astrophysics*, **397**, 789.

Vasyunin, A. I., Semenov, D., Henning, T., *et al.* 2008, *Astrophysical Journal*, **672**, 629.

Wakelam, V., Herbst, E., & Selsis, F. 2006, *Astronomy and Astrophysics*, **451**, 551.

Warin, S., Benayoun, J. J., & Viala, Y. P. 1996, *Astronomy and Astrophysics*, **308**, 535.

Watson, W. D. & Salpeter, E. E. 1972, *Astrophysical Journal*, **174**, 321.

Westley, M. S., Baragiola, R. A., Johnson, R. E., & Baratta, G. A. 1995, *Nature*, **373**, 405.

Willacy, K. 2007, *Astrophysical Journal*, **660**, 441.

Willacy, K., Langer, W., Allen, M., & Bryden, G. 2006, *Astrophysical Journal*, **644**, 1202.

Willacy, K. & Langer, W. D. 2000, *Astrophysical Journal*, **544**, 903.

Williams, D. A. 1972, *Astrophysics Letters*, **10**, L17.

Williams, D. A. 1993, in *Dust and Chemistry in Astronomy*, ed. T. Miller & D. Williams, Institute of Physics Publishing, 143–170.

Woodall, J., Agúndez, M., Markwick-Kemper, A. J., & Millar, T. J. 2007, *Astronomy and Astrophysics*, **466**, 1197.

Wooden, D., Desch, S., Harker, D., Gail, H.-P., & Keller, L. 2007, in *Protostars and Planets V*, ed. B. Reipurth, D. Jewitt, & K. Keil, University of Arizona Press, 815–833.

Young, E. D., Galy, A., & Nagahara, H. 2002, *Geochimica et Cosmochimica Acta*, **66**, 1095.

Yu, Y., Hewins, R. H., Clayton, R. N., & Mayeda, T. K. 1995, *Geochimica et Cosmochimica Acta*, **59**, 2095.

Yurimoto, H. & Kuramoto, K. 2004, *Science*, **305**, 1763.

Yurimoto, H., Kuramoto, K., Krot, A. N., *et al.* 2006, in *Prostostars and Planets V*, ed. B. Reipurth, D. Jewitt, & K. Keil, University of Arizona Press, 849–862.

Zasowski, G., Markwick-Kemper, F., Watson, D. M., *et al.* 2007, *Astrophysical Journal*, **694**, 459.

Zhu, X. K., Guo, Y., O'Nions, R. K., Young, E. D., & Ash, R. D. 2001, *Nature*, **412**, 311.

5

Laboratory studies of simple dust analogs in astrophysical environments

John R. Brucato and Joseph A. Nuth III

Abstract Laboratory techniques seek to understand and to place limits upon chemical and physical processes that occur in space. Dust can be modified by long-term exposure to high-energy cosmic rays, thus rendering crystalline material amorphous. It can be heated to high temperatures, thus making amorphous material crystalline. Dust may be coated by organic molecules, changing its spectral properties, or may act as a catalyst in the synthesis of both simple and complex molecules. We describe experimental studies to understand such processes and report studies that focus on the properties of simple oxide grains. We give an overview of the synthesis and characterization techniques most often utilized to study the properties of solids in the laboratory and have concentrated on those techniques that have been most useful for the interpretation of astrophysical data. We also discuss silicate catalysis as an important mechanism that may drive the formation of complex molecular compounds relevant for prebiotic chemistry.

Laboratory astrophysics is not an oxymoron. Laboratory studies are one of the very few means to understand the chemical and physical processes that occur in the outlandish environments (by terrestrial standards) observed by astronomers. Such processes often occur under conditions that are far removed from what is considered normal by most chemists and physicists, and the application of a terrestrially honed chemical intuition to processes in astrophysical systems is likely to yield incorrect interpretations of the observations unless the environmental effects are carefully considered. The design of experiments that address processes active in astrophysical environments must also account for differences between the laboratory and reality. Often one finds that it is neither necessary or even desirable to carry out experiments under typical astrophysical conditions. Few funding agencies will support experiments lasting ten million (or even ten!) years to duplicate processes in the interstellar medium, and few institutions will hire or promote a researcher who must wait tens of years before the publication of their next paper. However, by carrying out the experiment at higher pressure or by irradiating at higher fluxes, it is often possible to reduce the timescale of experiments from several million years to several days or weeks without changing the final product. Nevertheless, laboratory experiments do not necessitate the exact replication of all the possible chemical and physical processes that occur in complex astrophysical environments. Usually, laboratory studies are performed to understand processes occurring at atomic or molecular scale or to determine basic physical and chemical quantities (e.g. activation energies, reaction cross-sections, etc.) that are universally applicable.

The first decade of the twenty-first century is a revolutionary period in the study of extraterrestrial material. On January 15, 2006 samples of dust collected from the Wild 2 cometary coma by NASA's Stardust mission were successfully delivered to Earth. Thousands of particles originally in the protosolar nebula have been collected from a Kuiper Belt comet and are at the disposal of analysts worldwide. Beside these very special samples, extraterrestrial material is continuously collected from the near-Earth environment both as interplanetary dust particles (IDPs) and meteorites. Wide-ranging analytical investigations performed directly on these extraterrestrial samples, carrying information on the origin and evolution of the Solar System, have important advantages over analogs. However, a stringent limitation exists owing to the relatively small amount of material collected and the lack of precise information regarding their actual physical or thermal history. Furthermore, IDPs, micrometeorites and samples returned from Solar System bodies require dedicated collection efforts or even space missions, with severe time and economic constraints.

Interplanetary dust particles occur in space as more or less fluffy aggregates with variable proportions of silicate, oxide, pure metal, pure carbon, and organic components, which could be amorphous or crystalline; they are structurally mixed, and complex. Furthermore, recent laboratory examination of cometary dust samples collected by the Stardust mission, showed that cometary dust is highly heterogeneous and very fluffy, consisting of silicate fragments, organic and possibly volatile-rich particles that represent an un-equilibrated assortment of solar and presolar materials (Brownlee *et al.* 2006; Keller *et al.* 2006; Sandford *et al.* 2006). The presence of high-temperature minerals (forsterite and CAIs), formed in the hottest regions of the solar nebula (McKeegan *et al.* 2006; Zolensky *et al.* 2006), provided dramatic evidence confirming predictions of extensive radial mixing at early stages of the solar nebula (Nuth 1999; Nuth *et al.* 2000; Hill *et al.* 2001; Bockelée-Morvan *et al.* 2002). Astromaterials have complex natures that are the outcome of various astronomical environments they crossed. Thus, we are very interested in the formation and properties of solid particles that form in the outflows of stars, spend hundreds of millions of years in the interstellar medium, become part of a newly forming star or planetary system, and possibly undergo moderate to severe metamorphic reactions due to heating or reactions with water or ambient nebular gas.

The use of dust analogs offers several advantages with respect to the use of actual extraterrestrial materials: (a) there are no constraints to the type of measurements to be performed, so that characterization of such materials can be as thorough as required; (b) there is no limitation in the available abundance, so that tests can be repeated and destructive processes can be applied on subsets of the same material; (c) finally, since analog materials are generally simpler, and their chemical compositions are constrained by the initial choices of reactant, they can serve to ensure that the analytical technique in question is performing nominally. Having acknowledged that few laboratory experiments can be done under natural astrophysical conditions, we will nevertheless argue below that well-conceived experimental data will often provide insights into astrophysical processes that are impossible to obtain through models, or even through observations, of natural systems. This is not to diminish the importance of either modeling or observation as vital components contributing to our understanding of natural phenomena but, rather, to emphasize the fact that experiments provide practical information that is otherwise unavailable. This is especially true for complex chemical phenomena, such as the formation and metamorphism of refractory grains, under a wide range of astrophysical conditions. Experimental data obtained in our laboratories have been quite surprising in a number of areas, ranging from the chemical composition of the condensates themselves to the evolution of their spectral properties as a function of temperature and time. None of this information could have been predicted from first principles and would not have been deemed credible even if it had been.

There are several groups engaged in the experimental study of astrophysical grains and their simple (or more complex) analogs. A summary of the work of the research group at Naples, Italy can be found in the review by Brucato *et al.* (2002). This group forms grains via laser vaporization of natural minerals and synthetic chemical mixtures pressed into pellets. They record the infrared spectra of their condensates as a function of composition and degree of thermal annealing, as a function of both temperature and time (see Brucato *et al.* 1999). The research group at Jena, Germany has measured the optical constants of a wide range of natural minerals and synthetic glasses. They have also studied the infrared spectral evolution of glassy silicates (e.g. Fabian *et al.* 2000) and have applied their measurements of crystalline spectra to observations obtained by the Infrared Space Observatory (ISO) satellite (Jäger *et al.* 1998). The group at the Goddard Space Flight Center makes smokes via evaporation–condensation and by combustion, then either follows the spectral changes of these materials during annealing or hydration

(Nuth *et al.* 2002) or uses these materials as catalysts for the production of organics from H_2, N_2, and CO (Hill & Nuth 2003). There are also several collaborative research groups in Japan that measure the spectra of minerals in the far infrared (Koike *et al.* 1993), study the isotopic fractionation during evaporation of minerals (Nagahara *et al.* 1988), and observe the transformation of amorphous grains and grain mixtures into more highly ordered systems using state-of-the-art transmission electron microscopes (Kaito *et al.* 2003).

This is a very rapidly developing field of study and the interpretations of our separate experiments are not always in agreement. Most of the analog studies reported to date are attempts to understand the properties of oxide grains. This is both because there are considerable data already available in the literature on the properties of carbonaceous materials (Mennella *et al.* 1999; Dartois *et al.* 2004, 2005; Mennella 2005) and because it is difficult to carry out appropriate analog experiments on carbonaceous grains that add new information to this extensive database. We have not covered all of the laboratory techniques that have been used to study the properties of solids, but we have focused on those that have been most useful for the interpretation of astrophysical data.

5.1 Dust-analog synthesis

To model experimentally the formation and properties of solid particles that form in the outflows of stars under realistic conditions appropriate to a circumstellar outflow, we would need a system large enough that the walls of the chamber are unimportant and the chamber itself needs to be pumped down to pressures that are several orders of magnitude less than the partial pressure of the least abundant reactant. In a typical stellar outflow SiO is present at $\sim 10^{-6}$ the abundance of hydrogen, while hydrogen is present at 10^{10} particles per cm^3. In practical terms, this means that the experimental system must be capable of achieving a vacuum better than 10^{-15} atm and operate at about 10^{-9} atm or less. These conditions are barely achievable in terrestrial laboratories. However, for wall effects to be unimportant, the chamber radius must be several times the mean free path of the gas: at 10^{-9} atm the mean free path is ~ 100 m.

A spherical vacuum chamber several hundred meters in radius, capable of achieving a vacuum less than 10^{-15} atm and operating at 10^{-9} atm might be possible, but would be very expensive. Heating the system to more than 2000 K only adds spice to the problem. If the system were built and, in the one-meter cube center, SiO molecules were introduced at 10^4 cm^{-3}, then condensed to make 10 nm grains ($\sim 10^5$ SiO molecules per grain) that are allowed to settle onto a 10 m square collection plate, and if this experiment were repeated 10 times, then the collector would be covered with 10 nm scale grains at a concentration of 1 grain per cm^2. Collecting this sample, measuring its spectral properties, and using such materials as the starting point for additional experiments would be somewhat challenging. As a matter of fact, conditions in the relatively high pressure environment of a circumstellar outflow are easier to duplicate in the laboratory than many higher temperature and much lower pressure astrophysical environments such as the regions surrounding Eta Carinae or the Red Rectangle, where laboratory experiments under ambient natural conditions would be nearly impossible using today's technology.

5.1.1 *Vapor condensation, electrical discharge, and laser pyrolysis*

Solid grains form in circumstellar outflows via condensation from a high temperature gas. It is therefore not surprising that many analog experiments start with simple grains condensed from a high-temperature vapor. Vapors are produced in a variety of ways, ranging from the evaporation of simple metals from hot crucibles at controlled temperatures (Day & Donn 1978; Nuth & Donn 1981), to condensation in high-energy, but less-controlled environments such as in laser plumes (Stephens & Russell 1979; Brucato *et al.* 2002), plasmas produced via microwave discharge (Tanabé *et al.* 1988; Kimura & Nuth 2007), or electron bombardment (Meyer 1971). Each of these techniques produces interesting analog materials, but only in very small (milligram) quantities. The combustion of refractory precursors such as silane and pentacarbonyl iron highly diluted in hydrogen and mixed with oxygen or nitrous oxide at controlled temperatures in a tube furnace can produce gram quantities of analog oxides that are useful for many additional studies (Nuth & Hill 2004). The results from these studies will be discussed below.

High-voltage discharges in gases, driven by Tesla coils or microwave discharges, create a high-energy plasma that can break bonds and initiate reactions to produce solids from appropriate precursor molecules such as silane, siloxanes, or various volatile hydrocarbons. Grains produced in electrical discharges are a subset of vapor-phase condensates discussed above. While SiO molecules are produced in electrical discharges containing silane (Kimura & Nuth 2007) or various siloxanes (Tanabé *et al.* 1988), the effects of the ions or transient molecules such as ozone are poorly understood. As an example, it is likely that oxygen transfer from the non-mass dependently fractionated ozone to the growing silicate grains results in the production of similarly fractionated solid oxide grains (Kimura & Nuth 2007). However, the solids themselves are quite similar in their infrared spectra to grains produced via other vapor condensation techniques (Kimura & Nuth 2007). A wide range of carbonaceous morphologies can be produced in these discharges depending on the input gas and the characteristics of the discharge, though the technique has not been widely used to produce astrophysical analogs.

Laser pyrolysis is based on the interaction of a high-power laser beam with the target surface. It is basically a vapor-deposition process where condensation occurs in an expanding plasma plume caused by laser vaporization of the target. The physical processes involved in the vaporization of solids depend both on the laser and material properties. Laser power density, the absorption properties of the target material at the laser wavelength, and the target's thermal properties all drive the vaporization process. Materials with low thermal diffusivity and high absorbance reach a very high local thermal energy and, thus, high peak temperatures are achieved in a small area on the target surface. When temperatures rise sufficiently above the boiling point, evaporation occurs and material is removed as both vapor and solid/liquid ejecta. The interaction of the laser beam with the expanding material in the plume or a higher-powered laser with a sample surface can generate more energetic species to form a plasma. The plasma close to the surface emits radiation, which is more efficiently absorbed by the solid. At high power density (above 10^7 W cm^{-2}) the vapor becomes ionized and absorbs a fraction of the laser energy. The temperature conditions for evaporation are usually reached in a short time, which depends on the laser power density, thermal diffusivity, and heat of sublimation of the material. Generally, these quantities vary with temperature. One of the advantages of laser pyrolysis is the homogeneous evaporation of the target material. Since atomic diffusivities are several orders of magnitude smaller than thermal diffusivities, during the vaporization process fractionation of the material is usually not observed. For this reason the technique is useful for congruent evaporation. This generally results in a condensed sample with the same chemical composition as the target, even if segregation of a small fraction of the product is sometimes observed (Brucato *et al.* 1999).

A Nd–YAG solid-state pulsed laser has its fundamental laser wavelength output at 1064 nm. Laser beam energy may vary according to the instrument's technical configuration. However, the power emitted per laser pulse is typically of the order of 10^{8-10} W cm^{-2}. Sets of doubling crystals are frequently used to get the II and IV laser harmonics at 532 and 266 nm, respectively, and a typical optical set-up allows one or more of these wavelengths to reach the target. The optical properties of silicates used as target materials typically vary with wavelength in such a way that the absorption coefficient of the bulk target is lower by orders of magnitude at 1064 nm with respect to 266 nm. Therefore, a laser wavelength in the ultraviolet, rather than in the infrared, will vaporize silicates more efficiently. Targets may be bulk, flat polished slices or pressed powder. The flat surface optimizes the interaction between the incident laser beam and the target, having high output during the first part of the vaporization process. In order to produce condensates, a target is mounted inside a vaporization chamber, designed to work in different gas environments. The pressure inside the chamber is usually maintained at a few millibar and the gas used depends on the experimental conditions required. Quick quenching of the hot atoms by multiple collisions with cold atoms or molecules of gas present inside the evaporation chamber results in rapid energy loss that favors supersaturation of the vapor. These conditions result in condensation. The rapid freezing of the mutual interatomic positions, which are chaotic at the very high evaporation temperature, causes the formation of amorphous material, even if a few crystalline particles are produced simultaneously. In order to prevent the formation of pure iron grains during the vaporization of iron-bearing silicates and to ensure that the chemical composition of the condensed grains is maintained similar to that of the target, an oxygen atmosphere is required. Finally, the condensates are collected on different types of substrates depending on the analyses to be performed and on the type of sample processing planned.

5.1.2 Melt solidification, sol–gel reactions, Czochralski, and floating-zone methods

It is well known that a liquid can be cooled below its equilibrium freezing temperature. Crystallization during melt solidification is hampered due to an activation barrier caused by the surface energy of the crystal nuclei (surface energy is due to the unfilled or distorted chemical bonds on a solid surface). High cooling rates are usually needed to produce glasses, notably cooling rates of the order of thousands of $K\ s^{-1}$ might be needed. The supercooled liquid passes through a transition to a glass at a temperature that is typically two thirds of the melting temperature. At this transition some degrees of freedom are frozen-in. Since the transition is out of equilibrium, the properties of the resulting glass depend on its thermal history. Supercooled liquid below the melting point therefore becomes metastable with respect to the crystalline solid state. Glasses are, thus, out-of-equilibrium solids with many of the liquid's degrees of freedom frozen in. Solidification from a melt is often too fast to produce crystals and amorphous silicates usually result (Dorschner *et al.* 1995). Slower cooling allows some crystals to grow. Crystals can be achieved, for example, by immersing the sample container in a larger volume of liquid in a dewar, or by wrapping it in thermal insulating material. A variation of the simple cooling method uses thermal convection. Where applicable, it can be used to cool from a high-temperature melt to room temperature, or from a room-temperature liquid to a lower temperature. The latter is a specialized technique, usually carried out *in situ* on a diffractometer, with monitoring of the crystal growth by optical and X-ray methods; the sample is contained in a sealed capillary tube, and selective heating may be applied by an infrared laser to develop a single crystal. A more complete discussion of these topics can be found in Byrne (1965).

Sol–Gel reactions were developed in the 1950s as a method to synthesize pure silicate mineral grains such as forsterite or clinopyroxene. These techniques begin with sodium metasilicate or sodium orthosilicate, both of which are highly soluble in water. Adding magnesium chloride or iron (II) chloride to such a solution forms precipitates such as magnesium metasilicate or iron-orthosilicate gels, respectively. Freeze drying such gels yields amorphous powders that can be annealed to form fine-grained crystalline materials. Whereas making magnesium silicate grains is relatively easy, iron-bearing silicates are very sensitive to oxidation, requiring careful handling to prevent the formation of iron(III) during the freeze drying. The major difference between grains produced via the sol–gel technique and vapor-phase condensation is that sol–gel grains contain fully oxidized SiO_4^{-4} or $(SiO_3^{-2})_n$ structural units whereas many vapor-phase condensates form from SiO molecules and contain both under-oxidized metal and silicon atoms. Thompson *et al.* (2003) used the sol–gel technique to produce starting materials for a series of detailed thermal annealing experiments using X-ray diffraction to follow the changes in structure from amorphous grains to crystalline minerals (Thompson & Tang 2001).

Synthesis in the laboratory of large crystal minerals is often used not only to produce dust by grinding methods, but also to measure their optical constants in a wide spectral range (from ultraviolet to far-infrared). The Czochralski technique is mainly used to prepare large single crystals of silicon, which are subsequently sliced into wafers for use in electronic devices. Single crystals up to a few inches in diameter may be prepared from a melt by this technique, which involves contacting the melt with a seed crystal under an inert atmosphere and controlled conditions of temperature and withdrawal of the crystal from the melt. This technique has also been used to prepare in the laboratory highly pure synthetic crystals of forsterite and enstatite (Takei & Kobayashi 1974). The sizes of the bulk silicate crystals range from a few up to tens of millimeters.

Silicate crystals may also be prepared by a floating-zone technique (e.g. as can metals of high melting point, Takei 1978). A pure polycrystalline rod is gripped at the top and bottom in water-cooled grips and rotated in an inert gas or vacuum. A small melt zone, produced by either a water-cooled radio-frequency coil or by electron bombardment from a circular filament, is passed up its length. High purity is possible because the specimen has no contact with any source of contamination and also because there is a zone-refining action. Methods involving grain growth in the solid state depend upon the annealing of deformed samples. In the strain–anneal technique, a fine-grained polycrystalline metal is critically strained, approximately 1–2% elongation in tension, and then annealed in a moving-gradient furnace with a peak temperature set below the melting point or transformation temperature. Light straining produces very few nuclei for crystallization; during annealing, one favored nucleus grows more rapidly than the other potential nuclei, which the growing crystal consumes.

5.2 Characterization techniques

5.2.1 Spectroscopy

Spectroscopic measurement is a particularly favored analytical technique because spectra can be compared in a direct way to interpret the chemical and mineralogical composition of dust in various astronomical environments. Depending upon the different spectral regions under analysis and depending on the optical properties of the material, one must use different techniques. In regions of strong absorption, such as in the phonon band range (mid-infrared) or the ultraviolet, direct absorption measurements require very low column densities of material, which can only be achieved with thin films or diluted powder samples.

The resolution of the spectroscopic technique applied should be comparable to the resolution of the astrophysical observations, since artificial degradation of high spectroscopic resolution data sets are somewhat model dependent. However, higher-resolution data are also useful in that future missions could be designed to discriminate specifically between alternative materials whose high-resolution spectra are distinct, but whose lower-resolution spectra are virtually identical. This is a common problem for many silicate minerals and is also the case for some carbonaceous grains. It is always useful to cover as wide a range in wavelength as possible for each analog material characterized.

Although the emission spectra of some analog silicate grains have been published (Lynch *et al.* 1989) the most typical form of such data is the transmission or absorption spectrum of the analog grains embedded in KBr, CsI, polyethylene, or other materials. There has been some controversy in the past over the effect of the neutral medium in which the samples are embedded on the detailed spectrum of SiC crystals. However, as a general rule, there has been little controversy over the use of KBr, CsI, or polyethylene as embedding materials for silicates or metal oxides. Once samples are embedded in a matrix (or deposited onto a transparent surface such as a window) the spectra are recorded using normal spectroscopic techniques and instrumentation.

Thin films of silicates have been produced by pressing powders in a diamond anvil cell (Hofmeister 1997), by cutting grain samples to submicron thick slices with an ultra-microtome (Bradley *et al.* 1999), by electron-beam evaporation (Djouadi *et al.* 2005), and by laser deposition in a vacuum (Brucato *et al.* 2004). On one hand, powders produced in a laboratory are directly measured in transmittance when they are embedded in a matrix of transparent materials (e.g. KBr or polyethylene). On the other hand, reflectance measurements do not require the use of matrices; powders of selected-size grains are directly measured with an appropriate optical accessory. Through measurements in both transmittance and reflectance, it is possible to evaluate the optical constants of a material. These are certainly the physical parameters that best describe the nature of the interactions between electromagnetic waves and matter. In astronomical applications it is a very demanding task to construct models that use spectra and/or optical constants obtained in appropriate geometries. Even if this procedure is not straightforward, theoretical models such as those by Hapke (1993) or Shkuratov *et al.* (1999) have achieved a satisfactory degree of confidence and are widely used for astronomical spectral interpretations.

Raman spectroscopy can provide valuable information on the chemical species present in a sample and can probe the structural properties of materials; in particular, the structure of carbonaceous materials. Carbonaceous materials can vary widely in chemical composition and structure (diamond, graphite, glassy carbon, hydrogenated amorphous carbon, etc.). Depending on the degree of order of graphitic (sp^2 hybridization) materials, one or two first-order Raman bands are observed. According to the momentum conservation selection rule, the first-order Raman band at about $1582 \, cm^{-1}$ is known as the G (graphitic) line (Tuinstra & Koenig 1970). Spectra of microcrystalline graphite and disordered carbon show an additional band at $1360 \, cm^{-1}$. This D (disordered) line is attributed to phonons active in small crystallites or on the boundaries of larger crystallites (Fig. 5.1). In amorphous carbon and hydrogenated amorphous carbon, both G and D bands are present. These bands are quite a bit broader than those observed in disordered graphite: the broader the bands, the more disordered the amorphous carbon is. In amorphous carbon obtained by sputtering and deposition, the two bands cannot be distinguished; the G line is usually found at lower Raman shifts ($1560 \, cm^{-1}$) and the D line becomes a shoulder of the G line. The widths of the G and D bands are related to the bond-angle disorder and to the relative fraction of crystallites versus the amorphous matrix. Annealing experiments carried out on very disordered hydrogenated amorphous carbon films have indeed shown that the I_D/I_G line intensity ratio vs. the annealing temperature initially increases, then develops a maximum at some temperature between 800 and 900 °C, and finally decreases down to zero at high annealing temperature when graphitization occurs (Dillon *et al.* 1984). Under the hypothesis that annealing induces a monotonic increase in the average size of the sp^2 cluster, the initial increase of the I_D/I_G ratio can be interpreted as evidence that the sp^2 clusters grow in size and/or in number. When effects of momentum conservation become important (momentum is conserved in large crystals), the I_D/I_G ratio starts to decrease. A similar evolution of I_D/I_G has been observed during annealing of hydrogenated amorphous carbon grains (Baratta *et al.* 2004; Dartois *et al.* 2005).

In situ Raman spectra of amorphous carbon grains irradiated with 3 keV He$^+$ ions at different fluences are reported in Fig. 5.1 The amorphous carbon grains have been produced by arc discharge between two amorphous carbon electrodes in an inert argon atmosphere. Transmission electron microscopy (TEM) studies

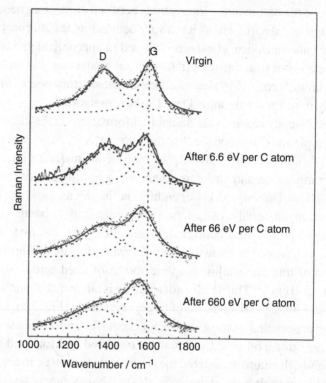

Figure 5.1 Raman spectra of amorphous carbon grains produced by arc discharge and deposited on an LiF substrate before and after irradiation at different fluences with 3 keV He$^+$ ions. The continuous lines are theoretical fits to the data given by the sum of two line profiles (dashed lines) representing the G and D lines (Baratta *et al.* 2004).

show that they are made up of spheroids with average radii of 5 nm arranged in a fluffy structure. Infrared spectroscopy has shown that amorphous carbon grains contain hydrogen (CH$_3$ and CH$_2$, H/C$\sim$4%) and a small amount of oxygen (CO, O/C$\leq$0.5%). The decrease in the I_D/I_G line intensity ratio versus the ion fluence is due to progressive disorder induced by ion irradiation (Baratta *et al.* 2004).

5.2.2 Electron microscopy

Most primary condensates are extremely small, ranging from 5 nm to 50 nm in diameter. Adequate characterization of such grains must rely on very high spatial resolution techniques such as transmission electron microscopy (TEM) or analytical electron microscopy (AEM). In the former technique, the emphasis is on obtaining very clear pictures of the morphology, homogeneity, elemental and mineralogical composition, shape, size, and degree of coagulation of the grains at nearly atomic-level resolution. The method allows for the direct imaging of the sample by which one is able to distinguish between amorphous, partly amorphous, and crystalline (mono- or polycrystalline) phases. The TEM observations of carbon soot samples reveal the presence of complex mixed structures such as amorphous grains organized in chain-like aggregates, minor amounts of poorly graphitized carbon and graphitic carbon ribbons, multi-walled onions, and fullerenes (Rotundi *et al.* 2006).

In the AEM technique, the emphasis is on obtaining chemical and/or structural information at the highest spatial resolution possible. Electron diffraction is used to determine the crystal structure of grains under the beam, and X-ray fluorescence to obtain their composition. However, AEM strives to analyze each separate grain independent from interference from any other. This can be especially difficult for small grains in epoxy mounts where the mount thickness could allow grains to lie below one another and the electron beam can produce signals from any particle in its path. Eliminating such interferences can be extremely important in understanding the properties of the samples. An example of this will be discussed below. Scanning electron microscopy (SEM) is a much lower spatial-resolution technique than TEM or AEM and is generally suitable either for determination of the properties of large grains (micron scale) or to determine the average chemical and structural composition of large aggregates of very small grains. Scanning electron microscopy is also a useful preliminary technique to characterize the properties of samples and to select portions of samples for characterization by TEM or AEM as necessary, or to characterize the success of specific sample preparation techniques. We have found SEM to be very useful in determining the average chemical composition of specific experimental runs producing gram-level quantities of 10 to 20 nm scale grains. Synthetic samples produced by vapor condensation are shown by SEM to have a very fluffy chain-like texture, formed by grains with spheroidal shapes. The size distributions of the condensed grains for a variety of chemical compositions (olivine, pyroxene, and SiC) follow a log-normal law, with a tail at sizes larger than the mean grain size, which is of the order of few tens of nanometers. This tail consists of both single grains that have grown faster than most as well as of grains that have first coagulated then fused (or partially fused) into single particles. While the condensed carbon-based samples synthesized by the arc discharge technique form a complex mixture of solid carbon phases, the analogs produced at the same pressure and atmospheric composition using laser ablation consist only of carbon with a chain-like texture. This confirms that chain-like carbon is the only texture strictly related to the condensation process and that it is not affected by auto-annealing (Rotundi *et al.* 1998, 2002).

5.3 Dust processing

5.3.1 Thermal annealing

If grains condense in a stellar outflow they should also be annealed to some extent since they remain in close proximity to the star and can be heated directly by absorption of energetic stellar photons or by contact with the hot gas. While tiny grains (nanometer scale) can be heated to quite high temperatures for short times (tens of seconds) by single ultraviolet photons, the effect on larger grains will be modest, heating them to temperatures just above the ambient gas kinetic temperature. However, even modest heating can have an effect on the structure of newly condensed grains. Laboratory studies of the annealing process have been carried out by several research groups, most commonly by placing samples of freshly condensed grain materials into a furnace under vacuum for carefully controlled times at specific temperatures. Although quantitative measurement of specific changes in the grains as a function of annealing time and temperature can be tricky, several different techniques have been used for such studies ranging from Fourier-transform infrared (FTIR) spectroscopy to X-ray diffraction, to simply imaging the grains as they annealed on a hot stage in a transmission electron microscope.

Hallenbeck *et al.* (1998) carried out a study of the spectral changes that occurred during the thermal annealing of amorphous magnesium silicate smokes (condensates) over a relatively narrow temperature range ($1026\,K \leq T \leq 1053\,K$) and obtained two interesting results. First, they observed a continuous change in the spectral properties of the grains as a function of annealing time up to a point that they identified as the stall (see Fig. 5.2). When the stall was reached, the spectral properties of the grains stopped changing for a period of time on the order of the total time that the grains had been annealed to that point. After this hiatus, the grains again began to exhibit roughly continuous changes in their spectral properties until they became crystalline. The stall spectrum is seen in Fig. 5.2 at 10.5 and 48 hr and is quite different from the 9.7 μm astronomical silicate feature seen in the interstellar medium: it is both broader and has peaks at 9.8 μm and at 10.8–11 μm. The stall spectrum is very similar to that of the crystalline olivine observed in comets P/Halley (Bregman *et al.* 1987; Campins & Ryan 1989), Bradfield 1987 XXIX (Hanner *et al.* 1990), Levy 1990 XX (Lynch *et al.* 1992), and Mueller 1993a (Hanner *et al.* 1994a). An excellent review of the 8 to 13 μm region of cometary spectra is provided by Hanner *et al.* (1994b) and is also discussed in Chapters 6 and 8.

Thompson and Tang (2001) started with amorphous magnesium silicate samples made by the sol–gel technique and used very-high-resolution X-ray diffraction to study the change in silicate structure as a function of annealing time and temperature. They confirmed the presence of the stall in their samples which is interesting because sol–gel samples are more highly ordered than are condensates. While grains

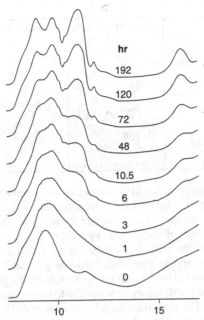

Figure 5.2 Infrared spectra from 8–15 μm of magnesium silicate smoke samples annealed in vacuum at 1027 K. Spectra of grains annealed for ∼ 10.5 hr are mainly identical to those annealed for up to 48 hr (Hallenbeck *et al.* 1998). These represent the "stall" spectrum.

prepared via the sol–gel technique begin with each silicon atom fully coordinated by four oxygen atoms (using sodium orthosilicate as the starting material), condensates begin with many unsaturated silicon atoms as condensation begins with SiO molecules as well as atomic magnesium or iron (see Fig. 5.3). Confirmation of a stall in sol–gel-produced grains proved that this phenomenon was not the result of some chemical rearrangement of the silicon oxidation state, but must have some other cause such as the formation of small polycrystalline regions.

Kaito *et al.* (2003) performed a very simple, yet elegant experiment. They placed amorphous magnesium silicate condensates onto a heated stage in a transmission electron microscope and recorded the changes in the physical structure of the grain as a function of temperature and time. They found that the initial stage of annealing produced a layer of polycrystalline material on the grain surface. The polycrystalline layer completely covered the surface at a time commensurate with the beginning of the stall. During the stall phase, no crystals were observed in the interiors of the grains, though the tiny crystals on the grain surfaces were growing due to a process called Ostwald ripening, where smaller grains lose mass to larger ones powered by an overall reduction in the surface energy of the system that results in an overall increase in the size of the crystals (Ostwald 1896). At the end of the stall phase the crystallization began to penetrate to the centers of the grains and the

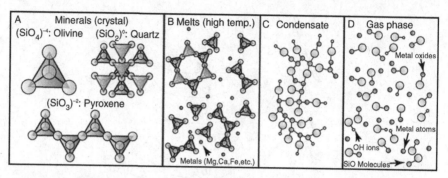

Figure 5.3 The amount of order in silicates can vary dramatically. A. The crystalline backbone structures for olivine, pyroxene, and quartz. The charge of the silicon tetrahedra is neutralized by metal cations in olivine and pyroxene. B. Silicate melts contain a mix of unaligned crystalline structures with metal cations randomly distributed in the melt. C. Chaotic condensates have not formed silicate tetrahedra; rather, they appear more like a frozen gas state. These materials are typically under-oxygenated and contain more metals than a glass. Annealing supplies the chaotic silicate with the energy needed to rearrange into the more stable silicate tetrahedra. D. The gas phase largely consists of SiO. Metals are typically present as atoms or simple monoxides while excess oxygen can be found as OH (Nuth *et al.* 2002).

average size of the crystals in the interior began to increase. Based on these studies, we can conclude that grains in comets and protoplanetary disks exhibiting a stall spectrum probably consist of an amorphous magnesium silicate core covered by a polycrystalline crust. Further annealing will slowly reduce the number of individual crystals, increasing their average size, until a single crystal is eventually produced.

Hallenbeck *et al.* (2000) used the spectral changes they had measured for thermally annealed amorphous magnesium silicate grains to construct a Silicate Evolution Index (SEI) that can be used to calculate the spectral properties of such materials based on an assumed thermal history. As an example, if a model predicted that a grain condensed at a particular temperature in an outflow and the grains remained in the gas as it flowed outward and cooled, then the final spectrum of the grains could be predicted by first calculating the SEI, then matching this to the appropriate standard spectrum. Harker & Desch (2002) used the SEI to demonstrate that small grains heated by strong shocks in the primitive solar nebula could be annealed to crystallinity on very short timescales out to nebular distances of about 10 to 12 AU.

5.3.2 Ion irradiation

Stars belonging to the asymptotic giant branch (AGB, Gehrz *et al.* 1989) and supernovae (Jones *et al.* 1996) are considered the principal sites of cosmic dust formation and, once formed, particles are injected into the interstellar medium (ISM) by expanding stellar winds. The AGB stars are considered to be ubiquitous in our galaxy (Habing 1996) and they account for 50% of the total stellar mass loss. Crystalline silicates are observed in the outflows around high mass-loss rate AGB stars (see Chapter 2). It was estimated that AGB stars could inject about 10–20% of the observed crystalline silicates into the ISM (Suh 2002). Contrary to what might be expected, spectroscopic analysis showed that an upper limit of only a few percent of crystals are present in the dense and the diffuse ISM (Li & Draine 2001; Demyk *et al.* 1999). Selective destruction, low production rates, or interstellar dust dilution mechanisms have been suggested to justify the absence or non-observation of crystalline silicates in the ISM (about 2.2% of crystalline silicates was estimated as the upper limit present in the ISM by Kemper *et al.* 2005).

An efficient way to produce amorphous grains from crystalline silicates is ion irradiation by energetic particles. Laboratory irradiation experiments on natural silicates were first performed by Day (1977). Protons with an energy of 1.5 MeV and fluences similar to those expected for cosmic rays were used in his laboratory. Day did not find evidence of structural modifications for crystalline forsterite, but the spectrum changed dramatically, most likely owing to changes in the structure of the surface layers of the crystal. More recently, low-energy (4–400 keV) ions (H, He, C, Ar) at similar fluences were used in experiments that demonstrated that the ion irradiation of crystalline silicates rather efficiently leads to their amorphization (Demyk *et al.* 2001; Carrez *et al.* 2002; Jäger *et al.* 2003; Brucato *et al.* 2004; Demyk *et al.* 2004; Davoisne *et al.* 2008). Further laboratory experiments have shown that single-crystal synthetic forsterite (Mg_2SiO_4) is amorphized when irradiated by heavy ions at high energy (10 MeV Xe ions) and at sufficiently high fluences (10^{13} ions per cm^2; Bringa *et al.* 2007). This process thus explains the absence of crystalline silicates in the ISM and it argues that the crystal-to-amorphous transition is affected significantly by various physical parameters of the impinging ions, as e.g. the mass, charge, and the kinetic energy.

5.4 Grain-growth studies

5.4.1 Efficiency of crystalline grain growth in stars

The interstellar grain population is overwhelmingly amorphous; no evidence for the presence of crystalline materials has ever been detected (Kemper *et al.* 2005). Some primitive meteorites contain grains that unambiguously formed prior to the formation of the Solar System in astrophysical environments such as the stellar winds from dying stars, in novae, and even in supernovae (see Chapter 2). Interestingly enough, the largest species of presolar material consists of crystalline graphite and

SiC grains that can range up to 25 μm in diameter. In addition, corundum (Al_2O_3) and spinel ($MgAl_2O_4$) crystals, formed in oxygen-rich systems, are also present, though these grains are generally smaller in size (~ 1 μm or so in diameter). These materials are presumably present in the interstellar medium as well, but are too rare to detect. A more intriguing question is, however, how did such large crystalline grains ever form?

Bernatowicz *et al.* (1996) and Sharp & Wasserburg (1995) both demonstrated that large graphite grains probably formed under near-equilibrium conditions in the atmospheres of low-mass AGB stars. Both studies calculated that, even if every appropriate atom that struck the growing grain actually stuck to its surface and became incorporated into the growing crystal, the largest graphite crystals would still require nearly a full year to reach their observed diameters. A new study on the efficiency of crystalline grain growth from a metallic vapor (Michael *et al.* 2003) has shown that only a few atoms of every 10^5 that strike a growing crystalline surface from the ambient gas phase will actually become incorporated into the crystal. Taken together, these studies imply that the growth of large graphite and SiC crystals, and of the smaller corundum and spinel crystals, must have occurred under near-equilibrium conditions that persisted for at least many years, if not for hundreds of thousands of years. The next big question is, "How does a growing grain stay suspended under near-equilibrium conditions in a stellar atmosphere for such long times?"

Nuth *et al.* (2006) showed that for particle sizes ranging from a few hundred Ångstroms up to several tens of microns in diameter the force exerted by radiation pressure in some red giant and AGB stars exceeds the force of gravity and thus offers the potential for graphite, SiC, corundum, and spinel grains to grow to the size range observed in primitive meteorites (e.g. up to ~ 25 μm). In the highest-mass AGB stars radiation pressure on growing grains greatly exceeds the force of gravity and thus ejects a grain from the star before it can grow larger than a few tens of nanometers. Only in very-low-mass AGB stars (less than 3 solar masses) does the radiative force balance the gravitational force to such a fine degree that the net acceleration on individual particles, ranging from a few nanometers up to about 25 μm, produces particle velocities that are comparable to atmospheric turbulence. Their analysis shows that the large graphite, SiC, corundum, and spinel crystals found in primitive meteorites can only have formed in atmospheres of the lowest-mass red giant and AGB stars, where particle growth is able to occur on timescales of a hundred thousand years under near-equilibrium conditions. We note that this suggestion is contrary to the standard assumption that grains can only form in stellar winds and implies that there is a class of grains that can form in chemical equilibrium deep within a stellar atmosphere, just above the photosphere.

5.4.2 *Metastable eutectic condensation and magnesium-rich*
minerals in astrophysics

Image, crystallographic, and chemical analyses of a wide range of astrophysical grain analogs are available in the literature. These include $FeO-Fe_2O_3-SiO_2$ smokes (Rietmeijer *et al.* 1999); Mg–Fe–SiO condensates (Rietmeijer *et al.* 1999); Mg–SiO condensates (Nuth *et al.* 2002), annealed Mg–SiO condensates (Rietmeijer *et al.* 2002); $Al_2O_3-SiO_2$ condensates (Rietmeijer & Karner 1999); and $Al_2O_3-Fe_3O_4-SiO_2$ vapors (Rietmeijer *et al.* 2008). In each of the systems above, we found evidence for the operation of complex chemical processes during the extremely short duration of the vapor-growth phase. This is in contrast to our initial expectation that, due to the rapid condensation and growth of the grains, their compositions would cluster about the average vapor-phase condensable composition. Some of the most striking evidence for chemical control of particle growth comes from experiments on Mg–SiO, Fe–SiO, and Mg–Fe–SiO smokes. Elemental analyses of individual grains condensed from Mg–SiO vapors clustered around five distinct compositions: pure SiO_x (the pure grains are always tridymite, SiO_2) and MgO grains, low-silica MgO grains, as well as serpentine and smectite dehydroxylate grains at $Mg_3Si_2O_7$ and $Mg_3Si_4O_{11}$, respectively (Nuth *et al.* 2002). In a similar fashion, analyses of Fe–SiO condensates also found condensate compositions clustered at the pure oxide (FeO_y, SiO_x) end-members and at intermediate compositions including the greenalite ($Fe_3Si_2O_7$) and iron saponite ($Fe_3Si_4O_{11}$) dehydroxylate compositions, and a distinct clustering at a low-iron ferrosilica composition (Rietmeijer *et al.* 1999). For a more detailed discussion of metastable eutectics see Rietmeijer & Nuth (2000).

To the extent that all experiments are somewhat flawed, the finding of pure oxide grains may well be, at least in part, an experimental artifact due to incomplete vapor mixing at the timescale of the experiment and silicon saturation of the vapor. The critical result of these experiments is that kinetically controlled condensation of silicate vapors leads to apparent chemical ordering in the condensed solids. Although fundamentally identical arguments will apply to the other binary systems that are discussed in this section, below we offer a few words on metastable eutectic solids using the $MgO-SiO_2$ phase diagram (Ehlers, 1972) as an example. If condensation had proceeded at thermodynamic equilibrium, the compositions of the condensed Mg–SiO solids would have been (1) MgO = 2 wt% and SiO_2 = 98 wt% at 1986 K, (2) MgO = 36 wt% and SiO_2 = 64 wt% at 1816 K, and (3) MgO = 63 wt% and SiO_2 = 37 wt% at 2123 K. These compositions are at the three stable eutectics in this phase diagram. But that is not what was found when the compositions of the individual Mg–SiO grains that condensed from the $Mg-SiO-H_2-O_2$ vapor were analyzed (Rietmeijer *et al.* 2002). The observed compositions were

(a) MgO = 25 wt%, (b) MgO = 45 wt%, and (c) MgO = 85 wt%. Composition (a) is between the eutectic compositions (1) and (2) while composition (b) lies between eutectic compositions (2) and (3). The third condensate's composition (c) was not constrained by the existing phase diagram. Thus, Rietmeijer *et al.* (2002) proposed a revision in the MgO-rich part of this diagram that was never experimentally defined. The observed grain compositions (a–c) are defined by the intersections of metastable extensions of the liquidus from two adjacent eutectic points (see Rietmeijer & Nuth 2000). Such an intersection is a metastable eutectic. A common example of such extensions would be superheated or supercooled water, that is, still-liquid water at 1 bar and 400 K or 1 bar and 265 K, as might occasionally be found in a microwave or a freezer, respectively. From an astrophysical perspective, a more interesting system is what was observed following the condensation of Fe–Mg–SiO vapor.

Figure 5.4 is a ternary diagram showing the compositions of individual particles condensed from mixed Fe–Mg–SiO vapors. The dot is the average composition of the condensable vapor phase as determined by SEM analysis of the smoke found on the collector: this represents the average of many thousands of individual, mostly 20 nm-sized grains. The open squares represent the individual grain compositions determined by AEM. Note that these grains roughly cluster at the pure end-member compositions (FeO_y, MgO, and SiO_x) and at the mixed metastable

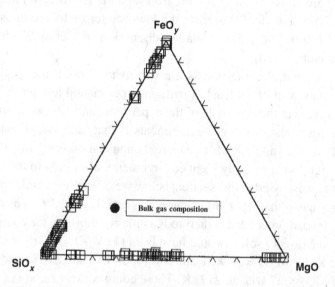

Figure 5.4 Ternary diagram showing the composition of individual 10–20 nm grains condensed from a mixed Fe–Mg–SiO–H_2–O_2 vapor. The large spot in the interior of the diagram represents the approximate composition of the vapor phase as measured via SEM analysis of the bulk smoke (modified after Rietmeijer *et al.* 1999).

eutectic grain compositions that were found in the Mg–SiO and Fe–SiO vapor condensates discussed above. However, what is most noticeable is that there were no mixed Fe–Mg–silicate grains found anywhere in this sample despite the very rapid nucleation and growth of the particles. We note that there are no eutectic compositions possible in the FeO–MgO binary phase diagram and therefore there are no metastable eutectics along this axis to direct the compositions of growing grains. With absent metastable eutectics along the MgO–FeO axis, grain growth is confined to the pure FeO–SiO and MgO–SiO axes producing pure iron silicate and magnesium silicate grains, as well as the end-member oxides FeO_y, MgO and SiO_x. The observation that kinetically controlled vapor-phase condensation alone does not lead to the formation of iron–magnesium silicates places important constraints on the common presence of such silicates in interplanetary dust particles (IDPs) and has interesting consequences when one notes that only pure magnesium silicate minerals have been observed in comets and stellar sources.

Condensation produces separate populations of amorphous iron silicate and magnesium silicate grains. Hallenbeck *et al.* (1998) demonstrated that magnesium silicates anneal at the same rate as iron silicates if the iron silicates are approximately 300 K hotter than the magnesium silicates. Nuth & Johnson (2006) thereby argued that shocks could produce both crystalline magnesium silicate and crystalline iron silicate whereas thermal annealing of a mixed small grain population might produce crystalline iron silicates, but would do so on a timescale sufficiently long that all of the grains would evaporate. At temperatures sufficiently high to produce crystalline magnesium silicates while still leaving the iron silicates as amorphous dust, the lifetimes of the grains against evaporation are more than long enough to ensure survival of the bulk of the grain population. They therefore conclude that the crystalline magnesium silicate minerals observed in comets and protostars are the products of thermally annealing a population of pure amorphous magnesium and iron silicate condensates. A similar process would account for the production of crystalline magnesium silicate minerals observed in the winds of high-mass-loss AGB stars.

5.4.3 Prediction of large-scale nebular circulation

Campins & Ryan (1989) first proposed the presence of crystalline silicates in Comet Halley, based on the presence of an 11.2 μm shoulder within the broad 10 μm silicate stretching mode. This shoulder seemed consistent with the presence of the mineral forsterite (Mg_2SiO_4). Until that time, silicates in comets were thought to be primarily unprocessed, amorphous interstellar grains coated by water ice containing various impurities (Greenberg 1983). The ISO was used to observe crystalline silicates in the dust around Comet Hale-Bopp (e.g. Crovisier *et al.* 1997,

2000). As the peaks observed near 33 and 69 μm are due to phonon transitions in the crystalline silicate lattice, the ISO observations are an unambiguous indication of the presence of crystalline silicates in cometary dust. More recently, crystalline olivine and orthopyroxene have been identified in Comet C/2001 Q4 (NEAT) via 10 μm observations (Wooden *et al.* 2004). There is considerable controversy concerning the origin of these crystalline grains.

As much as 2 to 5% of the interstellar silicate grain population could be crystalline, yet still lie below our current limits of detection (Li & Draine 2001; Kemper *et al.* 2005). As said before, it could equally well be true that any crystalline silicates injected into the interstellar medium become amorphous after long exposure to galactic cosmic rays. Finally, studies on the spectral evolution of silicate grains in protoplanetary disks indicate that the degree of crystallinity is not simply correlated with age (see Chapter 8 for more details). These observations, argue that processes within the protostellar nebula itself are responsible for the production of crystalline silicates.

Nuth (1999) previously argued that the presence of crystalline silicate grains in comets must be the result of thermal annealing of amorphous silicates within the inner regions of the primitive solar nebula, followed by transport of the new crystalline solids out beyond the nebular snowline in winds flowing well above and below the disk out to distances where the grains and ices they carried could be incorporated into newly forming comets (Nuth *et al.* 2000; Hill *et al.* 2001). One source of such winds might be the magneto-hydrodynamic interactions that Shu has proposed to produce chondrules in the primitive nebula (Shu *et al.* 1996), or they may be driven by interactions between such forces and nebular turbulence, or even by terms typically left out of model calculations for simplicity (Prinn 1990; Stevenson 1990). Models of transport in the solar nebula (Bockelée-Morvan *et al.* 2002; Boss 2004; Dullemond *et al.* 2006; Ciesla 2007) now show that materials readily move both inwards and outwards in the nebula prior to the formation of the giant planets. In contrast to simple thermal annealing, Harker & Desch (2002) have suggested that shock waves in the outer solar nebula could anneal the amorphous silicates to crystalline *in situ* prior to their incorporation into comets, thereby eliminating the need for large-scale nebular transport processes. However, if the Weidenschilling (1997) model for comet formation is correct, most comets began to form at distances much greater than 10 AU from the Sun. Therefore, grains annealed by shocks would still need to be transported out to distances where comets begin to form (as shown in Fig. 5.5; Nuth 2001). This hypothesis, based solely on observations of crystalline magnesium-rich minerals in comets and laboratory studies of their production, was spectacularly confirmed by the Stardust mission to Comet Wild 2 (Zolensky *et al.* 2006). Grains from this comet, which should never have been inside the orbit of Jupiter during its formative stages, contain fragments

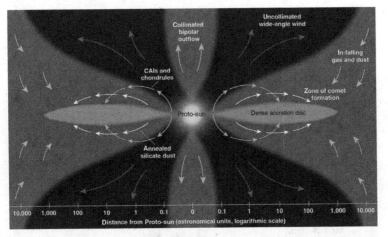

Figure 5.5 Winds in the solar nebula might be one of the possible processes responsible for the mixing of "hot" and "cold" components found in both meteorites and comets. Meteorites contain calcium–aluminum-rich inclusions (CAIs, formed at about 2000 K) and chondrules (formed at about 1650 K), which may have been created near the proto-Sun and then blown (gray arrows) several astronomical units away, into the region of the asteroids between Mars and Jupiter, where they were embedded in a matrix of temperature-sensitive, carbon-based "cold" components. The hot component in comets, tiny grains of annealed silicate dust (olivine) is vaporized at about 1600 K, suggesting that it never reached the innermost region of the disk before it was transported (white arrows) out beyond the orbit of Pluto, where it was mixed with ices and some unheated silicate dust ("cold" components). Vigorous convection in the accretion disk may have contributed to the transport of many materials and has been dramatically confirmed by the Stardust mission (Nuth 2001).

of calcium–aluminum inclusions (CAIs) and chondrules that could only form at the very high temperatures possible in the innermost regions of the solar nebula. If Comet Wild 2 was never inside the orbit of Jupiter as it formed, then such materials must have been transported out to the growing comet. As noted below, such large-scale circulation can have a major influence on the organic chemistry of protostellar nebulae, including the solar nebula.

5.5 Grain-catalysis studies

5.5.1 Amorphous dust grains as catalysts for the reduction of CO and N_2

Llorca & Casanova (2000) demonstrated that Fischer–Tropsch-type (FTT) catalytic reactions (see Fig. 5.6) occur under low pressures typical of the primitive solar nebula, converting CO and H_2 into hydrocarbons. The Haber–Bosch-type (HBT) reaction converts N_2 and H_2 into reduced nitrogen compounds such as NH_3. New

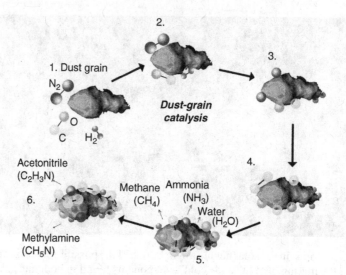

Figure 5.6 A simplified and somewhat speculative representation of the possible reaction steps on or near the catalytic surface associated with individual and joint FTT and HBT syntheses in a protostellar environment. A possible sequence is as follows: (1) A warm fresh dust grain (e.g. iron silicate) and common nebular gases (H_2, CO, N_2) exist, 0.1–1 AU from a protostar. (2) One or more of the gases are adsorbed onto the grain surfaces. (3) Molecules dissociate on the surface of the catalyst. (4) Increasingly longer carbon chains and other reactive atoms accumulate on the grain surface. (5) Organic products (CH_4, etc.), H_2O, and NH_3 form on the grain surface and are released into the nebula. (6) As the grains become increasingly covered with organic products and NH_3, a new suite of reactions are initiated, resulting in the synthesis of CN-bonded molecules [e.g. methylamine (CH_3NH_2) and acetonitrile (CH_3CN)]. Like the other (FTT and HBT) synthesis products, they too are released into the nebular environment (Hill and Nuth 2003).

experiments were designed to test the relative efficiency of various potential catalytic materials and also to produce mixtures of inorganic solids and organics that could serve as analogs of primitive asteroidal material for later experiments.

Grains in protostellar nebulae are exposed to the ambient gas for hundreds or even tens of thousands of years at pressures ranging from 10^{-3} to 10^{-4} atm or less. We do not have such times available for laboratory experiments, although we can reproduce the total number of collisions a grain might experience with components of the ambient gas by running experiments for shorter times at higher pressures. In the laboratory, experiments last from about three days at temperatures of 873 K to more than a month at 573 K. If an average experiment lasts a week (6×10^5 s), and if the only consideration for simulating the effects of the reaction is the total number of collisions of potential reactants with the catalytic surface, then one can simulate two centuries (6×10^9 s) of exposure to an ambient gas at 10^{-4} atm by running experiments at ~ 1 atm total pressure. This scaling assumes that the reactants strongly bind to the surface of the catalyst and that this is the rate-controlling step. Carbon monoxide and nitrogen must be strongly bound in order to weaken their bonds and thus increase their rates of reaction with H_2; Kress & Tielens (2001) deduced that this is the rate-controlling step. If the rate-controlling step is the reaction of H_2 with the CO or N_2 bound on the grain surface, then the important consideration is the number of collisions of H_2 with the CO or N_2 bound on the catalyst, rather than the number of collisions of CO or N_2 with the catalyst, and reaction time can once again be scaled to the number of collisions.

The experiments themselves were also very simple; see Fig. 5.7 (Hill & Nuth 2003). As the experiment proceeds they monitored progress by using periodic

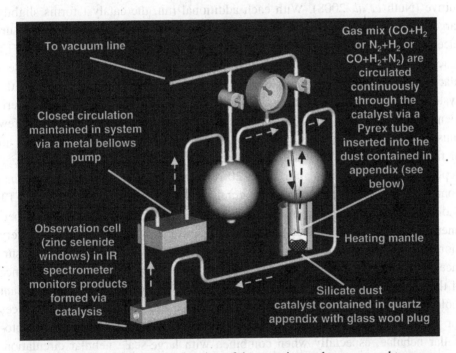

Figure 5.7 Three-dimensional drawing of the experimental system used to assess the catalytic properties of the amorphous iron silicate smokes. The (smoke) catalyst is contained in the bottom of a quartz finger (attached to a 2 L Pyrex bulb) that can be heated to a controlled temperature. A Pyrex tube brings reactive gas to the bottom of the finger. The gas then passes through the catalyst into the upper reservoir of the bulb and flows through a copper tube at room temperature to a glass-walled observation cell (with ZnSe windows) in an FTIR spectrometer. From there, a closed-cycle metal bellows pump returns the sample via a second 2 L bulb and the Pyrex tube to the bottom of the catalyst finger to start the cycle over again (Hill and Nuth, 2003).

FTIR spectra to follow the loss of CO and the formation of methane, water, and carbon dioxide. They also monitored smaller spectral features due to ammonia and *N*-methyl methylene imine. Once the CO had been reduced to about 10% of its starting concentration they took a final infrared spectrum, turned off the heater, cooled the system to room temperature, evacuated it to less than ~ 0.1 torr, then refilled the system with fresh gas and began a second run. Note, that they did not use a fresh batch of catalyst for this second run as they wanted to follow the changing properties of the catalyst as a function of exposure to the gas mixture and as a function of temperature. By making 15 runs with the same catalyst, they simulated ~ 3000 years of exposure of grains to nebular gas and built up a substantial coating of macromolecular carbon, nitrogen, and hydrogen.

For a typical, textbook catalyst, the results of these experiments were counterintuitive (Nuth *et al.* 2008). With each additional run, the catalyst forms slightly larger clumps, thus reducing surface area. The active metal atoms at the surface become more oxidized due to reaction with water generated by the FTT reaction. Some reactive sites on the catalyst become coated by the macromolecular carbon generated in previous runs and some catalyst simply is lost to the system. Each of these factors should slow reaction rates in subsequent experimental runs, yet they observed an increased rate of reaction after the first few runs, followed by a steady rate of reaction thereafter. One model can explain all of these observations, including the large mass fraction of carbon and nitrogen deposited onto the grains after about 20 runs; namely, the macromolecular carbonaceous coating produced on the grains via these reactions is a better FTT and HBT catalyst than the inorganic sites it covers. Furthermore, these experiments seem to indicate that this same organic catalyst has formed on every inorganic surface tested, though at different rates. In other words, every solid surface in a protostellar nebula could grow this material as a natural consequence of the exposure of grain surfaces to CO, N_2, and H_2, three of the most abundant molecules in the Universe. This hypothesis has several interesting consequences for the production of organic molecules via FTT and HBT reactions in protostellar nebulae, especially when combined with large-scale nebular circulation. The combination of a self-perpetuating catalyst that forms naturally on any grain surface, together with widespread outward transport of materials formed in the hot, dense inner regions of protostellar nebulae, makes these nebulae gigantic organic chemical factories that turn abundant CO, N_2, and H_2 into complex hydrocarbons and seed organic material throughout the nebula. If circumstances that favor the chemical evolution of simple organisms can occur in planetesimals formed from such materials, then primitive organisms might take advantage of this natural bounty.

5.5.2 Catalytic effects of dust on synthesis of biomolecules

Silicates could have played an important role in driving the formation of complex molecular compounds relevant for prebiotic chemistry. Due to the low efficiency of formation of complex molecules in the gas phase, in fact, it is not feasible for an active gas-phase chemistry to take place at low temperatures with condensate molecules strictly in contact with dust (Gould & Salpeter 1963; Allen & Robinson 1977; Tielens & Hagen 1982). Laboratory experiments have shown that different chemical–physical mechanisms might be responsible for the richness of molecules observed in space. Surface catalysis at low temperatures by dust is considered necessary to justify the presence of e.g. H_2, H_2O, or CO_2, as demonstrated experimentally (Vidali *et al.* 2005; Perets *et al.* 2007), according to the Langmuir–Hinchelwood mechanism (Langmuir 1918). However, to describe the presence of more complex molecules, such as CH_3OH, radicals, even organic refractory material, irradiation processes due to ions and ultraviolet photons are required (Strazzulla & Johnson 1991; Moore & Hudson 1998; Greenberg *et al.* 1999; Brucato *et al.* 2006b). Therefore, chemical compounds synthesized in the presence of dust grains may become even more complex in high-energy environments or if such materials are combined with fluids such as might exist in the interior of a carbon-rich asteroid.

It was estimated that between 10^7 and 10^9 kg yr^{-1} of carbon contained in organic compounds was delivered to the Earth in the first billion years as IDPs (Chyba & Sagan 1992). Carbonaceous chondrites, some of the most primitive objects of the Solar System, may contain up to 3% of organic matter and their possible role as sources of organic prebiotic material on Earth has long been emphasized. A minor part (10–30%) of the organic matter in the carbonaceous chondrites is made of a complex mixture of soluble molecules, some of which are similar to those found in biochemical systems. More than 70 different amino acids were identified in the Murchison meteorite with a total concentration of approximately 60 parts per million (Cronin *et al.* 1988; Martins *et al.* 2008). Nucleobases were also identified in the Murchison, Murray, and Orgueil meteorites at a level of several hundred parts per billion, only guanine could be identified in a set of Antarctic meteorites (Botta *et al.* 2002).

The stable isotopes of C and N in this organic matter suggest an origin in the ISM (Alexander *et al.* 1998; Busemann *et al.* 2006), but significant processing could have occurred in the solar nebula and in meteorite parent bodies. Although aqueous alteration in many parent bodies involved relatively oxidizing conditions and thus led to loss of organic material (e.g. conversion to CO_2 and carbonates, Naraoka *et al.* 2004), a few meteorites, particularly CM meteorites like Murchison, seemed to have been altered by reducing fluids (Browning & Bourcier 1998). Moreover, Shock & Schulte (1990) made thermodynamic arguments for amino acid synthesis by aqueous alteration of polycyclic aromatic hydrocarbons (PAHs), a common organic species in the ISM and primitive meteorites, and via Strecker synthesis by reaction of ketones or aldehydes with HCN and NH_3 (Schulte & Shock 1992).

Synthesis of organic molecules occurring in space have been performed in the laboratory following different strategies. Mainly simple nitrogen-bearing compounds such as hydrogen cyanide (HCN), isocyanic acid (HNCO), formamide (NH_2COH), ammonium cyanide (NH_4CN), or mixtures of H_2O, CO_2, CO, NH_3, CH_4, CH_2O, CH_3OH, etc. (Grim & Greenberg 1987; Demyk *et al.* 1998; Hudson *et al.* 2001; Moore & Hudson 2003; Palumbo *et al.* 2000, 2004; Brucato *et al.* 2006a,b) have been considered as potential astrobiological precursors. However, since carbonaceous chondrite meteorites have been shown to contain up to a few weight percent of water (Weidenschilling 1981) highly porous and permeable, water-rich parent asteroids would have been sites of hydrothermal circulation during their history. Water in the interior of asteroids would be liquified and mobilized by the heat from decaying ^{26}Al and ^{60}Fe, at least in the early Solar System. Additional internal heat would also be provided by serpentinization (Jewitt 2004) and possibly by impacts. With these physical conditions a rich chemistry might occur inside meteorite parent bodies, which is mainly driven by the catalytic effects of dust, temperature, and by previous exposure of such material to ultraviolet photons and irradiation by energetic ions.

Prebiotic reactions have thus been studied in the laboratory in conditions simulating the environments found on the early Earth and in space. Silicate cosmic dust analogs were used as catalysts, in particular olivine $(Mg,Fe)_2SiO_4$. This is because olivine is continuously confirmed by observations to be ubiquitous in space. The chemistry studied was that of formamide (NH_2COH). This very simple molecule, containing only one carbon atom, was observed to be present in different space environments, as gas in the ISM (Millar 2004) or in the long period Comet Hale–Bopp (Bockelée-Morvan *et al.* 2000) and, tentatively, in the solid phase on grains around the young stellar object W33A (Schutte *et al.* 1999). Formamide is confirmed to be a promising route to understand the first chemical steps that brought simple carbon bearing molecules towards largely complex mixtures of biomacromolecules. A large suite of pyrimidines (including cytosine and uracil) and purines have in fact been synthesized at 430 K using only formamide in the presence of cosmic dust analog silicates with an efficiency an order of magnitude larger than for terrestrial olivine, but also with chemical specificity dependent on the fraction of metal (Mg and Fe) present inside the mineral structure (Saladino *et al.* 2005). Another important aspect is to stabilize the biomolecules against the degradation effects that strongly limit their lifetime. Cosmic dust analog olivine has shown a high protective factor against degradation of polynucleotides, having a dual effect: to synthesize and protect biomolecules at the same time (Saladino *et al.* 2005).

The formation in space environments of solid formamide on grain surfaces was theoretically calculated starting with the reaction of CO with H. The resulting radical HCO reacts with atomic O, N, and C and ultimately leads to formation of NH_2COH. Grain-surface chemistry calculations predict abundances of the order of 1% for a wide range of space environmental conditions (Tielens & Hagen 1982) where low-energy cosmic rays irradiate grains. In cold dense clouds, ions deposit about 30 eV per molecule in 10^8 yrs, which is a factor of 10 less than in the diffuse medium (Moore 1999).

The first results obtained in the laboratory irradiating icy formamide, pure and mixed with cosmic dust analogs (CDAs) at 20 K, show that the presence of dust grains reduces the overall amount of synthesized species (Brucato *et al.* 2006b). This could be an effect related to the larger surface area of CDAs exposed to proton irradiation, which might favor a larger ejection of formamide into the gas phase. Olivine CDA acts chemically on the synthesis process of molecular species. In particular, NH_3 and CN^- molecules are absent when olivine CDA is present. In the latter case, this could be related to the charge exchange due to the presence of iron and magnesium metals in CDA and to the acid–base reaction of ammonia with HNCO. But CDA olivine also shows a specificity in driving the chemistry in the solid phase under ion irradiation. More important is the case of ammonium cyanate for its considerable theoretical importance in prebiotic chemistry. Thermal annealing of ammonium cyanate in aqueous solution forms urea $(NH_2)_2CO$ through dissociation into ammonia and isocyanic acid (Warner and Stitt 1933). Urea was also synthesized at 10 K by ultraviolet irradiation of isocyanic acid and tentatively detected in a protostellar object (Raunier *et al.* 2004). As observed by Brucato *et al.* (2006a), ammonium cyanate synthesized by proton irradiation of pure formamide at 20 K is stable up to room temperature. Therefore ammonium cyanate and formamide could have an important linking role between astrophysical and terrestrial environmental conditions. Finally, CDA $MgFeSiO_4$ is a selective catalyst preventing formation of NH_3 and CN^- molecules and changing the relative abundances of $NH_4^+OCN^-$, CO_2, HNCO, and CO. Thus, CDAs significantly influence the chemistry of formation and evolution of ices in space under radiation processing at low temperatures.

5.6 Conclusion

Laboratory production of dust analogs is a highly specialized field that attempts to reproduce a subset of the basic properties of the solids that we expect to exist in a particular astrophysical environment. The synthesis of dust in the outflows of evolved stars, however, is a much more complex process that occurs under non-equilibrium, continuously changing conditions and with the participation of many more chemical elements than can be accommodated by any practical experimental system. Even under the best of conditions it is difficult to control the rapid nucleation and growth of dust grains that occur in laboratory experiments where parameters such as temperature, pressure, and concentration are used to constrain the final characteristics of dust analogs.

Dust also plays an active role in astrophysics, both as a proxy and recorder of conditions that it has experienced, and as a catalyst for chemical reactions in many low-density astrophysical environments. Full chemical and structural characterization of the analog grains serves the dual purpose of ensuring that the experimental conditions used are appropriate and of understanding how the processing conditions affect the properties of the dust. The nature of both natural and synthetic dust is directly related to its prior processing history and can sometimes serve as an indicator of the metamorphic history of the body in which it resided. Furthermore, astrophysical observations require the measurement of accurate optical properties of appropriate materials in the laboratory to ensure reasonable interpretation with respect to chemical composition, structure, temperature, and the presence of several types of grains. We are aware that we have not covered all of the laboratory techniques that have been used to study the properties of solids, but we chose instead to focus on those efforts that have been most useful for the interpretation of astrophysical data. We have emphasized systematic studies that concentrate on the formation and evolution of the initial small-scale grains that by aggregation and further processing control the properties of comets, asteroids, and protoplanets.

Two of the first laboratory astrophysicists to recognize the potential to study chemical processes in space, rather than simply measure the spectral properties of potentially important gases and solids, were Mayo Greenberg and Bertram Donn, who began their efforts to understand ices and grains in the late 1950s. They recognized that they should better understand the formation of dust in circumstellar outflows, the chemical reactions in ice-coated grain mantles, or the catalytic role of solids in the generation of both simple molecules such as H_2, or the more complex organic materials, which might promote the chemical evolution of life. They did this by performing relatively simple experiments to delineate the roles of important variables. These experiments have since become more elaborate and sophisticated, but are still far from duplicating the conditions found in nature. Nevertheless, the results of these experiments have provided extraordinary insights into the way we view the processes that shape our Universe. Continuing increases in our ability to control the conditions in our experiments and to characterize their results should lead to further insight into the observations that flow from a remarkable and growing array of ground- and space-based telescopes. Improved observational and laboratory data will lead to more detailed understanding of specific astrophysical phenomena such as the birth and death of stars, chemical processes in the ISM, or even the origin of life.

References

Alexander, C. M. O., Russell, S. S., Arden, J. W., *et al.* 1998, *Meteoritics and Planetary Science*, **33**, 603.

Allen, M. & Robinson, G. W. 1977, *Astrophysical Journal*, **214**, 955.

Baratta, G. A., Mennella, V., Brucato, J. R., *et al.* 2004, *Journal of Raman Spectroscopy*, **35**, 487.

Bernatowicz, T. J., Cowsik, R., Gibbons, P. C., *et al.* 1996, *Astrophysical Journal*, **472**, 760.

Bockelée-Morvan, D., Gautier, D., Hersant, F., Huré, J.-M., & Robert, F. 2002, *Astronomy and Astrophysics*, **384**, 1107.

Bockelée-Morvan, D., Lis, D. C., Wink, J. E., *et al.* 2000, *Astronomy and Astrophysics*, **353**, 1101.

Boss, A. P. 2004, *Astrophysical Journal*, **616**, 1265.

Botta, O., Glavin, D. P., Kminek, G., & Bada, J. L. 2002, *Origins of Life and Evolution of the Biosphere*, **32**, 143.

Bradley, J. P., Keller, L. P., Snow, T. P., *et al.* 1999, *Science*, **285**, 1716.

Bregman, J. D., Witteborn, F. C., Allamandola, L. J., *et al.* 1987, *Astronomy and Astrophysics*, **187**, 616.

Bringa, E. M., Kucheyev, S. O., Loeffler, M. J., *et al.* 2007, *Astrophysical Journal*, **662**, 372.

Browning, L. & Bourcier, W. 1998, *Meteoritics and Planetary Science*, **33**, 1213.

Brownlee, D., Tsou, P., Aléon, J., *et al.* 2006, *Science*, **314**, 1711.

Brucato, J. R., Baratta, G. A., & Strazzulla, G. 2006a, *Astronomy and Astrophysics*, **455**, 395.

Brucato, J. R., Colangeli, L., Mennella, V., Palumbo, P., & Bussoletti, E. 1999, *Astronomy and Astrophysics*, **348**, 1012.

Brucato, J. R., Mennella, V., Colangeli, L., Rotundi, A., & Palumbo, P. 2002, *Planetary and Space Science*, **50**, 829.

Brucato, J. R., Strazzulla, G., Baratta, G., & Colangeli, L. 2004, *Astronomy and Astrophysics*, **413**, 395.

Brucato, J. R., Strazzulla, G., Baratta, G. A., Rotundi, A., & Colangeli, L. 2006b, *Origins of Life and Evolution of the Biosphere*, **36**, 451.

Busemann, H., Young, A. F., Alexander, C. M. O. D., *et al.* 2006, *Science*, **312**, 727.

Byrne, J. G. 1965, *Recovery, Recrystallization and Grain Growth*, Macmillan.

Campins, H. & Ryan, E. V. 1989, *Astrophysical Journal*, **341**, 1059.

Carrez, P., Demyk, K., Cordier, P., *et al.* 2002, *Meteoritics and Planetary Science*, **37**, 1599.

Chyba, C. F. & Sagan, C. 1992, *Nature*, **355**, 125.

Ciesla, F. J. 2007, *Science*, **318**, 613.

Cronin, J. R., Pizzarello, S., & Cruikshank, D. P. 1988, in *Meteorites and the Early Solar System*, ed. D. S. Lauretta & H. Y. McSween, University of Arizona Press, 819–857.

Crovisier, J., Brooke, T. Y., Leech, K., *et al.* 2000, *Thermal Emission Spectroscopy and Analysis of Dust, Disks, and Regoliths*, **196**, 109.

Crovisier, J., Leech, K., Bockelée-Morvan, D., *et al.* 1997, *Science*, **275**, 1904.

Dartois, E., Muñoz Caro, G. M., Deboffle, D., & d'Hendecourt, L. 2004, *Astronomy and Astrophysics*, **423**, L33.

Dartois, E., Muñoz Caro, G. M., Deboffle, D., Montagnac, G., & D'Hendecourt, L. 2005, *Astronomy and Astrophysics*, **432**, 895.

Davoisne, C., Leroux, H., Frère, M., *et al.* 2008, *Astronomy and Astrophysics*, **482**, 541.

Day, K. L. 1977, *Monthly Notices of the Royal Astronomical Society*, **178**, 49.

Day, K. L. & Donn, B. 1978, *Science*, **202**, 307.

Demyk, K., Carrez, P., Leroux, H., *et al.* 2001, *Astronomy and Astrophysics*, **368**, L38.
Demyk, K., Dartois, E., D'Hendecourt, L., *et al.* 1998, *Astronomy and Astrophysics*, **339**, 553.
Demyk, K., D'Hendecourt, L., Leroux, H., Jones, A. P., & Borg, J. 2004, *Astronomy and Astrophysics*, **420**, 233.
Demyk, K., Jones, A. P., Dartois, E., Cox, P., & D'Hendecourt, L. 1999, *Astronomy and Astrophysics*, **349**, 267.
Dillon, R. O., Woollam, J. A., & Katkanant, V. 1984, *Physics Review B*, **29**, 3482.
Djouadi, Z., D'Hendecourt, L., Leroux, H., *et al.* 2005, *Astronomy and Astrophysics*, **440**, 179.
Dorschner, J., Begemann, B., Henning, T., Jaeger, C., & Mutschke, H. 1995, *Astronomy and Astrophysics*, **300**, 503.
Dullemond, C. P., Apai, D., & Walch, S. 2006, *Astrophysical Journal*, **640**, L67.
Ehlers, G. E. 1972, *The Interpretation of Geological Phase Diagrams*, W. H. Freeman.
Fabian, D., Jäger, C., Henning, T., Dorschner, J., & Mutschke, H. 2000, *Astronomy and Astrophysics*, **364**, 282.
Gehrz, R. D., Ney, E. P., Piscitelli, J., Rosenthal, E., & Tokunaga, A. T. 1989, *Icarus*, **80**, 280.
Gould, R. J. & Salpeter, E. E. 1963, *Astrophysical Journal*, **138**, 393.
Greenberg, J. M. 1983, *Advances in Space Research*, **3**, 19.
Greenberg, J. M., Schutte, W. A., & Li, A. 1999, *Advances in Space Research*, **23**, 289.
Grim, R. J. A. & Greenberg, J. M. 1987, *Astronomy and Astrophysics*, **181**, 155.
Habing, H. J. 1996, *Astronomy and Astrophysics Review*, **7**, 97.
Hallenbeck, S. L., Nuth, J. A. III, & Daukantas, P. L. 1998, *Icarus*, **131**, 198.
Hallenbeck, S. L., Nuth, J. A. III, & Nelson, R. N. 2000, *Astrophysical Journal*, **535**, 247.
Hanner, M. S., Hackwell, J. A., Russell, R. W., & Lynch, D. K. 1994a, *Icarus*, **112**, 490.
Hanner, M. S., Lynch, D. K., & Russell, R. W. 1994b, *Astrophysical Journal*, **425**, 274.
Hanner, M. S., Newburn, R. L., Gehrz, R. D., *et al.* 1990, *Astrophysical Journal*, **348**, 312.
Hapke, B. 1993, *Theory of Reflectance and Emittance Spectroscopy*, Cambridge University Press.
Harker, D. E. & Desch, S. J. 2002, *Astrophysical Journal*, **565**, L109.
Hill, H. G. M., Grady, C. A., Nuth, J. A., III, Hallenbeck, S. L., & Sitko, M. L. 2001, *Proceedings of the National Academy of Science*, **98**, 2182.
Hill, H. G. M. & Nuth, J. A., III 2003, *Astrobiology*, **3**, 291.
Hofmeister, A. M. 1997, *Physics and Chemistry of Minerals*, **24**, 535.
Hudson, R. L., Moore, M. H., & Gerakines, P. A. 2001, *Astrophysical Journal*, **550**, 1140.
Jäger, C., Dorschner, J., Mutschke, H., Posch, T., & Henning, T. 2003, *Astronomy and Astrophysics*, **408**, 193.
Jäger, C., Molster, F. J., Dorschner, J., *et al.* 1998, *Astronomy and Astrophysics*, **339**, 904.
Jewitt, D. C. 2004, in *Comets II*, ed. M. C. Festou, H. U. Keller & H. A. Weaver, University of Arizona Press, 659–676.
Jones, A. P., Tielens, A. G. G. M., & Hollenbach, D. J. 1996, *Astrophysical Journal*, **469**, 740.
Kaito, C., Ojima, Y., Kamitsuji, K., *et al.* 2003, *Meteoritics and Planetary Science*, **38**, 49.
Keller, L. P., Bajt, S., Baratta, G. A., *et al.* 2006, *Science*, **314**, 1728.
Kemper, F., Vriend, W. J., & Tielens, A. G. G. M. 2005, *Astrophysical Journal*, **633**, 534.
Kimura, Y. & Nuth, J. A., III, 2007, *Astrophysical Journal*, **664**, 1253.
Koike, C., Shibai, H., & Tuchiyama, A. 1993, *Monthly Notices of the Royal Astronomical Society*, **264**, 654.
Kress, M. E. & Tielers, A. G. G. M. 2001, *Meteoritics and Planetary Science*, **36**, 75.

Langmuir, I. 1918, *American Chemical Society*, **40**, 1360.

Li, A. & Draine, B. T. 2001, *Astrophysical Journal*, **554**, 778.

Llorca, J. & Casanova, I. 2000, *Meteoritics and Planetary Science*, **35**, 841.

Lynch, D. K., Russell, R. W., Campins, H., Witteborn, F. C., & Bregman, J. D. 1989, *Icarus*, **82**, 379.

Lynch, D. K., Russell, R. W., Hackwell, J. A., Hanner, M. S., & Hammel, H. B. 1992, *Icarus*, **100**, 197.

Martins, Z., Botta, O., Fogel, M. L., *et al.* 2008, *Earth and Planetary Science Letters*, **270**, 130.

McKeegan, K. D., Aléon, J., Bradley, J., *et al.* 2006, *Science*, **314**, 1724.

Mennella, V. 2005, *Journal of Physics Conference Series*, **6**, 197.

Mennella, V., Colangeli, L., Brucato, J. R., *et al.* 1999, *Advances in Space Research*, **24**, 439.

Meyer, C. J. 1971, *Geochimica et Cosmochimica Acta*, **35**, 551.

Michael, B. P., Nuth, J. A., III, & Lilleleht, L. U. 2003, *Astrophysical Journal*, **590**, 579.

Millar, T. J. 2004, in *Astrobiology: Future Perspectives*, ed. P. Ehrenfreund, W. Irvine, T. Dwen *et al.*, Kluwer, 17–32.

Moore, M. H. 1999, in *Solid Interstellar Matter: The ISO Revolution*, ed. L. D'Hendecourt, C. Joblin, & A. Jones, Springer, 199.

Moore, M. H. & Hudson, R. L. 1998, *Icarus*, **135**, 518.

Moore, M. H. & Hudson, R. L. 2003, *Icarus*, **161**, 486.

Nagahara, H., Kushiro, I., Mori, H. & Mysen, B. O. 1988, *Nature*, **331**, 516.

Naraoka, H., Mita, H., Komiya, M., *et al.* 2004, *Meteoritics and Planetary Science*, **39**, 401.

Nuth, J. A., III 1999, *Lunar and Planetary Institute Conference Abstracts*, **30**, 1726.

Nuth, J. A., III 2001, *American Scientist*, **89**, 228.

Nuth, J. A., III & Donn, B. 1981, *Astrophysical Journal*, **247**, 925.

Nuth, J. A., III Hill, H. G. M., & Kletetschka, G. 2000, *Nature*, **406**, 275.

Nuth, J. A., III & Hill, H. G. M. 2004, *Meteoritics and Planetary Science*, **39**, 1957.

Nuth, J. A., III & Johnson, N. M. 2006, *Icarus*, **180**, 243.

Nuth, J. A., III, Johnson, N. M., & Manning, S. 2008, *Astrophysical Journal*, **673**, L225.

Nuth, J. A., III, Rietmeijer, F. J. M., & Hill, H. G. M. 2002, *Meteoritics and Planetary Science*, **37**, 1579.

Nuth, J. A., III, Wilkinson, G. M., Johnson, N. M., & Dwyer, M. 2006, *Astrophysical Journal*, **644**, 1164.

Ostwald, W. 1896, *Lehrbuch der Allgemeinen Chemie*, Engelmann, vol. 2.

Palumbo, M. E., Ferini, G., & Baratta, G. A. 2004, *Advances in Space Research*, **33**, 49.

Palumbo, M. E., Strazzulla, G., Pendleton, Y. J., & Tielens, A. G. G. M. 2000, *Astrophysical Journal*, **534**, 801.

Perets, H. B., Lederhendler, A., Biham, O., *et al.* 2007, *Astrophysical Journal*, **661**, L163.

Prinn, R. G. 1990, *Astrophysical Journal*, **348**, 725.

Raunier, S., Chiavassa, T., Duvernay, F., *et al.* 2004, *Astronomy and Astrophysics*, **416**, 165.

Rietmeijer, F. J. M. & Karner, J. M. 1999, *Journal of Chemical Physics*, **110**, 4554.

Rietmeijer, F. J. M. & Nuth, J. A., III 2000, *EOS Transactions*, **81**, 409.

Rietmeijer, F. J. M., Nuth, J. A., III, & Karner, J. M. 1999, *Astrophysical Journal*, **527**, 395.

Rietmeijer, F. J. M., Hallenbeck, S. L., Nuth, J. A., III, & Karner, J. M. 2002, *Icarus*, **156**, 269.

Rietmeijer, F. J. M., Pun, A., Kimura, Y., & Nuth, J. A., III, 2008, *Icarus*, **195**, 493.

Rotundi, A., Brucato, J. R., Colangeli, L., *et al.* 2002, *Meteoritics and Planetary Science*, **37**, 1623.

Rotundi, A., Rietmeijer, F. J. M., & Borg, J. 2006, *Natural C_{60} and Large Fullerenes: A Matter of Detection and Astrophysical Implications*, Springer.

Rotundi, A., Rietmeijer, F. J. M., Colangeli, L., *et al.* 1998, *Astronomy and Astrophysics*, **329**, 1087.

Saladino, R., Crestini, C., Neri, V., *et al.* 2005, *ChemBioChem*, **6**, 1368.

Sandford, S. A., Aléon, J., Alexander, C. M. O., *et al.* 2006, *Science*, **314**, 1720.

Schulte, M. D. & Shock, E. L. 1992, *Meteoritics*, **27**, 286.

Schutte, W. A., Boogert, A. C. A., Tielens, A. G. G. M., *et al.* 1999, *Astronomy and Astrophysics*, **343**, 966.

Sharp, C. M. & Wasserburg, G. J. 1995, *Geochimica et Cosmochimica Acta*, **59**, 1633.

Shkuratov, Y., Starukhina, L., Hoffmann, H., & Arnold, G. 1999, *Icarus*, **137**, 235.

Shock, E. L. & Schulte, M. D. 1990, *Geochimica et Cosmochimica Acta*, **54**, 3159.

Shu, F. H., Shang, H., & Lee, T. 1996, *Science*, **271**, 1545.

Stephens, J. R. & Russell, R. W. 1979, *Astrophysical Journal*, **228**, 780.

Stevenson, D. J. 1990, *Astrophysical Journal*, **348**, 730.

Strazzulla, G. & Johnson, R. E. 1991, in *Comets in the Post-Halley Era*, ed. R. L. Newburn, Jr., M. Neugebauer, & J. Rahe, Astrophysics and Space Science Library, vol. 167, 243–275.

Suh, K.-W. 2002, *Monthly Notices of the Royal Astronomical Society*, **332**, 513.

Takei, H. 1978, *Journal of Crystal Growth*, **43**, 463.

Takei, H. & Kobayashi, T. 1974, *Journal of Crystal Growth*, **23**, 121.

Tanabé, T., Onaka, T., Kamijo, F., Sakata, A., & Wada, S. 1988, in *Experiments on Cosmic Dust Analogues*, ed. E. Bussoletti, C. Fusco, & G. Longo, Astrophysics and Space Science Library, vol. 149, 175–180.

Thompson, S. P., Fonti, S., Verrienti, C., *et al.* 2003, *Meteoritics and Planetary Science*, **38**, 457.

Thompson, S. P. & Tang, C. C. 2001, *Astronomy and Astrophysics*, **368**, 721.

Tielens, A. G. G. M. & Hagen, W. 1982, *Astronomy and Astrophysics*, **114**, 245.

Tuinstra, F. & Koenig, J. L. 1970, *Journal of Chemical Physics*, **53**, 1126.

Vidali, G., Roser, J., Manicó, G., *et al.* 2005, *Journal of Physics Conference Series*, **6**, 36.

Warner, J. C. and Sitt, F. B. 1933, *Journal of the American Chemical Society*, **55**, 4807.

Weidenschilling, S. J. 1981, *Lunar and Planetary Institute Conference Abstracts*, **12**, 1170.

Weidenschilling, S. J. 1997, *Icarus*, **127**, 290.

Wooden, D. H., Woodward, C. E., & Harker, D. E. 2004, *Astrophysical Journal*, **612**, L77.

Zolensky, M. E., Zega, T. J., Yano, H., *et al.* 2006, *Science*, **314**, 1735.

6

Dust composition in protoplanetary disks

Michiel Min and George Flynn

Abstract This chapter discusses the composition of protoplanetary dust as derived from laboratory analysis of Solar System dust and from infrared remote sensing of protoplanetary dust around young stars. The advantages, disadvantages and limitations of different laboratory and remote sensing techniques used to derive compositional information are discussed in some detail. Also, an overview is given of the current state of knowledge of both the chemical and mineralogical composition of the dust. Finally, we briefly touch upon some of the implications of the findings for our understanding of the formation and processing history of the grains.

In this chapter we focus on the composition of protoplanetary and Solar System cosmic dust. This composition is important for at least two reasons. First, it provides us with a view on the origin of planets, asteroids, and comets. Second, and maybe even more importantly, the composition of dust can be used as a tracer of dynamical processes taking place in the protoplanetary disk. Certain dust species can only be formed under special circumstances, at certain temperatures or densities. Finding these species outside of their formation area allows us to trace disk dynamics.

So what do we mean by dust composition? We will address in this chapter both the chemical and the mineralogical composition, since these are both important in terms of tracing the thermal history of the grains. The mineralogical composition, i.e. the lattice structure or absence thereof, is one of the key characteristics used in current remote sensing of cometary and protoplanetary dust. (Note that Appendix 1 provides a list of minerals common to astrophysical settings as a reference point.)

In this chapter we will present information obtained by laboratory analysis of Solar System dust grains as well as observational constraints on the composition of dust around other stars. Each method has both advantages and disadvantages. The laboratory studies provide us with a very detailed view on the composition of the dust but on a generally limited sample in space and time. The comparison of the mineralogy and elemental composition of the dust in primitive objects that formed in different regions of the Solar System (e.g. by comparing Asteroid Belt material with samples from the comet-forming regions) with thermodynamic dust condensation models constrains the conditions in these different regions. The comparisons are also important for understanding whether there was significant mixing of dust between different regions. On the other hand, observations of dust around other stars provide us with a wider range in spatial scales and times, but the price for this is a great loss of detail. Therefore, the two methods complement each other, as we will try to show in this chapter.

6.1 Modeling the dust composition

6.1.1 Condensation sequences

The early solar nebula included dust grains that formed around other stars and dust grains that condensed in the solar nebula itself. Many of those that formed around other stars may have experienced significant processing in the interstellar medium before entering the solar nebula. Some of these circumstellar/interstellar grains survived nebular processing and can be identified in extraterrestrial materials. The study of these circumstellar and interstellar grains provides constraints on this component of the solar nebula.

The grains that formed in the solar nebula are believed to have been produced by direct condensation as the nebular gas cooled (see also Chapter 4). If the thermodynamic properties of the reactants and mineral products are known, it is possible to model the condensation of solids by allowing a gas of any initial composition to cool. Larimer (1967) pioneered the application of this technique to the condensation of minerals in the solar nebula. A limitation of it is that minerals whose thermodynamic properties have not been determined cannot be included in the modeling, and computer limitations generally require that the number of minerals that are included in any model must be restricted.

A detailed comparison of the predictions of these models with direct observations of the minerals present in dust from the primitive asteroids and comets provides a critical test of the accuracy of the presumed starting composition. For example, the effect of raising the C/O ratio in the initial composition results in some of the oxide and silicate minerals being replaced by carbides and carbonates (as discussed by Lodders 2003; Ebel 2006; and references therein).

Gail (2002, 2004) has set up a series of large-scale simulations predicting the abundances of minerals formed in the early Solar System as based on assumptions on the input dust materials from the interstellar medium (ISM), and condensation, annealing, and mixing of material in the solar nebula (also see Chapters 2, 4, and 5). These models predict the abundances of the most important dust components. Some of the general trends observed in the spectra of protoplanetary disks can be understood in terms of these predictions (see e.g. van Boekel *et al.* 2004, 2005).

6.1.2 Depletions in the gas phase

Spectroscopy allows one to see not only the elements in the gas phase, but also those in the dust grains. Using bright X-ray binaries as background sources, Ueda *et al.* (2005) were able to determine the total abundances of the elements in the diffuse ISM and found that most of them are approximately solar, with the exception of oxygen. From the details of the X-ray absorption spectra it could be determined to some extent what fraction of the elements were in the solid phase and also to a lesser extent the lattice structure could be determined. From this, they found the silicates to be predominantly rich in magnesium and poor in iron, in agreement with the infrared absorption spectroscopic study of the diffuse ISM by Min *et al.* (2007).

For 41 elements the mean elemental composition of the most primitive meteorites, CI carbonaceous chondrites (see Chapter 1), agrees to within 15% with the elemental composition of the solar photosphere (Lodders 2003). Thus, for all but the most volatile elements, like H and He, the bulk elemental composition of the solar nebula is usually taken to be the composition of these CI meteorites (Anders & Grevesse 1989). The important dust-forming elements Si, Fe, Mg, and S have solar abundances which lie very close together (Si:Fe:Mg:S = $1:0.87:1.05:0.44$). The abundances of many elements, including Mg, Si, P, S, Mn, Cr, Fe, Ni, Zn, and Ge, can be measured in the gas phase of interstellar clouds using high-resolution spectroscopy (Savage & Sembach 1996). Taking the CI composition as representative of the bulk composition of these interstellar clouds allows the composition of the solid phase, or dust grains, to be inferred by subtracting the gas composition from the presumed bulk composition (Savage & Sembach 1996). The major elements that limit the amount of silicate that can form, Mg, Fe, and Si, are each highly depleted in the gas phase. Thus, the grains that formed are expected to have approximately the CI ratios of these elements. However, many of the more volatile elements, e.g. S, are only slightly depleted in the gas phase in the diffuse ISM, indicating that the element/Si ratio for these elements should be low compared to the CI ratio in the interstellar grains. The properties of interstellar dust grains are reviewed in detail in Chapter 2 and in Draine (2003). We should note here the significant uncertainties in the elemental abundances in the ISM and emphasize that the assumption that they are solar is fundamental to all these conclusions.

6.2 Laboratory studies of Solar System dust

The dust of the early Solar System accreted to form the planets, moons, asteroids, and comets that are seen today. However, most of the bodies of the present-day Solar System have experienced significant thermal or aqueous processing, which modified the original dust, overprinting the record of early solar nebular events. A few primitive bodies, including some meteorites, interplanetary dust particles (see Fig. 6.1), and comets, are relatively unmetamorphosed. The examination of these samples provides the opportunity to examine relatively well-preserved dust from the early Solar System.

6.2.1 Measurements of elemental composition

The chemical composition of grains or particles is generally obtained by X-ray fluorescence (XRF) spectroscopy using electron excitation beams in scanning electron microscopes (SEMs), transmission electron microscopes (TEMs), or electron microprobes. Generally, Bremsstrahlung radiation, from electron deceleration in the sample acts as a source of background continuum flux and limits these analyses to elements present in concentrations greater than 100 ppm. X-ray microprobes (XRMs) extend the detection limit to $\sim 10^6$ atoms of an element, which corresponds to an ~ 100 attogram concentration in a 30 picogram sample (Sutton *et al.* 2002). Element abundances are also determined on ultrathin samples by measuring

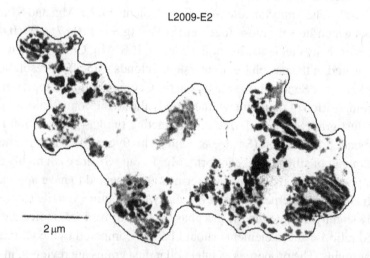

L2009-E2

2 μm

Figure 6.1 A TEM image of an ultramicrotome slice of an anhydrous, porous, chondritic IDP, L2009*E2. Dark areas are mineral grains, while the light areas are void spaces. The boundary of the particle is shown by the solid line. This particle is an aggregate of minerals of diverse compositions (TEM image from L.P. Keller).

the strong increase in X-ray absorption at the K-edge, or ionization energy, of a specific element (Manceau *et al.* 2002).

6.2.2 Identification of minerals

Minerals are identified by a variety of techniques. Some minerals can be identified on the basis of their elemental compositions, but diffraction or X-ray absorption near-edge structure (XANES) spectroscopy provides more compelling mineral identification. Diffraction patterns, produced either by electron or X-ray beams, provide a direct measurement of the spacing between planes of atoms in the crystal structure. When coupled with element analysis, diffraction provides mineralogical identification.

Application of XANES spectroscopy induces electron transitions from the ground state to unoccupied higher-energy electron states. Because these higher-energy states are a probe of the molecular bonding of an element, XANES spectroscopy identifies the oxidization state, the coordination number, and the nearest neighbors of an element (Manceau *et al.* 2002). Combining XRF and XANES analyses on a single spot can provide identification of the mineral. Low-energy XANES instruments – up to $\sim 2\,\mathrm{keV}$ – which access the K-lines of the light elements (generally C to Si) and the L-, M- etc. lines of heavier elements, have a spatial resolution better than 50 nm, while higher-energy XANES instruments – accessing the K-lines of heavier elements (Si to Sr) – currently have submicron resolution.

6.2.3 Isotopic analysis

The isotopic ratios of Solar System materials have been studied in detail (see also Chapter 4). Some large deviations from solar isotopic ratios indicate that the material formed in a circumstellar or interstellar environment outside our Solar System, and was not equilibrated during the formation of the Solar System. These grains are generally referred to as *presolar grains* (see Chapter 2 for detailed discussion). These isotopic ratios are generally measured using secondary ion mass spectrometry (SIMS, see Appendix 2), with the latest-generation instruments having submicron spatial resolution. By combining nano-SIMS isotopic analysis with TEM mineralogical characterization (Messenger *et al.* 2005) or with carbon-XANES and micro-infrared spectroscopy (Keller *et al.* 2004) the host of the isotopic anomaly can frequently be identified.

6.2.4 Infrared spectroscopy

Atoms in a molecule vibrate with well-defined frequencies that are characteristic of the particular bond or functional group. In infrared spectroscopy the sample is irradiated with infrared light of different wavelengths or photon energies. If the photon energy matches the energy quantum needed to excite a molecular vibration, the photon is absorbed. An infrared absorption spectrum determines the types and relative abundances of functional groups in a sample. Laboratory measurements of the infrared spectra of interplanetary dust material, including the grains in primitive interplanetary dust particles (IDPs) and meteorites (summarized in Flynn *et al.* 2002), can be used to identify the components of those grains or, if the mineralogy is known, to assign features in the astronomical spectra to specific minerals or compounds. The infrared spectra of some IDPs are an excellent match to infrared spectra of comets or protoplanetary disks (Sandford & Walker 1985).

6.3 Dust composition in Solar System samples

6.3.1 Composition of chondrites

Chondritic meteorites have elemental compositions similar to the composition of the solar photosphere (except for the volatiles), with the CI1 carbonaceous chondrite meteorites being the closest match to the photospheric abundance pattern (Anders & Grevesse 1989; Lodders 2003). The CI1 carbonaceous chondrites experienced the most complete aqueous alteration of any type of meteorite, most likely in processing that occurred after they were incorporated into their asteroidal parent bodies (Zolensky & McSween 1988). Other types of chondrites experienced less aqueous processing, but they experienced more severe thermal processing after incorporation into their parent bodies. All known meteorites experienced aqueous processing, thermal processing, or, in some cases, both. Nonetheless, the mineralogical content of the CV3 and CO3 carbonaceous chondrites, which experienced little or no aqueous processing and are among the least severely thermally processed meteorites, show significant agreement with the mineralogy predicted by solar nebula condensation models, suggesting that these meteorites have not been severely altered subsequent to their assembly by accretion of the original Solar System dust. The classification of meteorites and the elemental and mineralogical properties of meteorites are described in detail by Sears & Dodd (1988).

6.3.2 Presolar grains in meteorites

Some of the chondritic meteorites contain grains (including crystalline and amorphous silicates, diamonds, silicon carbide, graphite, metal oxides, and metal nitrides) that have been identified as presolar based on non-solar isotopic ratios (Zinner 1988; Anders & Zinner 1993; Bernatowicz *et al.* 2006), particularly for

C, N, and O. Many of these presolar grains can be associated with specific sources, such as asymptotic giant branch (AGB) stars, red giant stars, supernovae, or novae, by comparing their particular isotopic patterns with models of the isotopic ratios in these sources (Chapter 2; Bernatowicz 2006; Meyer & Zinner *et al.* 2006). Some of these grains are so well preserved that original crystal-growth features on their surfaces are intact (Bernatowicz *et al.* 2003). Some meteorites also contain organic matter that is identified as presolar based on non-solar H and N isotopic compositions (Zinner 1988; Messenger & Walker 1997). Chapter 2 provides an extensive discussion of the origins and properties of presolar grains.

6.3.3 Primitive interplanetary dust particles

The zodiacal cloud of dust in our Solar System is produced by impacts onto asteroid surfaces and by dust emission from comets. Some of these particles are perturbed into Earth-intersecting orbits by Poynting–Robertson drag, which occurs when particles interact with solar radiation. Small dust particles, generally in the range from a few micrometers to more than 50 μm in size, experience a relatively gentle deceleration over distances of many kilometers in the Earth's upper atmosphere. These interplanetary dust particles (IDPs) are collected by NASA from the Earth's stratosphere using a high-altitude aircraft.

The majority of IDPs are fine-grained, porous aggregates (Fig. 6.2) dominated by submicron anhydrous minerals (i.e. that contain no water), but some are hydrous particles (Bradley 1988). Some IDPs are severely heated during deceleration in the Earth's atmosphere, but many of the anhydrous, porous IDPs show no evidence of heating, either during atmospheric deceleration or on their parent body. These fine-grained porous IDPs, which are generally anhydrous, are different in elemental composition and in the size scale of mineralogical diversity from any material found in meteorites. Their high porosity suggests that they would break up on atmospheric deceleration if they were comparable in size to meteorites. These IDPs are some of the most primitive samples of Solar System dust available for laboratory study, preserving samples of the condensates that formed in the solar nebula as well as presolar grains (Ishii *et al.* 2008). The hydrous IDPs, on the other hand, contain secondary minerals, most likely produced by later processing on an asteroidal parent body. Rietmeijer (1998) has summarized the mineralogy of both types of IDPs. The fine-grained, porous IDPs (mostly anhydrous) have a generally chondritic elemental composition (Schramm *et al.* 1989), although carbon and some moderately volatile elements may be enriched over the CI concentrations (Flynn *et al.* 1996). These particles are aggregates of mostly submicron crystalline silicate, amorphous silicate, Fe–Ni sulfides, and minor refractory minerals, all held together by an organic-rich, carbonaceous matrix. Some minerals and organic matter in IDPs

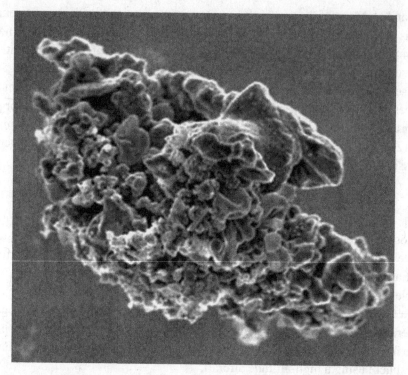

Figure 6.2 An SEM image of an ∼10 micrometer fine-grained, anhydrous IDP. The surface texture results from the aggregation of many individual submicron grains into an aggregate particle (SEM image from S. Messenger).

exhibit non-solar isotopic ratios, which can be used to identify presolar components (Messenger *et al.* 2003).

6.3.4 Grains from Comet 81P/Wild 2

NASA's Stardust spacecraft passed through the dust coma of Comet 81P/Wild 2, collecting dust grains by impacting at ∼6 km s^{-1} velocities into ∼0.1 m^2 of low-density silica aerogel capture cells and ∼150 cm^2 of aluminum foil (Brownlee *et al.* 2006; Hörz *et al.* 2006). Many of these dust particles are associated with jets of material observed in images of Wild 2, indicating they were released from the interior of the comet only hours before collection. However, even that short time may have been sufficient for the sublimation of water ice from the dust, and the 6 km s^{-1} collection generated sufficient heat to mobilize some organic compounds (Sandford *et al.* 2006). The Wild 2 dust was delivered to Earth in January 2006. Particles striking the aerogel produced elongated cavities called *tracks* as they decelerated, frequently leaving material along most of the length of a track. The bulk contents of the major rock-forming elements in 23 tracks were approximately CI, although an enrichment of some moderately volatile elements was observed (Flynn *et al.* 2006a). However, the bulk contents of O and Si could not be determined on the tracks because the capture medium was SiO_2, and bulk Al and Mg contents could not be determined because of X-ray absorption effects. Measurements of the debris in craters produced by the impact of Wild 2 particles onto Al-foils placed between the aerogel capture cells indicates the Mg/Si ratio of the Wild 2 grains was consistent with CI (Flynn *et al.* 2006a).

The Stardust Mineralogy/Petrology Preliminary Examination Team (PET) examined Wild 2 material from 52 of these tracks, finding it to consist of mixtures of grains and compacted or melted aerogel. Twenty-Five of the tracks, chosen at random from those of average length, were studied in detail. Eight of these tracks were dominated by olivine, seven by low-Ca pyroxene, three by roughly equal amounts of olivine and pyroxene, five by iron sulfides, one by sodium silicate minerals, and one by refractory minerals similar to Ca–Al inclusions in primitive meteorites like the Allende CV3 carbonaceous chondrite (Zolensky *et al.* 2006). The PET found a high content of crystalline silicate (Zolensky *et al.* 2006), but much of the fine-grained (1 μm) material was destroyed or altered on capture (Brownlee *et al.* 2006). The abundance of amorphous silicate in the Wild 2 particles is difficult to determine because many of the extracted particles are mixtures of Wild 2 material and compressed or melted aerogel (Zolensky *et al.* 2006), itself an amorphous silicate.

6.3.5 Deep Impact results

A projectile from NASA's Deep Impact spacecraft struck Comet 9P/Tempel 1, producing a dust plume that was examined using the infrared spectrograph on the Spitzer Space Telescope. Lisse *et al.* (2006) inferred a dust-to-gas ratio ≥ 1.3, and reported that the spectrum was best fitted by a mixture of the laboratory standard spectra of amorphous silicate ($MgFeSiO_4$ with a lower amount of $MgFeSi_2O_6$), olivine (with about a 4 : 1 ratio of Mg_2SiO_4 to Fe_2SiO_4), pyroxene (with comparable amounts of $CaMgSi_2O_6$ and $Mg_2Si_2O_6$), smectite [$Na_{0.33}Fe_2(SiAl)_4O_{10}(OH)_2 \cdot H_2O$], carbonate ($FeCO_3$ and $MgCO_3$), and niningerite ($Mg_{0.5}Fe_{0.5}S$), with minor amounts of polycyclic aromatic hydrocarbons, amorphous carbon, water ice and gas-phase water. The signal intensity in infrared emission spectroscopy is roughly proportional to the surface areas of the emitting minerals, although corrections for emission efficiency as a function of grain size must be made (as discussed in Section 6.4.1). Lisse *et al.* (2006) reported that $\sim 8\%$ of the surface area of all the silicate emission was from the phyllosilicate. They reported carbonate and phyllosilicate, both surprising because these minerals are generally believed to form in the presence of liquid water. The high content of crystalline silicate, particularly the Mg-rich olivine, suggests formation in a region with a relatively high temperature (between 1100 K and 1400 K). These authors also reported strong similarities between the dust spectrum of the Tempel 1 ejecta and that of the dusty disk of the young star HD 100546, suggesting a similar origin for the comet and the disk material. They found no Fe-rich olivines and few crystalline pyroxenes in Hale–Bopp and HD 100546, and suggested the silicate in Hale–Bopp is less processed than in Tempel 1, indicating an earlier age of formation for the former. They also compared the Tempel 1 results with those reported by the Stardust PET for the Wild 2 grains and results obtained in 1986 by observations of Comet 1P/Halley.

6.4 Remote sensing of dust around young stars and in comets

The composition of dust grains can be traced remotely by using the radiation reflected, emitted, and absorbed by the dust grains. In particular, vibrational, stretching, and bending modes in the lattice of the grain material translate into emission and absorption resonances (see e.g. Fabian *et al.* 2001; Min *et al.* 2003, 2006). Most of these resonances occur at infrared wavelengths. The location, shape, and strength of these features can be used to identify certain grain materials and derive grain sizes, shapes, and structures. Another tracer of dust grain properties that can be used in remote sensing applications is the degree of polarization of the light scattered by the dust grains. This diagnostic is frequently applied to the dust in cometary comae. Although the degree of polarization contains valuable information on the dust grain size and structure, its direct diagnostic value to the composition of the grains is limited. However, the information can be used in order to constrain grain sizes and structure and thereby make the analysis of the composition more robust. In the following we will focus on remote sensing using infrared spectroscopy.

In the laboratory the complex refractive index as a function of wavelength can be measured for different materials (see e.g. Dorschner *et al.* 1995; Fabian *et al.* 2001). This refractive index contains resonances corresponding to the vibrational resonances in the lattice of the material. The refractive index is used in combination with a model for the dust grain to solve Maxwell's equations and obtain an absorption and scattering efficiency. How the resonances in the refractive index translate into resonances in the absorption and scattering spectrum of the grain depends on the grain properties. Figure 6.3 illustrates the effects of grain shape on the spectral appearance of the absorption efficiency of the grain. Details on methods employed to obtain the optical properties of dust grains can be found in Appendix 3. Here we only emphasize that the choice of the dust model to go from refractive index to optical properties is crucial in obtaining reliable results. For example, it is well

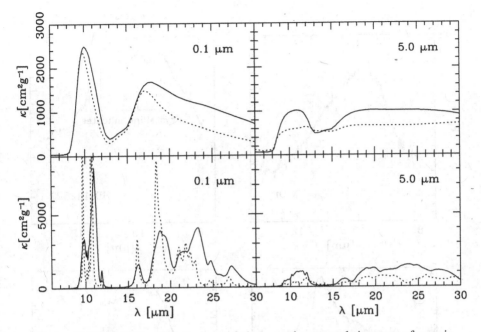

Figure 6.3 The effects of grain size and shape on the spectral signature of cosmic dust grains of various compositions. The upper panels show the mass absorption coefficient for amorphous silciates and the lower panels for grains composed of the crystalline silicate forsterite. The left panels show the computations for grains of 0.1 μm in size, the right panels for grains of 5 μm in size. The solid lines show the spectral signature for irregularly shaped grains while the dotted lines display the ones for homogeneous spherical particles. The refractive index as functions of the wavelength are taken from Servoin & Piriou (1973) for the forsterite data and from Dorschner *et al.* (1995) for the amorphous silicate data. The computations for the irregularly shaped grains have been performed using the highly irregular Gaussian random field particles (see Min *et al.* 2007).

established that using homogeneous spherical particles gives results very different from astronomical observations (see e.g. Min *et al.* 2003; Fabian *et al.* 2001). Examples of the effects of the processing of dust grains on their spectral appearance is given in Fig. 6.4 where the infrared spectrum is shown for each of four protoplanetary disks. In a perfect crystal the lattice vibrations only occur at specific wavelengths and thus show sharp resonances in the infrared spectrum. When the lattice is broken, i.e. an amorphous silicate is considered, the number of possible oscillations increases and the spectrum displays one broad feature. Thus, the appearance of narrow spectral structure indicates the presence of crystalline silicates. The broadening of the feature as the grains grow in size can be understood in

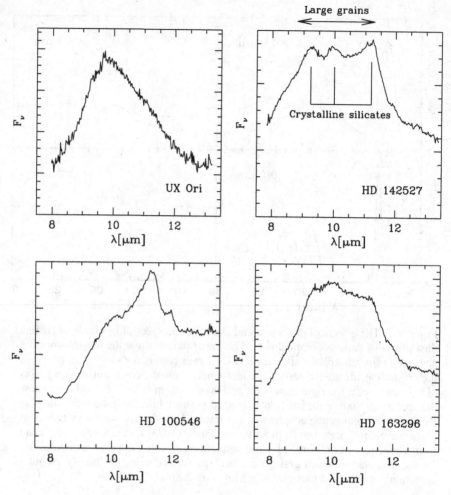

Figure 6.4 The effects of grain processing on the spectral appearance of proto-planetary disks in the infrared. The 10 μm features of several young stars are shown with various degrees of grain growth and crystallization. As indicated in the upper-right panel, the broadening of the feature indicates grain growth, while the appearance of sharp resonances indicates the presence of crystalline silicates. The spectra are taken from van Boekel *et al.* (2005).

a highly simplified way as the grains becoming optically thick, causing the features to saturate.

Observing emission spectra from protoplanetary disks and comets in the infrared is a difficult task from the ground. The Earth's atmosphere is largely opaque in the infrared with the exception of a few windows. The clearest window, and therefore the most frequently applied one, is centered around 10 μm. Luckily, this window covers just the wavelength range enabling us to investigate the so-called 10 μm silicate feature. Since silicates are one of the most abundant cosmic dust species and the 10 μm silicate feature offers great diagnostic value to the size and composition of the silicate grains, this window has been frequently used. The possibility to view a much broader wavelength range came with the infrared space telescopes; first ISO (Infrared Space Observatory) offering a view from roughly 2 to 200 μm, and later the Spitzer Space Telescope offering spectra in the 5 to 40 μm range. This wavelength range offers a richness of additional features and opens the possibility to study cosmic dust species that have no distinct resonances in the 10 μm window.

6.4.1 Limitations

There are two important intrinsic limitations of infrared spectroscopy for the study of cosmic dust which are discussed below.

The first limitation is that only a certain size range of grains is detectable, as determined by the size of the grain with respect to the wavelength of observation. The so-called size parameter, $x = 2\pi r/\lambda$, with r the radius of the grain and λ the wavelength of radiation, has to be smaller than unity in order to have a clear and distinct spectral signature. For larger grains the spectral signature becomes less pronounced and the emission efficiency rapidly decreases (see Fig. 6.3 and Chapter 7). We note here that Min *et al.* (2004) have shown that some spectral signature will always remain. This spectral signature can in principle easily be detected with the sensitivity of current-day telescopes in optically thin environments, where the emission from the large grains dominates and is not swamped by the much more efficient radiation from smaller grains. In practice, however, in most cases the small grains do dominate the emission while in other cases the spectroscopic signature of the large grains is destroyed by efficient scattering and radiative transfer effects. Thus, in general, when observing from the ground using the 10 μm region, information is only available for grains smaller than ∼ 2 μm. When studying the broader wavelength range provided by for instance the Spitzer Space Telescope up to 40 μm, we can access grains up to ∼ 6 μm in size. This limit on the size range available reflects on the derived composition when there is a difference in typical grain size between various materials. For example, in the study of the composition of the dust in the coma of Comet Hale–Bopp the small-grain component has a very high fraction of crystalline silicates (∼ 30–70% has been reported; see e.g. Brucato *et al.* 1999; Wooden *et al.* 1999). However, when considering a broader wavelength range it turns out that the crystallinity of the large-grain component is negligible, resulting in an overall crystallinity which is significantly lower (on the order of ∼ 10%; Min *et al.* 2005). This counter-intuitive result of two separate size distributions can most likely be understood as a result of the underlying simplifying assumption that different materials are in separate grains, while in reality they are most likely in one mixed agglomerate. When realistic, mixed aggregates are considered, a natural outcome is that the materials with small abundances appear spectroscopically to be in smaller grains (Min *et al.* 2008). This is because the small-abundance components act like "raisins" in a "bread" of the dominant material. This picture is consistent with the results from IDP studies in which it is found that forsterite and enstatite grains are predominantly submicron sized, while the IDP as a whole can be several microns in diameter. An overall complicating factor when deriving the size of grains from their infrared spectrum is that there is a degeneracy between grain size and structure. A solid, small grain displays a spectral signature which can hardly be distinguished from that of a larger fluffy aggregate (Min *et al.* 2006). Roughly, what the remote sensing observations are measuring is the thermally emitting effective cross-section of the grains in the beam or field of view of the instrument. The conversion of this measurement into an estimate of the number and mass of dust particles requires a physical model for the dust particles, which includes a detailed description of their shape, structure, temperature, and composition. In many cases the selection of the physical model is not easily uniquely determined.

To summarize, there are two important questions that one has to ask. (i) What is the structure of the grains expected to be present in the environment observed? (ii) For this grain structure, and given the observed wavelength coverage, what is the grain size range accessible? The results of any analysis from infrared spectroscopy on the compositional or geometric characteristics of dust grains should be viewed in the light of these two basic questions.

The second important limitation in studying infrared spectra of protoplanetary disks is the fact that the radiation we observe in the infrared originates in the hot surface layer of the disk. The cold mid-plane material, accounting for the bulk of the dust mass, can only be observed at long (mm) wavelengths, where the disk becomes transparent. Unfortunately, at these wavelengths, only very few solid-state resonances can be found to trace the composition of the dust. Typical chemical bond strengths in solids are such that the energies required to vibrate the atoms around the equilibrium lie in the 1–50 μm (i.e. 0.2–1 eV) regime. Only unusually weak bonds, like the hydrogen bonds found in water ice, cause resonances at longer wavelengths. Differential settling of the larger and more compact grains as discussed in Chapter 3, and trickling of molecular cloud material on the surface layer might cause a difference between the composition of the surface layer and the mid-plane of the disk. In addition, radial mixing of high-temperature material formed close to the star toward the cold outer regions appears to be most effective close to the mid-plane (Ciesla 2007b). In general, this could also cause the mid-plane material to have a different composition than the upper layers, although radial mixing timescales are generally thought to be longer than those for vertical mixing (see also Chapter 3). This would cause the mixed-out materials to be transported to the surface layers.

In addition to the intrinsic limitations, there are practical ones. For example, remote sensing studies rely heavily on the available spectroscopic laboratory data (discussed in Chapter 5). Another important limitation is that when using a single dish 10 m class telescope most protoplanetary disks are only marginally spatially resolved in the infrared, while many are unresolved. Using infrared *space* telescopes it is yet impossible to obtain spatial information since for the typical distances of the most nearby protoplanetary disks, the spatial resolution of these telescopes is several hundreds of astronomical units. Therefore, usually the infrared spectrum observed is the integrated contribution from all distances from the star. Interferometry allows us to zoom in on the inner disk regions. Using this technique van Boekel *et al.* (2004) have shown that there are in general radial gradients in the composition of the dust. They find that the silicate dust in the innermost regions is much more crystalline (up to 100%) than the dust in the outer regions. In addition, in one source, a hint of a gradient in the crystalline pyroxene over crystalline olivine ratio was found in that the outer regions contain more pyroxene-type material. These gradients have to be taken into account when analyzing spatially unresolved infrared spectra. One possibility of disentangling the spatial distribution of the dust is to use a broad-wavelength coverage in order to extract the temperature distribution for all dust materials separately. For example, for crystalline silicates the underlying temperature distribution can be derived by comparing the strengths of the resonances at different wavelengths. In this way Bouwman *et al.* (2003) have found from the broad-wavelength ISO spectrum of the young Herbig star HD 100546 that the crystalline silicates in this source are "piled up" around 12 AU from the star, indicating a possible planet causing mixing in the disk and clearing most of the dust within this radius. This result is compatible with mid-infrared imaging of the disk by Grady *et al.* (2001).

To make things even more complicated, species without detectable emission features, like metallic iron, iron sulfide, or amorphous carbon, can dominate the radiative transfer in protoplanetary disks and thereby set the temperature structure. However, the abundances and size distributions of such grains are extremely difficult to constrain.

6.4.2 Analysis methods

Reading the above, one might get a pessimistic view of the possibilities of remote sensing techniques to derive the composition of cosmic dust in protoplanetary disks. However, when the limitations of the techniques are properly considered, it is possible to derive a wealth of information. The above limitations force one to consider not purely the absolute characteristics, but focus on the relative differences between various sources. Trends and correlations in the dust properties are the strongest tools to gain insight into the evolution of dust in protoplanetary disks.

The analysis of the observations can be done in a direct manner or by comparing the observed infrared spectra with predictions from models of varying complexity. Direct analysis methods use the observed spectrum directly to measure quantities like feature strengths, wavelength position, and spectral slope and use these measures to look for correlations between these parameters for different sources. The advantage of direct methods is that they are not sensitive to assumptions and shortcomings of theoretical models. Thus, any observed correlation stands and can be easily verified by different observers. The disadvantage is that the information on the physical causes for trends and correlations found is unclear. Furthermore, it is hard to compare the findings of these direct methods to information obtained from material in our Solar System. In order to disentangle the physics behind the observed correlations one can try to reproduce them by using complex radiative transfer models, thus constraining possible parameters in these models.

Another very interesting way to study differences in protoplanetary disk spectra is to compare the spectra in a direct, statistical manner. In this way Pascucci *et al.* (2008) showed for a large sample of Herbig, T Tauri, and brown dwarf disks that the strength of the continuum-subtracted silicate features is weaker (statistically significant levels) for brown dwarfs than for the members of the earlier-type countersample; the comparison also suggests that the shape of the silicate emission features around brown dwarfs show more dust processing than around T Tauri and Herbig Ae stars.

The second possibility, i.e. to compare directly with theoretical models, is not always straightforward. Since simulations of protoplanetary disks including dust physics, disk structure, and radiative transfer are computationally very demanding and most methods must involve several simplifying assumptions. The most important assumption which is frequently employed is that the radiation emerges from an optically thin region, allowing simply the addition of the contributions from different dust components (see e.g. Bouwman *et al.* 2001, 2003; van Boekel *et al.* 2005; Honda *et al.* 2006; Sargent *et al.* 2006; Bouwman *et al.* 2008). This assumption is justified to some extent since the radiation predominantly emerges from the optically thin upper atmosphere of the disk. The second important assumption employed in most analysis techniques in the current literature is that the dust has a homogeneous composition throughout the disk. Both these assumptions are currently being tested and the effects on the derived dust compositions are being analyzed using full radiative transfer methods to compute the spectra emerging from realistic disk configurations.

6.5 Composition of the dust

6.5.1 Silicate dust

Crystalline silicates

The major silicate minerals in most of the fine-grained, anhydrous IDPs and in the Wild 2 particles are olivine and pyroxene.

Olivine is a solid solution series ranging from the magnesium end-member called forsterite, Mg_2SiO_4 (denoted as Fo_{100}), to the iron end-member, fayalite, Fe_2SiO_4 (denoted as Fo_0). Analysis of individual olivine grains shows a wide range of compositions. Olivine grains in the fine-grained, anhydrous IDPs range in composition from almost pure forsterite to about Fo_{52} (Rietmeijer 1998). The olivine in Wild 2 ranges from submicron in size to 10 μm and spans the composition range from Fo_{100} to Fo_4 (Zolensky *et al.* 2006, 2007). There is a pronounced peak in the composition–frequency plot near Fo_{99}, almost pure forsterite (Zolensky *et al.* 2007). The Deep Impact results, which were obtained using the two end-member reference spectra, had a best fit with an $\sim 4 : 1$ forsterite to fayalite ratio (Lisse *et al.* 2006).

Pyroxene is a solid solution series from the magnesium end-member enstatite, $MgSiO_3$ (denoted En_{100}), to the iron end-member ferrosilite, $FeSiO_3$ (denoted En_0). In IDPs, low-Ca pyroxenes span the compositional range from En_{100} to En_{46} (Zolensky & Barrett 1994). Low-Fe, Ca-poor pyroxene that is enriched in Mn, up to 5.1 wt% MnO, has been found in some IDPs (Klöck *et al.* 1990). Both low-Ca and high-Ca pyroxene were identified in the Wild 2 grains. In some cases diffraction data were available, and the low-Ca pyroxene was identified as orthoenstatite (Zolensky *et al.* 2006, 2007). The composition of the low-Ca pyroxene has a pronounced peak in the composition–frequency plot at En_{95} (Zolensky *et al.* 2007). Calcium-rich pyroxenes, including diopside and augite, have been observed in a few IDPs (Rietmeijer 1989, 1998). The best fit to the Tempel 1 spectrum was with about $3:1:1$ ratios of ferrosilite, diopside, and orthoenstatite (Lisse *et al.* 2006).

Also from remote sensing of protoplanetary and cometary dust, the most abundant crystalline silicates are olivine and pyroxene. The crystalline olivines identified today in protoplanetary dust by infrared spectroscopy are highly magnesium-rich (Fo_{90}). Indeed, annealing experiments indicate that the product of annealing under certain conditions is a forsterite crystal with metallic iron inclusions (Davoisne *et al.* 2006). Including iron in the lattice of the crystal shifts the resonances to longer wavelengths than generally observed in protoplanetary disks (Koike *et al.* 2003). Here we should note that these conclusions rely on detailed measurements of the mineral spectra, and recent laboratory measurements by Tamanai *et al.* (2006) indicate that a small iron fraction might be allowed in the protoplanetary crystals (on a ~ 5–10% level).

Forsterite is much easier to detect than enstatite using infrared spectroscopy. However, Wooden *et al.* (1999) show that in fact enstatite might be dominant in mass over forsterite in cometary grains. Forsterite-to-enstatite ratios have been determined by van Boekel *et al.* (2005) for a sample of Herbig Ae stars, the ratio they find ranging from 0.1 to 1 for different sources. They show that in protoplanetary disks there seems to be a correlation between the mass fraction of crystalline dust and the forsterite-to-enstatite ratio, the enstatite fraction increasing with increasing crystallinity. However, it is difficult to obtain exact values on the forsterite fraction in Herbig stars because of a possible confusion with polycyclic aromatic hydrocarbon (PAH) emission. It would therefore be interesting to see if a similar correlation can be found in the lower-mass objects that suffer less from PAH contamination. In addition, van Boekel *et al.* (2004) show that the forsterite-to-enstatite ratio for at least one Herbig star seems to decrease with distance from the central star. Both these observations are in overall agreement with the prediction by Gail (2004) on the formation of forsterite and enstatite close to the central star and subsequent radial mixing of this high-temperature material to the outer disk regions. For a more complete picture of thermal processing of grains refer to Chapter 8.

Amorphous silicates

Amorphous silicates are common in anhydrous, porous IDPs, generally constituting $\sim$30 to 60 wt% of the particle (Keller & Messenger 2007). The major amorphous silicate in the anhydrous, porous IDPs is glass with embedded metals and sulfides (GEMS). These GEMS grains are rounded and range from 0.1 μm to $\sim$1 μm, with most being <0.5 μm in diameter (e.g. Bradley 1994 and see Chapter 7). A systematic survey of GEMS grains indicates that their average S/Si, Mg/Si, Ca/Si, and Fe/Si are $\sim$0.6×CI (Keller & Messenger 2007), where the CI composition is the solar nebula composition (Anders & Grevesse 1989). Mineralogic characterization indicates that GEMS grains are amorphous magnesiosilicates that contain finely dispersed nanocrystals of Fe–Ni metal and Fe sulfide. Bradley (1994) suggested that these GEMS grains had the expected characteristics of interstellar silicate grains. However, Martin (1995) pointed out that the sulfur content of the GEMS grains was much higher than expected for interstellar grains, because the sulfur in the gas phase in interstellar clouds shows only a small depletion from the presumed bulk value. Bradley *et al.* (1999) reported that the $\sim$10 μm silicate absorption feature of GEMS grains matched that of some interstellar silicates (Fig. 6.5), suggesting that GEMS grains share a similar formation process with the latter. Combined TEM and nano-SIMS measurements of individual GEMS grains indicate that only a few percent (< 5%) of the GEMS grains have non-solar O-isotopic compositions and are demonstrably interstellar grains (Keller & Messenger 2007). The origin

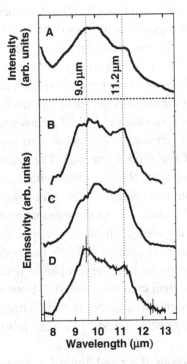

Figure 6.5 Comparison of the 10 μm Si–O stretching bands of a "GEMS-rich" IDP and astronomical silicates. (A) Chondritic IDP L2008V42A. Profile derived from transmittance spectrum. (B) Comet Halley (Campins & Ryan 1989). (C) Comet Hale–Bopp (Hayward *et al.* 2000). (D) Late-stage Herbig Ae/Be star HD 163296 (Sitko *et al.* 1999). The structure at 9.5 μm in (B), (C), and (D) is due to telluric O_3. Figure from Bradley *et al.* (1999).

of the remaining GEMS grains is uncertain, since interstellar material can have an O-isotopic content consistent with the solar composition.

Other amorphous silicates are also found in IDPs. The compositions range from ferromagnesian silica, with variable Mg/Fe ratios, containing Ca and Al (Bradley 1988) to Fe–Mg-bearing aluminosilica (Klöck & Stadermann 1994). Bradley (1988) reported pyroxene glass in one IDP.

From infrared spectroscopy it is very difficult to obtain the composition of the amorphous silicates. This is because the spectral signature observed is a combination of grain composition, shape, size, and structure, making it difficult to isolate the pure amorphous silicate signal. This, in combination with the relatively small spectral changes caused by the composition of the silicates, makes it hard to get a definitive answer in most cases. In the case of interstellar dust we have a unique opportunity: the grains are very small and (almost) all silicates are amorphous.

For these interstellar grains Min *et al.* (2007) determined the composition of the amorphous silicates to be highly magnesium-rich (Mg/(Fe+Mg) ≈ 0.9). In addition, (Mg+Fe)/Si ≈ 1.35. This is inconsistent with the derived composition for the glassy silicate material in GEMS grains [that have (Mg+Fe)/Si ≈ 0.7, and Mg/(Fe+Mg) ≈ 0.9–1; Keller & Messenger 2004], indicating that the bulk silicates in the diffuse ISM are not the same as the GEMS grains found in the Solar System.

Because the capture medium for the Wild 2 grains was an amorphous silicate, the Stardust preliminary examination was not able to measure the intrinsic amorphous silicate content of the Wild 2 material. The Tempel 1 emission spectrum was best fitted by including amorphous silicates with a predominantly olivine-type stoichiometry, while some amorphous silicate with pyroxene-type stoichiometry was also found. The total surface area of these amorphous phases was found to be about one fifth that of the crystalline silicate (Lisse *et al.* 2006). Wild 2 samples collected by Stardust were found to contain "GEMS-like" structures. Chi *et al.* (2007) report that the amorphous phase in these samples is often pure SiO_2, suggesting it is melted aerogel rather than cometary material. However, Keller & Messenger (2008) have identified amorphous silicates in Wild 2 tracks that exhibit chondritic Mg/Fe and S/Fe ratios, suggesting GEMS-like materials may have been abundant in Wild 2 particles.

Both the anhydrous, porous IDPs and Tempel 1 appear to have a higher crystalline to amorphous silicate ratio than is inferred from infrared spectroscopy of interstellar or most circumstellar grains. This could simply be a result of grain alteration in interstellar space, with radiation processing converting crystalline silicates to amorphous silicates.

6.5.2 Carbonaceous dust

Amorphous and graphite elemental carbon

Spherules of relatively well-ordered crystalline graphite, ranging from ∼1 to 7 µm in size, that are found in some primitive meteorites, particularly the hydrous meteorites Murchison and Orgueil, are identified as originating in AGB stars and supernova outflows (see Chapter 2 for details). These graphite grains have exotic carbon isotopic ratios, but approximately solar nitrogen isotopic ratios. Submicron-sized amorphous carbon globules in the Tagish Lake chondritic meteorite have been identified as presolar, having H and N isotopic ratios characteristic of cold molecular clouds and the outer proto-solar disk (Nakamura-Messenger *et al.* 2006).

Carbon grains have a very smooth continuum opacity in the infrared making their detection difficult, so proper removal of stellar photosphere and scattered light contributions is critical to determining their abundance properly. Also, it is very hard to distinguish the emissivity signature of metallic iron and iron sulfide grains, which also provide a largely featureless continuum flux. However, a source of opacity is definitely needed in order to explain the infrared spectra of comets and protoplanetary disks. Min *et al.* (2005) have shown that for Comet Hale–Bopp a mass fraction of $\sim 20\%$ of amorphous carbon grains can successfully reproduce the observed infrared spectrum and the polarization of scattered light at optical wavelengths. Also, for protoplanetary disks, similar carbon abundances were found to best reproduce available observations for protoplanetary disks. Unfortunately, it has not been possible to derive the exact carbon abundance in the Wild 2 particles and thus a detailed comparison of this number to that in IDPs is not possible. In the primitive meteorites and IDPs carbon is abundant (a few percent in the meteorites and $\sim$12 wt% in the IDPs; Thomas *et al.* 1994), but elemental carbon is generally rare, with the bulk of the carbon being present as organic compounds (Flynn *et al.* 2003).

Diamonds

Nano-diamonds, typically ranging from <1 nm to $\sim$10 nm in size, are found in meteorites and IDPs. Meteoritic nano-diamonds can be concentrated by acid dissolution of the other components of the meteorite. Although the bulk C^{12}/C^{13} ratio of the meteoritic nano-diamonds is solar, this acid concentrate exhibits an isotopically anomalous Xe component linked to a supernova origin (Lewis *et al.* 1987), which suggests the nano-diamonds are non-solar. However, the Xe concentration is so low that only one in 10^6 would contain a Xe atom, so it has not been compellingly established that all the meteoritic nano-diamonds are non-solar; or even that some minor, undetected component of the acid concentrated material is the Xe host. The meteoritic nano-diamonds are generally believed to have formed by condensation from a vapor (Daulton *et al.* 1996). However, they do not exhibit the exciton feature in their carbon XANES spectrum (Flynn *et al.* 2000), a feature detected in synthetic vapor-deposited nano-diamonds but not seen in shock diamonds. Nano-diamonds have also been identified by electron diffraction in three large IDPs, two of which are anhydrous and one hydrous (Dai *et al.* 2002).

Two infrared spectral features have been attributed to nano-diamonds; at 3.43 and 3.53 μm. A search for these features in a sample of 60 Herbig Ae/Be stars has resulted in only two significant detections of such diamonds, in HD 97048 and Elias 3–1 (see Acke & van den Ancker 2006). These two sources are in no real aspect different from the rest of the sample. Habart *et al.* (2004) show that the diamond emission originates from the inner (i.e. < 15 AU) region of the disk. Also, the interstellar 3.47 μm absorption feature is attributed to nano-diamonds (see Pirali *et al.* 2007).

Organic carbon

The anhydrous, porous IDPs are very carbon-rich, having a mean carbon content of 12 wt% (Thomas *et al.* 1994), which is about three times higher than the carbon content of the most carbon-rich meteorites. Most of the carbon in IDPs that have not been modified by severe heating during atmospheric deceleration is organic, in that it consists of compounds of carbon and hydrogen, frequently including oxygen, nitrogen, or other elements (Flynn *et al.* 2004). The 3.4 μm aliphatic C–H stretching absorption features indicate a significantly higher $C-H_2$-to-$C-H_3$ ratio than is found in typical interstellar dust spectra, suggesting longer aliphatic chains in the IDPs (Flynn *et al.* 2003). Because these IDPs are the least processed samples of extraterrestrial material available for laboratory study, Flynn *et al.* (2008a) suggest that the interstellar organic matter experienced more severe radiation processing than the organic matter in IDPs. The carbonyl (C=O) functional group was also detected, with a C=O-to-aliphatic-$C-H_2$ ratio higher than is seen in interstellar spectra. Carbon XANES spectra demonstrate that carbon rings are abundant, but the aromatic C–H infrared absorption was rarely detected, suggesting the carbon may be in large-mass PAHs, which have fewer sites for H-bonding than smaller PAHs. Laser desorption laser ionization mass spectrometry measurements show the mass distribution of PAHs in IDPs has broad peaks centered near 250 and 370 amu. This is significantly higher than the PAH mass distribution seen in the Allende CV3 meteorite, which may indicate more extensive alkylation in the IDPs than in Allende. In an IDP where the organic matter was separated from the silicate, the C:O ratio was ~2:1 and the C:N ratio was ~10:1, indicating a higher O and N content in the organic matter in IDPs than in the insoluble organic matter extracted from CI and CM meteorites (Flynn *et al.* 2006b), but comparable to the O:C and N:C ratios seen in organic matter in the Wild 2 grains (Sandford *et al.* 2006). This latter study shows a correlation between increasing O:C and N:C ratios and degree of primitiveness in meteoritic organic matter. An extension of this trend line suggests that some of the organic matter in IDPs and Wild 2 particles is significantly more primitive, i.e. less metamorphosed, than the organic matter in meteorites. Combined carbon-XANES, micro-Fourier transform infra-red spectroscopy, and nano-SIMS on D-rich and D-poor regions of IDPs suggest the carrier of the anomalous H (high D/H) in that particle is an aliphatic hydrocarbon (Keller *et al.* 2004).

A few Wild 2 grains that were extracted from the aerogel were analyzed for carbon by similar techniques to the IDPs, and some were found to contain a significant amount of organic matter, some of which exhibits non-solar D/H (Sandford *et al.* 2006). Infrared spectra show a high $C-H_2$-to-$C-H_3$ ratio, athough the exact ratio cannot be determined because of $C-H_3$ contamination in the aerogel.

Carbon-XANES spectroscopy demonstrates the presence of the carbon-ring and C=O functional groups, with high ratios of O:C and N:C (Sandford *et al.* 2006). Many of the characteristics of the Wild 2 organic matter are generally similar to those of the organic matter in fine-grained, anhydrous IDPs, but not similar to the organic matter in the diffuse interstellar medium (Sandford *et al.* 2006). Polycyclic aromatic hydrocarbons were also detected spectroscopically in the Tempel 1 ejecta (Lisse *et al.* 2006). The best-fit model used small PAHs of pyrene (a three ringed $C_{16}H_{10}$) averaged size.

Emission from PAHs is detected in some, but not in all, protoplanetary disks. Also, PAH emission occurs more frequently in disks around higher-mass stars (in $\sim 57\%$ of the Herbig stars; Acke & van den Ancker 2004) than from disks around lower-mass stars (for the T Tauri stars the detection rate is closer to 10%; Geers *et al.* 2006). A relation between the presence of PAHs and disk geometry is reported by Acke & van den Ancker (2004); disks with a flaring outer surface show significantly more PAH emission. From this the idea emerges that PAHs are present in all disks, but visible only when they are able to absorb sufficient ultraviolet radiation from the central star to be excited and emit infrared radition. This idea is reinforced by the discovery of weak PAH features in T Tauri stars (Geers *et al.* 2006). Spectroscopic features attributed to PAH molecules have also been detected in a variety of other astrophysical locations, including the diffuse interstellar medium. If the PAHs were as abundant in protoplanetary disks as they are in interstellar space, theory predicts they would be detected in all disks (Dullemond *et al.* 2007). The fact that cometary spectra and the spectra of most protoplanetary disks show no, or very little, PAH emission could indicate that PAHs are destroyed in the early phases of protoplanetary disk evolution.

Carbonates

Carbonates are common in hydrous meteorites and hydrous IDPs, where they are believed to have formed by parent-body aqueous processing. Since simple models of cometary evolution involve no aqueous processing, carbonates were generally presumed not to occur in comets. However, carbonates have also been detected by infrared spectroscopy in the dust shell around evolved stars and in protostars, where liquid water is not expected (Ceccarelli *et al.* 2002; Kemper *et al.* 2002). Indeed, Toppani *et al.* (2005) have performed experiments that indicate that carbonates can be formed by non-equilibrium condensation in circumstellar environments where water is present as vapor, not as liquid. Detections of carbonates in other exosolar systems are reported by Ceccarelli *et al.* (2002) and Chiavassa *et al.* (2005).

Although carbonates are rare in anhydrous IDPs, Rietmeijer (1990) reported a magnesium carbonate grain in an anhydrous, porous IDP, and Flynn *et al.* (2006c) have found carbonate, including magnesium carbonate. The Vega-1, Vega-2, and Giotto spacecraft flew through the coma of Comet 1P/Halley determining the elemental compositions of dust particles, which were separated into groups based on their element abundance patterns. One group of particles with high Mg but relatively low Si and a high C/S ratio, indicating they were not magnesium sulfides, were identified as magnesium carbonate grains (Fomenkova *et al.* 1992). A weak infrared emission feature at 6.8 μm, attributed to carbonate, was also observed in Comet Halley (Bregman *et al.* 1987) and significant abundances of both magnesium and iron carbonates were reported in the Tempel 1 dust (Lisse *et al.* 2006). Carbonate was quite rare in the Wild 2 particles, but calcite, dolomite, and very small ferroan magnesite grains have been reported (Flynn *et al.* 2008b).

6.5.3 Other dust components
Sulfides

Sulfides are the only mineral group found in all types of extraterrestrial materials studied in the laboratory (Zolensky *et al.* 2006). Iron–nickel sulfide grains are a significant component of the fine-grained, anhydrous IDPs, pyrrhotite, a low-Ni iron sulfide, and pentlandite, with a higher nickel content, are both found in these IDPs (Rietmeijer 1989). Rare grains of iron-rich zinc sulfide are also found (Rietmeijer 1989). Infrared spectra of Tempel 1 dust produced in the Deep Impact cratering event was best fitted by including the magnesium–iron sulfide niningerite (Lisse *et al.* 2006). Iron sulfide was also detected using impact ionization mass spectrometers that flew through the dust coma of Comet Halley on the Giotto and Vega spacecraft in 1986 (Jessberger *et al.* 1988). Sulfides were abundant in the Wild 2 grains analyzed by transmission electron microscopy after collection by Stardust (Zolensky *et al.* 2006). Almost pure FeS was the dominant sulfide in the Wild 2 samples, but rare grains of iron-nickel sulfide (pentlandite) were also found.

In comparison, in the emission spectra of protoplanetary disks a very broad resonance around 23 μm was detected (Keller *et al.* 2000). This coincides with a resonance detected in the laboratory spectra of iron sulfide. However, laboratory iron sulfide spectra also display strong, narrow spectral structure at wavelengths longward of 30 μm, not seen in the spectrum of protoplanetary disks, perhaps due to grain size effects. An increase in grain size will cause the strongest features, i.e. those at long wavelength, to be weakened with respect to the broad, weaker feature around 23 μm. Since there are little reliable laboratory data on iron sulfide it is, at present, hard to make true quantitative statements on this.

A question that needs answering is what the origin of iron sulfide is. As discussed above, the sulfur in our Solar System is all locked up in solid FeS. In a nebula of solar composition, iron sulfide condenses at about 650 K (Lauretta *et al.* 1996). However, in the diffuse interstellar medium the sulfur is all found in the gas phase (Sofia 2004). Somewhere during the formation of the Solar System, or in the parent molecular cloud, the FeS was formed. Finding the onset of this formation might hold important information on the processing history and dynamics of materials in protoplanetary disks.

Silicon carbide

Silicon carbide (SiC) is detected in primitive meteorites in the Solar System at a very low abundance (< 0.001%). This SiC is shown to be of presolar origin, i.e. formed in the outflows of evolved stars or supernovae (see Chapter 2). Indeed SiC is detected spectroscopically to be forming around carbon-rich AGB stars (see e.g. Speck *et al.* 1997). When considering the typical ratio of oxygen-rich to carbon-rich AGB stars, the expected abundance of SiC emitted into the diffuse interstellar medium corresponds to ~10% of all Si atoms being in SiC. However, the abundance of SiC in the primitive meteorites is around a few parts per million, indicating that most of it has been destroyed prior to, or during, the formation of the Solar System. In agreement with this, no detection of SiC is reported in the spectra of protoplanetary disks. The detection of SiC along one line of sight in the diffuse interstellar medium by Min *et al.* (2007) indicates that most of the SiC at least survived the diffuse interstellar medium to arrive in the molecular cloud from which the Solar System formed. If this is indeed true, the question remains as to whether the SiC was destroyed in the molecular cloud, or in the early solar nebula.

6.6 Processing history of grains as derived from the dust composition

From the results summarized above on the composition of the dust grains in proto-planetary disks the general picture that emerges for the processing of material in a protoplanetary disks is neither simple nor static. The high abundance of crystalline silicates, believed to form only in the inner region of the solar nebula, reported in both Wild 2 and Tempel 1, suggests significant transport of inner-disk material outward to the comet-formation region. In addition, these high-temperature materials are found in protoplanetary disks also at large distances from the central star. Scenarios for mixing of materials formed in the inner disk and transported outward could account for this. For example, radial transport based on turbulent mixing is found to be able to account for some high-temperature condensates at large distances (around several tens of astronomical units) from the central star (see Chapter 3 for details; and also Bockelée-Morvan *et al.* 2002; Gail 2004; and

Ciesla 2007). The X-wind scenario, suggested by Shu *et al.* (2001), allows even larger particles to be transported to the outer-disk regions. Other possibilities for forming these high-temperature condensates far out in the disk are shock heating or lightning (see Chapter 8; and also Pilipp *et al.* 1998; Desch & Cuzzi 2000; and Harker & Desch 2002). If these processes are the main causes of the appearance of crystalline silicates in the outer region of protoplanetary disks, it makes the use of crystallinity as an indicator of disk dynamics more complicated. In addition to these mechanisms, (proto)planets can drive shocks that anneal dust grains, or cause a collisional cascade of larger bodies, releasing material which has been altered by parent-body processing. If this "second generation" dust comes from differentiated bodies it is expected to be crystalline and have a high iron content. In our Solar System, modeling by Farinella & Davis (1996) indicates that most of the current 1 km to 10 km size bodies in the Kuiper Belt are collisional fragments of much larger bodies. This mechanism might allow us to probe the formation of planets by studying the amount of processing visible in the small-grain component. A different picture was proposed by Dullemond *et al.* (2006) who computed that the material falling onto the protoplanetary disk in the very beginning enters at quite close proximity to the central star, at high temperatures, and then spreads out into the protoplanetary disk.

In conclusion, (radial) mixing and the accompanying thermal processing of the grains are evident from both remote sensing and laboratory studies of protoplanetary dust. These mechanisms will be discussed extensively in Chapter 8. Further development of the remote sensing techniques requires accurate models of the optical properties and predictions on the aggregate shapes expected to form in disks. These will be discussed in Chapter 7.

References

Acke, B. & van den Ancker, M. E. 2004, *Astronomy and Astrophysics*, **426**, 151.
Acke, B. & van den Ancker, M. E. 2006, *Astronomy and Astrophysics*, **457**, 171.
Anders, E. & Grevesse, N. 1989, *Geochimica et Cosmochimica Acta*, **53**, 197.
Anders, E. & Zinner, E. 1993, *Meteoritics*, **28**, 490.
Bernatowicz, T. J., Messenger, S., Pravdivtseva, O., Swan, P., & Walker, R. M. 2003, *Geochimica et Cosmochimica Acta*, **67**, 4679.
Bernatowicz, T. J., Croat, T. K., & Daulton, T. L. 2006, *Meteorites and the Early Solar System II*, ed. D. S. Lauretta and H. V. McSween, University of Arizona Press, 109–126.
Bockelée-Morvan, D., Gautier, D., Hersant, F., Huré, J.-M., & Robert, F. 2002, *Astronomy and Astrophysics*, **384**, 1107.
Bouwman, J., de Koter, A., Dominik, C., & Waters, L. B. F. M. 2003, *Astronomy and Astrophysics*, **401**, 577.
Bouwman, J., Henning, T., Hillenbrand, L. A., *et al.* 2008, *Astrophysical Journal*, **683**, 479.

Bouwman, J., Meeus, G., de Koter, A., *et al.* 2001, *Astronomy and Astrophysics*, **375**, 950.

Bradley, J. P. 1988, *Geochimica et Cosmochimica Acta*, **52**, 889.

Bradley, J. P. 1994, *Science*, **265**, 925.

Bradley, J. P., Keller, L. P., Snow, T. P., *et al.* 1999, *Science*, **285**, 1716.

Bregman, J. D., Witteborn, F. C., Allamandola, L. J., *et al.* 1987, *Astronomy and Astrophysics*, **187**, 616.

Brownlee, D., Tsou, P., Aléon, J., *et al.* 2006, *Science*, **314**, 1711.

Brucato, J. R., Colangeli, L., Mennella, V., Palumbo, P., & Bussoletti, E. 1999, *Planetary Space Science*, **47**, 773.

Campins, H. & Ryan, E. V. 1989, *Astrophysical Journal*, **341**, 1059.

Ceccarelli, C., Caux, E., Tielens, A. G. G. M., *et al.* 2002, *Astronomy and Astrophysics*, **395**, L29.

Chi, M., Ishii, H., Toppani, A., *et al.* 2007, *Lunar and Planetary Institute Conference Abstracts*, **38**, 2010.

Chiavassa, A., Ceccarelli, C., Tielens, A. G. G. M., Caux, E., & Maret, S. 2005, *Astronomy and Astrophysics*, **432**, 547.

Ciesla, F. J. 2007, *Science*, **318**, 613.

Dai, Z. R., Bradley, J. P., Joswiak, D. J., Brownlee, D. E., & Genge, M. J. 2002, *Lunar and Planetary Institute Conference Abstracts*, **33**, 1321.

Daulton, T. L., Eisenhour, D. D., Bernatowicz, T. J., Lewis, R. S., & Buseck, P. R. 1996, *Geochimica et Cosmochimica Acta*, **60**, 4853.

Davoisne, C., Djouadi, Z., Leroux, H., *et al.* 2006, *Astronomy and Astrophysics*, **448**, L1.

Desch, S. J. & Cuzzi, J. N. 2000, *Icarus*, **143**, 87.

Dorschner, J., Begemann, B., Henning, T., Jaeger, C., & Mutschke, H. 1995, *Astronomy and Astrophysics*, **300**, 503.

Draine, B. T. 2003, *Annual Review of Astronomy and Astrophysics*, **41**, 241.

Dullemond, C. P., Apai, D., & Walch, S. 2006, *Astrophysical Journal*, **640**, L67.

Dullemond, C. P., Henning, T., Visser, R., *et al.* 2007, *Astronomy and Astrophysics*, **473**, 457.

Ebel, D. S. 2006, *Meteorites and the Early Solar System II*, ed. D. S. Lauretta & H. V. McSween, University of Arizona Press, 253.

Fabian, D., Henning, T., Jäger, C., *et al.* 2001, *Astronomy and Astrophysics*, **378**, 228.

Farinella, P. & Davis, D. R. 1996, *Science*, **273**, 938.

Flynn, G. J., Bajt, S., Sutton, S. R., *et al.* 1996, *Astromical Society of the Pacific Conference Series*, **104**, 291.

Flynn, G. J., Bleuet, P., Borg, J., *et al.* 2006a, *Science*, **314**, 1731.

Flynn, G. J., Henning, T., Keller, L. P., & Mutschke, H. 2002, in *Optics of Cosmic Dust*, ed. G. Videen & M. Kocifaj, Kluwer, 37–56.

Flynn, G. J., Keller, L. P., Feser, M., Wirick, S., & Jacobsen, C. 2003, *Geochimica et Cosmochimica Acta*, **67**, 4791.

Flynn, G. J., Keller, L. P., Hill, H., Jacobsen, C., & Wirick, S. 2000, *Lunar and Planetary Institute Conference Abstracts*, **31**, 1904.

Flynn, G. J., Keller, L. P., Jacobsen, C., & Wirick, S. 2004, *Advances in Space Research*, **33**, 57.

Flynn, G. J., Keller, L. P., Wirick, S., & Jacobsen, C. 2006b, in *X-ray Microscopy: Proceedings of the 8th International Conference*, ed. S. Aoki Y. Kagoshima & Y. Suzuki, IPAP Conference Series 7, 315–317.

Flynn, G. J., Keller, L. P., Wirick, S., & Jacobsen, C. 2006c, *Geochimica et Cosmochimica Acta Supplement*, **70**, 179.

Flynn, G. J., Keller, L. P., Wirick, S., & Jacobsen, C. 2008a, *Proceedings of the IAU Symposium*, **251**, 267.

Flynn, G. J., Leroux, H., Tomeoka, K., *et al.* 2008b, *Lunar and Planetary Science Insitute Conference Abstracts*, **39**, 1391.

Fomenkova, M. N., Kerridge, J. F., Marti, K., & McFadden, L.-A. 1992, *Science*, **258**, 266.

Gail, H.-P. 2002, *Astronomy and Astrophysics*, **390**, 253.

Gail, H.-P. 2004, *Astronomy and Astrophysics*, **413**, 571.

Geers, V. C., Augereau, J.-C., Pontoppidan, K. M., *et al.* 2006, *Astronomy and Astrophysics*, **459**, 545.

Grady, C. A., Polomski, E. F., Henning, T., *et al.* 2001, *Astronomical Journal*, **122**, 3396.

Habart, E., Testi, L., Natta, A., & Carbillet, M. 2004, *Astrophysical Journal*, **614**, L129.

Harker, D. E. & Desch, S. J. 2002, *Astrophysical Journal*, **565**, L109.

Hayward, T. L., Hanner, M. S., & Sekanina, Z. 2000, *Astrophysical Journal*, **538**, 428.

Honda, M., Kataza, H., Okamoto, Y. K., *et al.* 2006, *Astrophysical Journal*, **646**, 1024.

Hörz, F., Bastien, R., Borg, J., *et al.* 2006, *Science*, **314**, 1716.

Ishii, H. A., Bradley, J. P., Dai, Z. R., *et al.* 2008, *Science*, **319**, 447.

Jessberger, E. K., Christoforidis, A., & Kissel, J. 1988, *Nature*, **332**, 691.

Keller, L. P., Hony, S., Bradley, J. P., *et al.* 2002, *Nature*, **417**, 148.

Keller, L. P. & Messenger, S. 2004, *Lunar and Planetary Institute Conference Abstracts*, **35**, 1985.

Keller, L. P. & Messenger, S. 2007, in Workshop on the Chronology of Meteorites and the Early Solar System, November 5–7, Kauai, Hawaii. LPI Contribution No. 1374, 88.

Keller, L. P. & Messenger, S. 2008, *Meteoritics and Planetary Science Supplement*, **43**, 5227.

Keller, L. P., Messenger, S., Flynn, G. J., *et al.* 2004, *Geochimica et Cosmochimica Acta*, **68**, 2577.

Kemper, F., Jäger, C., Waters, L. B. F. M., *et al.* 2002, *Nature*, **415**, 295.

Klöck, W. & Stadermann, F. J. 1994, *American Institute of Physics Conference Proceedings*, **310**, 51.

Klöck, W., Thomas, K. L., McKay, D. S., & Zolensky, M. E. 1990, *Lunar and Planetary Institute Conference Abstracts*, **21**, 637.

Koike, C., Chihara, H., Tsuchiyama, A., *et al.* 2003, *Astronomy and Astrophysics*, **399**, 1101.

Larimer, J. W. 1967, *Geochimica et Cosmochimica Acta*, **31**, 1215.

Lauretta, D. S., Kremser, D. T., & Fegley, B. J. 1996, *Icarus*, **122**, 288.

Lewis, R. S., Ming, T., Wacker, J. F., Anders, E., & Steel, E. 1987, *Nature*, **326**, 160.

Lisse, C. M., VanCleve, J., Adams, A. C., *et al.* 2006, *Science*, **313**, 635.

Lodders, K. 2003, *Astrophysical Journal*, **591**, 1220.

Manceau, A., Marcus, M. A., & Tamura, N. 2002, *Reviews in Mineralogy and Geochemistry*, **49**, 341.

Martin, P. G. 1995, *Astrophysical Journal*, **445**, L63.

Messenger, S., Keller, L. P., & Lauretta, D. S. 2005, *Science*, **309**, 737.

Messenger, S., Stadermann, F. J., Floss, C., Nittler, L. R., & Mukhopadhyay, S. 2003, *Space Science Reviews*, **106**, 155.

Messenger, S. & Walker, R. M. 1997, *American Institute of Physics Conference Proceedings*, **402**, 545.

Meyer, B. S. & Zinner, E. 2006, in *Meteorites and the Early Solar System* II, ed. D. S. Lauretta & H. Y. McSween, University of Arizona Press, 69–108.

Min, M., Dominik, C., Hovenier, J. W., de Koter, A., & Waters, L. B. F. M. 2006, *Astronomy and Astrophysics*, **445**, 1005.

Min, M., Dominik, C., & Waters, L. B. F. M. 2004, *Astronomy and Astrophysics*, **413**, L35.

Min, M., Hovenier, J. W., & de Koter, A. 2003, *Astronomy and Astrophysics*, **404**, 35.

Min, M., Hovenier, J. W., de Koter, A., Waters, L. B. F. M., & Dominik, C. 2005, *Icarus*, **179**, 158.

Min, M., Hovenier, J. W., Waters, L. B. F. M., & de Koter, A. 2008, *Astronomy and Astrophysics*, **489**, 135.

Min, M., Waters, L. B. F. M., de Koter, A., et al. 2007, *Astronomy and Astrophysics*, **462**, 667.

Nakamura-Messenger, K., Messenger, S., Keller, L. P., Clemett, S. J., & Zolensky, M. E. 2006, *Science*, **314**, 1439.

Pascucci, I., Apai, D., Luhman, K., et al. 2009, *Astrophysics Journal*, **693**, 143.

Pilipp, W., Hartquist, T. W., Morfill, G. E., & Levy, E. H. 1998, *Astronomy and Astrophysics*, **331**, 121.

Pirali, O., Vervloet, M., Dahl, J. E., et al. 2007, *Astrophysical Journal*, **661**, 919.

Rietmeijer, F. J. M. 1989, *Lunar and Planetary Science Conference* Abstracts, **19**, 513.

Rietmeijer, F. J. M. 1990, *Meteoritics*, **25**, 209.

Rietmeijer, F. J. M. 1998, *Reviews in Mineralogy*, **36**, 2.

Sandford, S. A., Aléon, J., Alexander, C. M. O., et al. 2006, *Science*, **314**, 1720.

Sandford, S. A. & Walker, R. M. 1985, *Astrophysical Journal*, **291**, 838.

Sargent, B., Forrest, W. J., D'Alessio, P., et al. 2006, *Astrophysical Journal*, **645**, 395.

Savage, B. D. & Sembach, K. R. 1996, *Annual Review of Astronomy and Astrophysics*, **34**, 279.

Schramm, L. S., Brownlee, D. E., & Wheelock, M. M. 1989, *Meteoritics*, **24**, 99.

Sears, D. W. G. & Dodd, R. T. 1988, *Meteorites and the Early Solar System*, ed. J. F. Kerridge and M. S. Matthews, University of Arizona Press, 3–31.

Servoin, J. L. & Piriou, B. 1973, *Physica Status Solidi (b)*, **55**, 677.

Shu, F. H., Shang, H., Gounelle, M., Glassgold, A. E., & Lee, T. 2001, *Astrophysical Journal*, **548**, 1029.

Sitko, M. L., Grady, C. A., Lynch, D. K., Russell, R. W., & Hanner, M. S. 1999, *Astrophysical Journal*, **510**, 408.

Sofia, U. J. 2004, *Astronomical Society of the Pacific Conference Series*, **309**, 393.

Speck, A. K., Barlow, M. J., & Skinner, C. J. 1997, *Monthly Notices of the Royal Astronomical Society*, **288**, 431.

Sutton, S. R., Bertsch, P. M., Newville, M., et al. 2002, *Reviews in Mineralogy and Geochemistry*, **49**, 429.

Tamanai, A., Mutschke, H., Blum, J., & Meeus, G. 2006, *Astrophysical Journal*, **648**, L147.

Thomas, K. L., Keller, L. P., Blanford, G. E., & McKay, D. S. 1994, *American Institute of Physics Conference Proceedings*, **310**, 165.

Toppani, A., Robert, F., Libourel, G., et al. 2005, *Nature*, **437**, 1121.

Ueda, Y., Mitsuda, K., Murakami, H., & Matsushita, K. 2005, *Astrophysical Journal*, **620**, 274.

van Boekel, R., Min, M., Leinert, C., et al. 2004, *Nature*, **432**, 479.

van Boekel, R., Min, M., Waters, L. B. F. M., et al. 2005, *Astronomy and Astrophysics*, **437**, 189.

Wooden, D. H., Harker, D. E., Woodward, C. E., et al. 1999, *Astrophysical Journal*, **517**, 1034.

Zinner, E. 1988, in *Meteorites and the Early Solar System*, ed. J. F. Kerridge and M. S. Matthews, University of Arizona Press, 956–983.

Zolensky, M. & McSween, Jr., H. Y. 1988, in *Meteorites and the Early Solar System*, ed.
    J. F. Kerridge and M. S. Matthews, University of Arizona Press, 114–143.
Zolensky, M., Zega, T., Weisberg, M., *et al.* 2007, *Lunar and Planetary Institute
    Conference Abstracts*, **38**, 1481.
Zolensky, M. E. & Barrett, R. A. 1994, *Meteoritics*, **29**, 616.
Zolensky, M. E., Zega, T. J., Yano, H., *et al.* 2006, *Science*, **314**, 1735.

7

Dust particle size evolution

Klaus M. Pontoppidan and Adrian J. Brearley

Abstract This chapter describes how the growth of particles from submicron to centimeter sizes in protoplanetary disks is chronicled both by astronomical observations and microscopic imaging of pristine meteoritic material. A growing sample of planet-forming disks at a range of evolutionary stages is available for astronomical remote sensing studies, but limitations of spatial resolution as well as very high optical depths still hide the inner mid-plane of the disks – the exact region where planets are believed to form. Conversely, meteoritic studies currently exclusively sample material from the inner mid-plane of the solar nebula, and the fact that such material can be brought into a laboratory setting allows very detailed studies of its properties. The dust component in the matrices of chondritic meteorites carries a record of continuous processing and modification of fine-grained material during protoplanetary disk evolution that tends to obscure the earliest stages of dust formation, growth, and coagulation. We discuss how knowledge on dust particle size evolution gained from these two very different approaches can be combined to significantly enhance our understanding of the first stages in planet formation.

Solids in protoplanetary disks undergo a growth process resulting in extreme changes in the size distribution of dust particles over time. We know this because the particle size distribution of dust in the interstellar medium (ISM) is so different from that of planets, leftover planetesimals, and dust in our own Solar System; the formation of the Earth from submicron grains corresponds to a change of 12 orders of magnitude! The presence of a large number of extrasolar planets shows that the planet-formation process is relatively common. It is curious how difficult it is actually to measure and study this growth process in nearby protoplanetary systems and in the fossil record of the Solar System. To astronomers, much of the difficulty stems from a three-way degeneracy between the particle size distribution, the optical properties of the bulk solid, and radiative transfer effects. Cosmochemists studying meteoritic materials have difficulties identifying pristine grains in large enough numbers to constrain the early solar nebula as a whole. A crucial question is therefore: can the two disciplines, astrophysics and cosmochemistry, be synergized to improve our understanding of the early evolution of small dust particles in protoplanetary disks?

The focus of the chapter is on the smallest range of particle sizes (<1 cm) and thus the initial phases of dust coagulation in young gas-rich disks, since these are the ones that can be observed in extrasolar systems and have preserved some very early record in chondritic meteorites and interplanetary dust particles (IDPs).

The theoretical understanding of grain coagulation in protoplanetary disks is relatively advanced, and several excellent and current reviews of this theory exist (Henning *et al.* 2006; Dominik *et al.* 2007). Our knowledge of the initial growth of primordial dust particles to "pebble sizes" ($\sim$1 cm) is therefore severely limited by our ability to accurately observe and measure particle size distributions, composition, and evolutionary history. In this chapter, our main goal is to provide an overview of data relevant for dust coagulation processes as obtained from astronomical observations of protoplanetary disks, and to compare data obtained from studies of early Solar System materials. Of specific interest is the determination of particle size of the materials present in protoplanetary disks as an indicator of coagulation processes. We will also examine the key assumption at the core of this comparison; that the extraterrestrial sample suite (chondritic meteorites and IDPs) are to some degree representative of dust in extrasolar protoplanetary disks.

7.1 Dust coagulation in the Solar System and in extrasolar protoplanetary disks

There are clearly a number of major differences between astrophysics and cosmochemistry which must be considered when making comparisons. The astronomical data set is now based on a sample of hundreds of circumstellar disks that are presumably actively experiencing grain coagulation as a first step toward planet formation. For this reason, they are often referred to as putative "protoplanetary" disks. Conversely, the data from the Solar System are based on a single, ancient disk, but one we know eventually did form planets, and that can be studied in great detail. Other important differences include the observational methodologies, the scale of the observations (i.e. the volume of material sampled and the spatial resolution) and the location within the disk that is being sampled.

The spatial resolution of astronomical observations varies over several orders of magnitude from 1–100 AU, depending on the observing technique, whereas the extraterrestrial chondritic sample suite predominantly represents materials that were derived from a feeding zone significantly less than 1 AU in width, centered around 3.5–4.0 AU. The actual volume of protoplanetary disk material sampled by an individual chondrite is hard to define precisely. However, given that asteroid spectroscopy has shown that different asteroid classes occur distributed within certain zones through the Asteroid Belt, we can infer that specific asteroid types as the parent bodies of chondritic and other meteorites accreted within feeding zones that were fractions of an AU in width. The difference in scale between the two different types of observations therefore spans several orders of magnitude in both terms of spatial resolution and the mass of material which is being imaged.

Further, many of the astronomical data do not probe dust in the disk mid-planes, but rather dust in the surface layers of the disks, whereas chondrites sample material from the mid-plane. Comparisons therefore rely on assumptions of communication between the surface layer and mid-plane at a few astronomical units, such as vertical mixing.

The techniques used in astronomical and cosmochemical studies are quite different. Astronomical studies use the optical characteristics of dust to extract particle size distributions, whereas studies of chondrites involve direct measurement of grain sizes from electron microscope images. The particle size distributions measured by astronomical techniques are therefore averaged over a huge volume of material within the disk and the statistics are excellent. No comparable size distributions for the fine-grained ($< 5\,\mu m$) fraction of meteorites are available. This disparity in the data sets comes about because size measurements of large numbers of chondritic grains is both difficult and very time-consuming. Although characteristic grain sizes for different phases have been measured, a major gap in our knowledge of the characteristics of the fine-grained dust fraction of chondrites is the grain size distribution, based on a large number of measurements.

7.2 Nomenclature and definitions

The field of protoplanetary dust size distributions and properties can be confusing. We therefore begin by briefly reviewing the most common definitions and assumptions. Owing to the nature of astrophysical observations, extrasolar dust particles are usually described in terms of their optical properties. This is generally not the case for Solar System material, which is available to direct laboratory studies. This gives rise to potentially confusing differences in nomenclature.

7.2.1 What is considered evidence for protoplanetary dust coagulation?

It is non-trivial to define what is meant by "particle growth." For extrasolar protoplanetary disks, particle growth refers to all processes that contribute to increase the "typical" particle size in a given region in a disk. Grain coagulation is one type of particle growth in which cohesive grain units stick together to form loosely bound clusters. Particles may also grow by direct gas-phase deposition, called condensation. Particles formed by nucleation in a cooling gas or processed by melting are often called monomers in astrophysics. The distinction between monomers and clusters is made because it is much easier to fragment and break apart a cluster particle (into monomers) than a single monomer. While not impossible, it is difficult to distinguish between monomers and cluster grains from astronomical observations, in particular if the particle population includes both. Note that in the astrophysical literature "dust grain" therefore typically refers to both monomers and aggregates. However, to avoid confusion with the grains of cosmochemistry, which always refer to monomers, we use "particle" to describe free-floating entities that may be monomers or aggregates.

When searching for evidence for particle growth and coagulation in protoplanetary disks, the reference point is the particle size distribution of dust in the ISM prior to its incorporation into the disk. It is difficult to infer that the particle size distribution in a given protoplanetary disk has evolved through aggregation processes, or otherwise, without knowing this starting point. Thus, the pre- and protostellar size distribution and other properties of dust must be known. The particle size distribution in the local diffuse interstellar medium is relatively well understood for the size range $a = 0.0035$–$1.0\,\mu m$ and several models fitting to ultraviolet to infrared extinction curves have been constructed (e.g. Mathis *et al.* 1977; Kim *et al.* 1994; Weingartner & Draine 2001). The population of particles larger than a few microns does not strongly affect the shorter wavelength extinction curves, and extinction at long wavelengths ($\lambda \gtrsim 10\,\mu m$) is difficult to measure (Indebetouw *et al.* 2005). Thus, the upper cutoff in the size distribution is often inferred from the known elemental abundances of carbon and silicon. Any population of large particles unconstrained by direct observations of extinction and scattering must leave enough grain volume in small particles to reproduce a measured gas-to-extinction ratio, e.g. $A_V/N_H = 1.9 \times 10^{21}\,cm^2$ (where N_H is the number of H atoms and A_V is the visual extinction)[1] (Lyα absorption, Bohlin *et al.* 1978) for the diffuse ISM, or $A_J/N_H = 5.6 \times 10^{21}\,cm^2$ (where A_J is the J-band extinction) for dense clouds (X-ray absorption, Vuong *et al.* 2003). Even with these constraints, carbonaceous particles may include a significant population with sizes up to 10 μm (Weingartner & Draine 2001). Silicate particles seem to be constrained to somewhat smaller sizes of <1 μm.

---

[1] It is generally very difficult to measure total gas column densities through a cloud, while it is relatively straightforward to measure the extinction (essentially optical depth) due to dust at some wavelength. The extinction is related to the total dust column density via a dust model, including a particle size distribution. These dust models are typically not unique, hence the additional constraint from the elemental abundances of dust constituents, in particular Si and Fe. Therefore, the conversion factor between the two is an important number, and significant efforts have been directed toward its measurement.

To further complicate matters, it is expected that dust particles undergo significant coagulation at H_2 densities of 10^5–10^6 cm^{-3} in protostellar envelopes (Ossenkopf & Henning 1994), *prior* to their incorporation in a protoplanetary disk. While some or all of this coagulation may be undone when prestellar dust particles fall with high impact velocities on the young protoplanetary accretion disk as part of its formation, the presence of particles a few microns in size is not in itself ironclad evidence that coagulation has taken place. The presence of, say, millimeter-sized particles can be considered much stronger evidence.

7.2.2 Distribution functions

In astrophysics, particle sizes in protoplanetary disks are usually described in the context of a distribution, $dn = f(a)da$, where a is the particle radius and dn is the number density of particles with radii between a and $a + da$. Note that in cosmochemistry, grain size usually refers to the *largest diameter* of a monomer. The description is often associated with two assumptions. First, it is often assumed that the distribution function, $f(a)$, is close to being a power law in the ISM, except at very large or very small sizes (Weingartner & Draine 2001), and power laws are therefore often used in protoplanetary disk studies. Second, the definition of a particle radius assumes that the particles are spherical – a convenience for converting the size distribution to an opacity law.

Neither of these two assumptions are likely very representative of protoplanetary dust. Even interstellar dust size distributions likely deviate significantly from power laws (Kim *et al.* 1994). The initial product of grain coagulation, fractal-like aggregates of compact monomers, is far from compact and spherical, and such particles may interact very differently with light than compact grains with similar geometric cross-sections. The initial particle size distribution quickly diverges from a single power law in a protoplanetary disk due to competing growth and destruction processes as well as dynamical effects causing larger dust particles to settle and drift inwards (Wurm & Blum 1998; Dullemond & Dominik 2005). Many of the observational studies of protoplanetary dust are thus concerned with reconciling $f(a)$ as predicted by theory.

7.2.3 "Large" and "small" particles

The optical properties of spherical particles at wavelength λ change drastically around *size parameter*, $x = 2\pi a/\lambda \sim 1$; particles with $x \gg 1$ are in the "geometric" limit, and their absorption cross-sections scale with the cross-sectional area: $C_{abs} \sim a^2$. Particles with size parameters $\ll 1$ have absorption cross-sections that roughly scale as particle volume $C_{abs} \sim a^3$. The scattering properties of both large and small spherical particles are more complicated since not only the scattering cross-section is important, but also the directional dependence of the scattered irradiance and polarization.

When references are made to *large* and *small* particles (size parameters larger and smaller than unity, respectively) by observers, an implicit dependence on the wavelength of the observational data is being introduced. Thus, the "large" particles in a study of near-infrared scattering are the "small" particles of a study of millimeter emission. In this chapter, "large" and "small" particles will refer to whether they have size parameters larger or smaller than unity, and the use of such modifiers is always linked to an observed wavelength regime.

7.3 Coagulation basics

Dust coagulation refers to the formation of ever larger aggregates from inelastic collisions of individual compact dust grains (monomers) and previously formed aggregates. Coagulation processes are divided into monomer–cluster aggregation [usually called particle–cluster aggregation (PCA); "particle" refers to what we call monomer] and cluster–cluster aggregation (CCA), the first being relevant for dust populations dominated by monomers, while the second is relevant when most collisions occur between aggregates. To facilitate the coagulation process, the dust particles must have sufficiently large relative velocities to ensure that collisions are not rare, but not so large velocities that the collision partners rebound or even shatter into smaller fragments. Thus, the process of dust coagulation in protoplanetary disks is intimately connected to the dust–gas interactions that govern the relative velocities.

The most basic model of grain coagulation can be summarized as follows: submicron particles are small enough to be fully coupled to the gas. Their relative motion is Brownian, leading to small, non-destructive collisional velocities (~ 1 cm s^{-1}), and potentially turbulent, characterized by somewhat larger relative velocities (perhaps 10–100 cm s^{-1}, Wurm & Blum 1998). As particles grow larger they begin to move toward the mid-plane along the vertical component of the gravitational vector from the central star in a process known as settling. As this happens, the particles may sweep up smaller grains, still suspended in the gas, thus increasing their mass and cross-section. Larger cross-sections in turn lead to an increased rate of coagulation. For sticking probabilities near unity, this leads to an exponential growth, which ceases only when the particles reach the disk mid-plane as "pebbles" or "boulders." At this point, the particles will experience gas-drag forces because pressure support in the disk causes the gas to orbit at slightly sub-Keplerian velocities in the mid-plane. Large particles in the disk mid-plane, which tend to orbit at Keplerian velocities, will thus experience a head wind, causing them to migrate inwards on timescales as short as $100 \, \text{yr} \, \text{AU}^{-1}$ (Weidenschilling 1977). At higher altitudes, the direction of the gas pressure gradient may switch, leading to super-Keplerian gas velocities and outward migrating particles (Takeuchi & Lin 2002). Such radial drift will induce relative velocity differences depending on particle size and location in the disk, and will generally enhance coagulation (as well as provide a sink of boulder-sized particles as these may migrate all the way to the star where they are lost).

In addition to inducing relative velocities between dust particles, gas turbulence will also lift dust particles that have already settled toward the mid-plane and return them to the disk surface. Particles that have been lifted up will be allowed another pass at settling down to the disk mid-plane while sweeping up smaller dust particles in their path.

All these mechanisms contribute to the complicated coupling between grain coagulation and gas dynamics, and generally require a grain coagulation simulation to be solved along with an evolving gas disk model in order to gain a detailed understanding of the resulting particle size distributions and grain structures. However, even simple models conclude that the coagulation of submicron particles to centimeter- and meter-sizes is fast – much shorter than the lifetime of gaseous disks (Blum 2004; see also Chapters 3 and 9). This must be reconciled with the observation that small particles have generally not been completely depleted from disks in $<10^6 \, \text{yr}$ (Dullemond & Dominik 2005). The inference is that efficient fragmentation mechanisms exist that can replenish the population of micron-sized particles. Further information on dust coagulation can be found in Blum (2004) and Dominik *et al.* (2007).

7.4 Laboratory simulations of dust coagulation

Many parameters that go into dust coagulation models are difficult or too complex to estimate from purely theoretical considerations or from astronomical observations, and laboratory simulations of particle–particle collisions are needed to constrain them. Important parameters that can be determined from laboratory experiments include the relative particle sizes and velocities that allow grains to stick, that will compact loosely bound aggregates, and that will fragment aggregates. The shape and porosity of aggregates is important since the ratio of cross-section to particle mass may be very different than for compact spherical particles. For instance, fluffy, fractal aggregates will interact very differently with the gas and have quite different opacities than compact particles with the same mass, because of differences in cross-section. Indeed, experiments have confirmed that dust particles formed by ballistic aggregation will have a highly fluffy and fractal structure (Wurm & Blum 1998). The structure of such fluffy particles is often described by a power-law relation between the mass of the particle and the gyration radius, r

(or other size measure), $m \propto r^{D_f}$, where D_f is the fractal dimension of the particle. For compact (spherical) particles, $D_f = 3$, but note that the converse is not necessarily true, i.e. a fractal dimension of 3 does not imply compact particles, and it is possible to find particles with porosities of 90% or higher with $D_f = 3$. The other extreme, a linear string of monomers, results in $D_f = 1$. Experiments and numerical simulations show that for CCA, $D_f \sim 1.9$ (Wurm & Blum 1998), while for PCA, $D_f \to 3$ for aggregate sizes $\gtrsim 1000$ monomers (Mukai *et al.* 1992; Ossenkopf 1993). If collision velocities are sufficient, the dissipation of kinetic energy may restructure CCA aggregates to make them more compact and increase D_f to values close to 3 (Dominik & Tielens 1997).

Dependencies of grain composition, e.g. icy grains versus silicates or carbon, as well as their charge state, are also of interest, and for destructive collisions, the resulting fragment size distribution is important (Wurm *et al.* 2005). These are dependent on the microphysics of the adhesive forces between monomers (Chokshi *et al.* 1993; Blum & Schräpler 2004). A review of laboratory simulations of dust coagulation can be found in Blum & Wurm (2008).

7.5 Observational tracers of grain coagulation

Consider the typical observables of dust particles in an astronomical object; one may have access to a spectrum covering a wide range of wavelengths, and with luck some degree of spatial information resulting in a brightness distribution at each wavelength. The aim of the observer is to convert this information to a particle size distribution as a function of position in the disk. By their very nature, disks are at least two-dimensional objects, and although they almost certainly contain significant, or even strong three-dimensional structure, most particle size and dust coagulation studies do not consider them as such. Since it is expected that the particle size distribution is strongly dependent on location in the disk, searching for such differences within individual disks is an important test of theory. Thus, it is a general aim of observations of protoplanetary dust to determine the particle size distribution as a function of a radial (distance from the central star) and vertical (distance from the disk plane) coordinate $f(a, R, z)$, given a wavelength-dependent brightness distribution, sometimes including polarimetry. Further, a method that is as independent of the bulk solid properties as possible is needed. This is unfortunately a classical non-linear inverse problem, and will typically not have a unique, general solution, or even a set of solutions with finite measure. Much of the challenge observers face is to identify favorable geometries that produces a reasonably unique solution for a well-defined limited dust property, such as the presence of dust particles of a certain size. In the following, the various observational techniques that each yield a small part of the solution are discussed.

A major obstacle to solving the inverse problem is that a typical proto-planetary disk is optically thick at most wavelengths, particularly in the inner planet-forming regions. For a minimum-mass solar nebula with typical assumptions for disk and dust structure, the vertical optical depth only drops below unity at radii of 1–10 AU for wavelengths longer than 1 mm. Shorter wavelengths are limited to probing the uppermost layers of the disk only until much later evolutionary stages. This problem is discussed in Section 7.5.6.

Another important complication is that dust opacities are calculated from the dielectric properties of the bulk material in combination with the particle size/shape model. While the opacities of dust particles are very sensitive to their detailed shapes and sizes, these effects are difficult to separate from the underlying dielectric properties of the grain material (see Chapter 6).

7.5.1 The inverse problem

The radius distribution function, $dn/da = f(a)$, is not always an intuitive description when one is interested in understanding the opacities resulting from the dust model. While a size distribution could span many decades in particle radius, the corresponding opacity at a given wavelength is often dominated by particles spanning a certain subrange of radii. For instance, while the total *dust mass* in the classical MRN (Mathis *et al.* 1977) $dn/da \propto a^p$ ($p = -3.5$) distribution of the ISM is slightly dominated by particles with larger radii, the opposite is true for the total *opacity* associated with the distribution. This can be understood if the size distribution is expressed as a volume or cross-section distribution:

$$dV = \frac{4\pi}{3}a^4\frac{dn}{da}d\ln a \quad \text{or} \quad d\sigma_v = \pi a^3 Q_{abs}\frac{dn}{da}d\ln a \qquad (7.1)$$

Expressing the size distributions in this way is analogous to the way a spectral energy distribution νF_ν presents equal energy radiated in equal logarithmic frequency bins.

Examples of the dust cross-section distributions for astronomical silicate at different wavelengths and for different power-law particle size distributions are shown in Fig. 7.1. The 9.7 μm absorption cross-section distribution for particles with $p = -3.5$ actually peaks at ~ 1 μm, but is otherwise dominated by particles $\lesssim 3$ μm. Thus, for this distribution, observations at 9.7 μm are not very sensitive to large particles and their presence cannot be ruled out. For only a slightly shallower size distribution with $p = -3.0$, the absorption cross-section suddenly becomes dominated by large particles, and particles smaller than 1 μm will not contribute much to the total opacity. At 1 mm, the picture is different. Here, the absorption

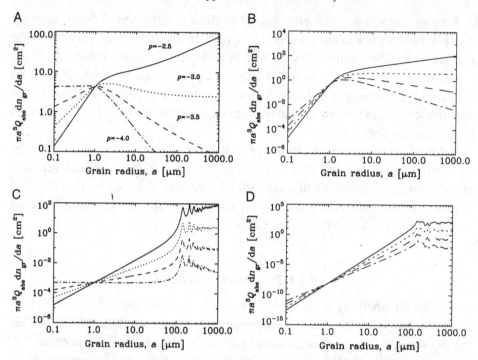

Figure 7.1 The 9.7 μm (top) and 1000 μm (bottom) opacity distributions for cor-responding power-law particle size distributions with exponents −2.5 (solid), −3.0 (dotted), −3.5 (dashed), −4.0 (dot-dashed). The left-hand panels show the absorption efficiency, while the right-hand panels show the scattering efficiency.

cross-section (or emissivity) is practically always dominated by the largest parti-cles.[2] Thus, dust size tracers at this wavelength range are sensitive to the upper cutoff in the particle size distribution. In this way, by using tracers at a wide range of wavelengths, small windows on the dust particle size distribution can be opened. In the following sections, each tracer will be discussed in turn.

7.5.2 Mid-infrared silicate features

The spectral profiles of solid-state resonances from small particles depends on both the shape and the size of the particles. This is particular true for the strong mid-infrared Si–O stretching and O–Si–O bending modes, characteristic of circumstellar silicate dust. In the dielectric spectrum, the stretching mode is roughly centered on 10 μm, while the bending mode is centered on 18 μm. Both resonances are very broad and strongly dependent on the structure of the molecules in which the SiO_x

---

[2] Up to a certain point: very large (meter-sized or larger) particles will no longer contribute to the opacity.

groups are located, i.e. crystallinity, size of the individual domains, and chemical composition (see also Chapters 5, 6, and 8).

Like all dielectric resonances, the silicate bands are strong for small particles ($x \lesssim 1$). In the transition region $x \sim 1$, the band shape changes rapidly as a function of particle size, while for much smaller particles, the band shape is independent of particle size (the Rayleigh limit). The observed shape of the 10 μm silicate band is therefore sensitive to the presence of particles of sizes $a \sim 2$ μm, while the 18 μm band is sensitive to particles twice that size. In short, larger particles will produce flatter silicate features with decreasing feature-to-continuum ratios [$(F_{9.7\,\mu m} - F_C)/F_C$, e.g. van Boekel *et al.* 2005].

Silicate dust in both the diffuse ISM as well as in dense protostellar envelopes has been observed to exhibit a nearly constant band profile dominated by small particles (Bouwman *et al.* 2001; Kemper *et al.* 2005). In sharp contrast, silicate bands observed in emission from protoplanetary disks exhibit a wide range of silicate band profiles, indicating a dominant presence of dust particles larger than the Rayleigh limit (Bouwman *et al.* 2001; van Boekel *et al.* 2005; Kessler-Silacci *et al.* 2006). The interpretation of this observation is that the characteristic size of the dust particles has grown either by a move to a shallower dust size distribution, or by the removal of smaller particles (cf. Fig. 7.1). Either way, the inference usually made is that small particles have been removed by a coagulation process not occuring in the ISM.

Mechanisms for removing small particles from the upper layers of disks other than grain coagulation exist. For instance, small (submicron) particles may effectively be carried away by a disk wind (Shu *et al.* 1996). Conversely, the lifetime of large particles in the upper layers is likely limited, as they will tend to settle to the mid-plane. Such competing effects may create complex behaviors in the area of $f(a, R, z)$ traced by silicate emission features.

While the shape of silicate features in emission from protoplanetary disks is often used to argue for general particle growth, it should be kept in mind that they trace only a very limited part of the total particle population. Apart from the fact that the silicate features only probe silicate grains, particles with radii larger than 10–20 μm do not produce silicate features at all, and, in protoplanetary disks, this population may very well dominate by mass. The emission features are only produced in the innermost regions and the uppermost layers of the disk. For a solar-type star, the 10 μm band is produced in the so-called surface layer of the disks within ~ 1 AU, although this radius increases with stellar luminosity (Kessler-Silacci *et al.* 2007).

The term "surface layer" is used interchangeably for somewhat different concepts in the literature. It might refer to the parts of the disk that are optically thin to (generally ultraviolet or visible) radiation from the central star, i.e. along a line of sight from the disk surface to the star. This is a physical definition because the surface layer will typically have very different properties than the mid-plane in terms of temperature, ionization fraction, chemistry, etc. However, the term may also refer to the part of the disk that is actually observed. In this case, the observed surface layer at a given wavelength is defined as the material at optical depths less than unity, as measured from infinity along the vertical axis.[3] The physical and observed surface layers may often be similar, but are not necessarily so. Because the opacity of the disk is dominated by dust at most wavelengths, the depth of the surface layer depends on the particle size distribution, which in turn may also depend on the radial and vertical coordinates. These interconnected dependencies often make it hard to estimate what fraction of the vertical column is traced by a given observational data set. For instance, at $10\,\mu m$ and at a radius of 1 AU, a typical disk is so optically thick that only the upper 0.1–1.0% of the disk (by mass) can be observed.

A large number of >100 mid-infrared spectra of protoplanetary disks from in particular the Spitzer Space Telescope are now available. The silicate emission features show a wide range of profiles indicating differences both in composition (see Chapter 6) and size distributions dominated by particles ranging from 1–10 μm (van Boekel *et al.* 2005; Furlan *et al.* 2006; Kessler-Silacci *et al.* 2006; Pascucci *et al.* 2009). While the observations show that the opacity-dominating particle sizes in the upper layers of disks are indeed larger than those of the ISM, the differences are really very mild compared to what might be expected from runaway grain coagulation to centimeter-sized particles. Bouwman *et al.* (2008) find from Spitzer spectroscopy that crystalline grains in protoplanetary disks are systematically smaller (submicron sized) than amorphous grains and argue that this suggests that the crystalline grains are not formed by annealing of amorphous grains, but by nucleation and condensation from the gas phase. The ubiquitous presence of silicate emission features is thus indicative of a *retention* or *replenishment* mechanism of particles in the 1–10 μm range in the upper layers of the inner regions of protoplanetary disks, rather than direct evidence for grain coagulation as a step to the path of planetesimal formation. Thus, an important constraint on grain coagulation models supplied by observations of mid-infrared silicate emission features is that a small ($a \sim$ a few microns) particle population generally persists on 10^6 year timescales.

7.5.3 Visible and infrared scattering

Light scattered off the disk surface is perhaps the best independent tracer of particle size distributions in disks. This is because particle size and structure not only affect the scattering efficiency, but also change the phase function and polarization of the scattered light. These extra sets of observables can often be exploited to

---

[3] Strictly speaking, the surface layer should be calculated along the line of sight to the observer, but for illustrative purposes, this is only an inconvenient geometrical difference.

obtain constraints on the nature of dust that is not easily available from thermal emission, which tends only to be weakly polarized and isotropic in comparison (although potentially important exceptions exist at millimeter wavelengths, such as in the case of spinning grains aligned in a magnetic field; see Tamura *et al.* 1999; Cho & Lazarian 2007). The scattering phase function of large particles is highly anisotropic. Light scattered on the near and far sides of a disk will in general result in a very asymmetric surface brightness if the particles are large. Thus, wavelength dependencies of scattering properties are often used to trace particle size populations, and search for $x \sim 1$.

The use of scattered light as a tracer of size distributions is not problem-free. Because scattered light from a protoplanetary disk is faint relative to the glare of the central star, studies of highly inclined, or edge-on systems use the disk itself as a natural coronograph (e.g. Burrows *et al.* 1996; Stapelfeldt *et al.* 1998; Wolf *et al.* 2003; and Figs. 1.5 and 2.11). Without this favorable geometry, only a few nearby optically thick disks have been imaged in scattered light, in particular using the Hubble Space Telescope, which allows accurate subtraction of the stellar point source (Krist *et al.* 2000; Kudo *et al.* 2008). Scattered light from the outer disk has been detected at wavelengths as long as $\sim 10\,\mu$m showing that the conclusions reached using the silicate emission from the inner regions (<10 AU) hold throughout the disk $\gtrsim 100\,\mu$m, at least in some cases (McCabe *et al.* 2003; Pontoppidan *et al.* 2007b); the optical properties of the surface layer are dominated by 1–10 μm particles.

7.5.4 Millimeter spectral slope

Since dust particles emit inefficiently at wavelengths much shorter than their size, it is necessary to observe at millimeter to centimeter wavelengths to probe particles that have experienced strong particle growth. The tracer of such large particles is the spectral slope β of the emissivity of dust, defined as $\kappa_\nu \sim \nu^{-\beta}$ at wavelengths $\gtrsim 500\,\mu$m. The disk spectral energy distribution (SED, see Chapter 3) is related to the opacity, $\tau_\nu = \kappa_\nu \Sigma(R)$ as:

$$4\pi d^2 \nu F_\nu = \int_0^\infty 8\pi^2 \nu B_\nu(T[R])(1 - e^{-\tau_\nu})R\,dR, \qquad (7.2)$$

where d is the distance to the star, F_ν is the flux at frequency ν and T is the temperature in the disk surface layer at radius R. For optically thick disks $1 - e^{-\tau_\nu} \simeq 1$, and the SED is independent of dust opacity. For optically thin disks, where $1 - e^{-\tau_\nu} \simeq \tau_\nu$, the SED depends on the functional form of the opacity, and the spectral slope $[\alpha = \mathrm{d}(\log F_\nu)/\mathrm{d}(\log \nu)]$ of the SED at long wavelengths (in the Rayleigh–Jeans tail) is $\alpha = 3 + \beta$. In the millimeter regime, several effects conspire to simplify matters. First, minimum-solar-mass disks begin to become optically thin above 500 μm over most of their radii. Second, at typical disk temperatures, the SEDs are well into the Rayleigh–Jeans limit, making Eq. (7.2) independent of temperature. A problem recognized by Beckwith *et al.* (1990) is that even at these long wavelengths, optical depth effects do exist, especially in the inner parts of disks, and such saturation also causes the spectral slope to become shallower. Because the total dust disk mass can be inferred from the millimeter fluxes, it can be determined whether optical depth effects apply and can be corrected for, but only if the disk size is known. Thus, the availability of millimeter interferometric imaging is usually required to determine β from a measured α.

Unfortunately, radiative transfer effects and particle size distribution are not the only parameters affecting the observed millimeter slope of disk SEDs. While commonly assumed, it is not certain that the dielectric function of the particle material produces a ν^2 slope even for small particles in the Rayleigh limit (Draine & Lee 1984). This concern can be partly alleviated by comparing the slope of the millimeter opacity in disks to that in the ISM. A value of β close to 2 in the ISM is consistent with many observations of dense clouds (Bianchi *et al.* 2003; e.g. Kramer *et al.* 2003), and recently, it was directly determined to be $(1.7 - 2.0)^{+0.29}_{-0.33}$ by Shirley *et al.* (2008).

Figure 7.2 illustrates the distribution of long-wavelength opacity indices, β, determined for disks spatially resolved in the millimeter wavelength regime, as compared to the 9.7 μm silicate emission-feature-to-continuum ratio discussed in Section 7.5.2. Clearly, the typical opacity slope in disks is very different from that of the ISM, and also spans a wide range from 0.0 to 1.6. These observations probe the disk mid-plane, as opposed to the surface probed in the infrared as discussed above, and the common conclusion is that dust in the mid-plane of protoplanetary disks generally has grown to at least millimeter or centimeter sizes. The one caveat is that it remains possible that chemical and compositional changes relative to ISM dust cause the imaginary part of the index of refraction to rise beyond 500 μm leading to shallow opacity slopes not related to large particle sizes.

7.5.5 Particle growth and the presence of very small grains and PAHs

Very small grains and macromolecules are known to be present in the surface layers of some disks as well as in the ISM. They are usually revealed by emission features due to polycyclic aromatic hydrocarbons (PAHs, see also Chapter 6). These particles, with frameworks of six to several thousand carbon atoms, are so small that they can be excited by single ultraviolet photons. Subsequently, they will non-thermally re-radiate the energy in discrete, but broad, bands stretching across the mid-infrared wavelength region. They are found in protoplanetary disks when the ultraviolet radiation field is sufficiently high (Habart *et al.* 2004), but

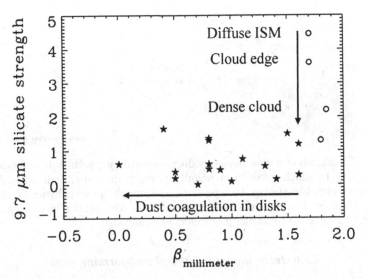

Figure 7.2 The relation between the particle growth in the disk mid-plane traced by the millimeter opacity index and that of the inner disk surface traced by the 9.7 μm silicate emission feature. The star symbols represent individual disks. Data points are from van Boekel *et al.* (2003), Natta *et al.* (2004), Furlan *et al.* (2006), Rodmann *et al.* (2006), and Lommen *et al.* (2007). Typical errors are 10–30% in both β and silicate band strength. Note also that differences in how the silicate band strengths were derived may introduce slight systematic offsets for the different data sets. The circle symbols represent dust opacity models calculated for the interstellar medium at a range of densities. From top to bottom the circles are: $R_V = 3.1$ and $R_V = 5.5$ from Weingartner & Draine (2001), a Spitzer-constrained dust opacity for dense clouds from Pontoppidan *et al.* (in preparation) and the particle growth simulation for protostellar envelopes [thin ice mantles, Ossenkopf & Henning (1994)].

in a few disks there is evidence of significant abundance enhancements (Geers *et al.* 2006; Perrin *et al.* 2006; Geers *et al.* 2007). Their presence, however, is not expected from coagulation theory. In terms of coagulation, PAHs behave more like molecules than grains. Their low mass causes their thermal random velocities to be high, while also maintaining a high sticking probability. This results in PAH "freeze-out" timescales of <1 Myr, even at H_2 densities as low as $10^5 \, cm^{-3}$. The PAHs are therefore removed almost completely in protostellar envelopes owing to efficient sticking on the surfaces of larger grains (Bernstein *et al.* 2005). This will hold for most of the protoplanetary disk, except perhaps for the uppermost layers (Dullemond *et al.* 2007). Their presence in the surface layers of disks in relatively large amounts suggests that they are being formed *in situ*, either by destruction of larger particles, or by gas-phase formation. If formed chemically from the gas phase, they may represent the first stage in grain condensation in protoplanetary disks.

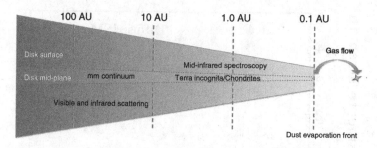

Figure 7.3 Sketch of a protoplanetary disk surrounding a solar-type star showing
the regions in which different techniques probe grain and particle sizes. More
luminous stars, like Herbig Ae-type stars, will stretch the scale by up to a factor
of 10, while less luminous stars, like brown dwarfs, will shrink it by a comparable
factor. See Table 8.1 for stellar parameters of young stars.

7.5.6 Terra incognita: the planet-forming zone

In review of the various methods to trace particle size distributions in protoplanetary
disks, a crucial problem emerges: the currently available dust tracers only probe
the uppermost layers of the inner disk and the mid-plane and upper layers of the
outer regions of the disk, as illustrated in Fig. 7.3. The mid-plane of the inner disk
($\lesssim 10$ AU), where most planet formation actually occurs, is a region that seems to
be out of reach of current observational facilities and methods. There are some cases
in which infrared tracers penetrate to the mid-plane, such as when grain growth has
proceeded to the point where the inner disks become optically thin, or if vertical
turbulent mixing is efficient enough to allow the surface dust composition to reflect
that deeper in the disk. A very important step to obtain an unambiguous tracer of the
dust in the inner-disk mid-plane will be the advent of the Atacama Large Millimeter
Array (ALMA). This facility will achieve a spatial resolution as high as ~ 10 milli-
arcsec for continuum observations of disks, or 1–2 AU for the nearest star-forming
clouds. This will potentially allow a direct determination of radial differences in
the particle size distribution of dust in the disk mid-plane, and will allow a direct
comparison with the surface-layer dust as traced by infrared observations. What
does, however, trace the inner-disk mid-plane in at least one protoplanetary disk is
the archaeological record available in primitive material in the Solar System. This
record will be the subject of the remainder of the chapter.

7.6 Chondritic meteorites

Infrared observations show that micron to submicron dust is abundant in the surfaces of protoplanetary disks. A significant component of this material appears
to be amorphous in character, although, closer to protostars, higher relative abundances of crystalline materials are present (van Boekel *et al.* 2004). Chondritic

meteorites and interplanetary dust particles (IDPs) also contain fine-grained material, but the relationship between this material and that detected in the infrared is unclear. Most cosmochemists argue that the components of chondrites formed close to the nebular mid-plane by thermal processing of interstellar dust and so may differ from the material on the surface of the disk. Therefore, a key question is how the infrared observations and information from the sample suite of chondrites can be related. For example, is there a significant proportion of interstellar (unprocessed) dust on the disk surface or was dust processed at the disk mid-plane transported to the disk surface by turbulence within the disk? Amorphous silicate material on the disk surface could be interstellar dust, but amorphous silicates, dominated by thermally processed nebular materials, are also present in primitive chondrites and IDPs.

Fine-grained materials in the matrices of chondritic meteorites consist of a complex mixture of material formed in different locations within the protoplanetary disk. Only a small amount of silicate interstellar dust is preserved in chondrites, which is consistent with the observations of surface-layer dust.

Below we review the characteristics of fine-grained Solar System dust as preserved in primitive chondrite matrices and draw inferences from these observations about the mechanisms of coagulation of nebular dust as well as the timing and location of coagulation.

7.6.1 Fine-grained materials from the early Solar System

Remote observations provide information about the processes of dust formation and coagulation in extrasolar protoplanetary disks. In contrast, the suite of fine-grained Solar System materials available for study in the laboratory consists of materials that formed 4.56 Gyr ago during the earliest stages of the formation of the Solar System.

In order to consider the processes of dust coagulation in the early Solar System, we first review the characteristics of this material. Of considerable importance is the fact that these samples – represented principally by chondritic meteorites, but also by IDPs and by samples from Comet Wild 2 collected by the Stardust mission – all come from parent bodies of different kinds. As a result, even the most primitive of these materials has been processed, both physically and chemically, to different degrees. The processes that affected Solar System dust may have occurred in different environments such as the solar nebula (e.g. evaporation/condensation, annealing) and asteroidal parent bodies (aqueous alteration and/or thermal processing, mild compaction to extensive lithification). A major challenge is to understand the effects of this secondary processing.

7.6.2 Chondritic meteorites

Chondritic meteorites provide the most voluminous sample of early Solar System materials available for study in the laboratory. These complex objects consist of a mixture of coarse-grained materials, such as chondrules and calcium–aluminum-rich inclusions (CAIs or refractory inclusions) embedded within a fine-grained dust-like matrix (see Chapter 1 and e.g. Scott & Krot 2005; Weisberg *et al.* 2006). Chondrules and CAIs are typically a few millimeters in diameter and formed by high-temperature processes such as melting, evaporation, and condensation. The origins of CAIs and chondrules is discussed in detail in Chapter 8. Here we focus on the matrix component of chondrites, defined as the fine-grained ($< 5\,\mu m$ in size), predominantly silicate material that is interstitial to macroscopic, whole, or fragmented entities such as chondrules and CAIs etc. (Scott *et al.* 1988). In the most primitive chondrites typical sizes (diameters) of matrix grains are tens to hundreds of nanometers. We also discuss so-called fine-grained rims (Fig. 7.4a), which are distinct occurrences of fine-grained matrix materials that coat chondrules, CAIs, etc. in many chondritic meteorites.

The detailed characteristics of chondritic meteorites are presented in several recent reviews (e.g. Brearley & Jones 1998; Krot *et al.* 2003; Weisberg *et al.* 2006). Of importance for the following discussion is that chondrites are classified based both on their chemical and petrographic characteristics (e.g. carbonaceous (C), ordinary (O), and enstatite (E) chondrites and their subgroups; see Chapter 1 and Fig. 1.1) as well as their degree of secondary processing (aqueous alteration – petrologic Types 1–2; and thermal metamorphism – petrologic Types 3–6). Secondary processes are significant, because they modify the primary nebular characteristics of chondrites and can obscure, or obliterate, the nebular record that they contain.

Petrologic Type 3 chondrites are the least modified by secondary processes, but still show evidence of thermal metamorphism (and to a lesser extent aqueous alteration). A progressive increase in the effects of metamorphism is indicated by the division of Type 3s into petrologic Subtypes 3.0 to 3.9. For petrologic Type 3.0–3.1 chondrites even more subtle changes due to metamorphism are present, meriting further subdivisions (e.g. 3.00, 3.01, 3.02, etc.; Grossman & Brearley 2005). Type 3.00 chondrites are the most pristine of all the chondrites, but even these may have undergone some modification that is too subtle to be recognized by current analytical techniques. To minimize the complicating effects of secondary processes, we focus on petrologic Type 3.2 chondrites and lower (3.0–3.2).

Interaction with minor amounts of aqueous fluids can modify the primary mineralogy and thermal metamorphism at low temperatures (above $\sim 520-570\,K$) drives recrystallization of the matrix to form coarser-grained materials. Unfortunately, the effects of thermal metamorphism on matrices have only been fully appreciated recently. Consequently, much of the older literature on matrices is

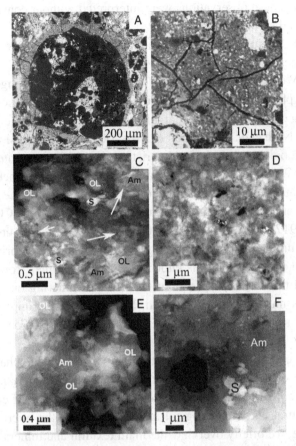

Figure 7.4 Electron microscope images showing the textural characteristics of matrices and fine-grained rims in primitive chondritic meteorites. (a) Backscattered electron (BSE) SEM image of a typical fine-grained rim around a chondrule in the pristine chondrite ALHA77307 (CO3.0). The contrast in the image is a function of atomic number, with Fe-rich grains appearing brighter in the image. (b) BSE image of a region within a fine-grained rim in ALHA77307 showing that the rim contains a small percentage of grains in the 1–10 μm range, typically iron, nickel sulfides and metal grains. The fine-grained silicate matrix with a grain size <1 μm is beyond the resolution of the SEM. (c) Dark-field scanning transmission electron microscope (DF-STEM) image of a region of a fine-grained rim in ALHA77307. The rim consists of a complex mixture of extremely fine-grained crystalline olivine (OL), pyroxene, and sulfide grains (S) intermixed with amorphous silicate material (Am), that often appears to occur as fluffy aggregates (arrowed). (d) BSE SEM image of a region of matrix within ALHA77307, showing the presence of irregularly shaped domains with different BSE contrast that are intermixed on a very fine scale. (e) DF-STEM image of a region of matrix in Acfer 094 showing that this material consists of a complex mixture of amorphous silicate material (Am) and fine-grained crystalline phases such as olivine (OL) and pyroxene. (f) DF-STEM image of a region of matrix in the EET 92042 CR chondrite showing the remarkably low abundance of crystalline phase (cf. Fig. 7.4e). The matrix consists largely of amorphous or nanocrystalline (Am) silicate material in which sulfide grains (S) are embedded. Figures (a) and (b) are from Brearley (1993).

contradictory, because observations were interpreted without consideration of the effects of thermal metamorphism.

The primary conclusion of most researchers up until the early 1990s was that FeO-rich olivines in chondrite matrices were nebular condensates (e.g. Palme & Fegley 1990; Weisberg & Prinz 1998). However, there is a growing consensus that these FeO-rich phases are, in fact, the result of parent-body metamorphic recrystallization (among other processes) of largely amorphous materials (e.g. Grossman & Brearley 2005; Nuth *et al.* 2005; Scott & Krot 2005). This conclusion has led to the recognition that only a handful of truly pristine chondritic meteorites of petrologic Type 3.1 or less exist in the world's meteorite collections. These meteorites include the un-equilibrated ordinary chondrites (UOCs), Semarkona (LL3.0), Bishunpur, and Chainpur; and the carbonaceous chondrites ALHA77307 (CO3.0), Y81020 (CO3), QUE 99177 (CR3.0), and MET 00426 (CR3.0), and the unique chondrite Acfer 094. A further important point is that few of these meteorites are pristine falls; indeed all the carbonaceous chondrites are finds either from Antarctica or hot deserts and therefore the effects of terrestrial weathering on the characteristics of matrices must also be considered. Nevertheless, the importance of these rare meteorites to our understanding of early Solar System processes cannot be overstated.

The study of chondrite matrices is both time-consuming and challenging, principally because of the fine-grained characteristics of the material. Transmission electron microscopy (TEM) has been the analytical technique of choice, because it provides textural and mineralogical information from the micron down to the nanoscale. Here, we briefly overview the main textural and mineralogical characteristics of matrices in very un-equilibrated chondrites, focusing principally on the carbonaceous chondrites.

7.6.3 Carbonaceous chondrites

The CO3.0 chondrite ALHA77307 and the unique carbonaceous chondrite Acfer 094 are two of the most primitive chondrites known (e.g. Brearley 1993; Newton *et al.* 1995; Greshake 1997; Grossman & Brearley 2005; Nguyen *et al.* 2007). The matrices and fine-grained rims ($\sim$ 40–50 vol%) of these two meteorites share a number of similarities, in being extremely fine-grained ($< 0.5\,\mu$m) and consisting of a complex, heterogeneous, un-equilibrated mixture of crystalline silicates, amorphous silicate material, oxides, sulfides, iron, nickel metal (Fig. 7.4b,c), and carbonaceous grains. Larger mineral grains (1–10 μm) occur embedded within this finer-grained material. The amorphous silicate material is particularly significant, because such material now appears to be an important signature of very pristine chondrites (see below and Chapters 6 and 8).

In ALHA77307, the textural characteristics of the matrix and rims suggest that they are composed of distinct aggregates or clusters of phases that have variable textural and mineralogical characteristics. They have been compacted together, obscuring their original outlines to a significant degree (Figure 7.4d). Within these distinct aggregates, crystalline phases are embedded in a groundmass of the amorphous Si–Mg–Fe-rich material. An extreme state of disequilibrium in ALHA77307 matrix and fine-grained rims is indicated by highly variable olivine and pyroxene compositions. This disequilibrium is also reflected in the variable oxidation state of iron over very short distances as indicated by the presence of magnetite (Fe_3O_4) and kamacite (FeNi metal) grains within less than 1 μm of each other. Collectively, these observations are indicative of a very low degree of parent-body processing (thermal metamorphism) for these meteorites.

Recent observations of the Antarctic CR chondrites, QUE 99177 and MET 00426 (CR3.0) demonstrate the widespread occurrence of amorphous materials in the matrices of primitive chondrites. Most CR chondrites are classified as petrologic Type 2 meteorites because they have undergone aqueous alteration (e.g. Weisberg *et al.* 1993). However, these two CR chondrites are pristine Type 3 chondrites, (e.g. Abreu & Brearley 2006; Abreu 2007), a classification that is supported by the evidence that QUE 99177 has the highest abundance of presolar silicate grains of any chondrite studied to date (Floss & Stadermann 2008). Compared with ALHA77307 and Acfer 094, the abundance of amorphous silicate material is higher and crystalline silicates (e.g. olivines and pyroxenes) are essentially absent from the fine-grained matrix. However, like ALHA77307 and Acfer 094, both CR chondrites contain abundant nanometer-sized sulfides (see Fig. 7.4f for an example) embedded within the amorphous materials. Although they have been affected by aqueous alteration, at least one CM2 chondrite provides further evidence of the widespread occurrence of amorphous material in primitive nebular dust. The weakly altered CM2 chondrite Y-791198 (Metzler *et al.* 1992) has fine-grained rims and a matrix that consist almost exclusively of silicate material that is either completely amorphous or nanocrystalline, in which nanophase iron, nickel sulfides are embedded (Chizmadia & Brearley 2008). The only anhydrous crystalline silicate phases are olivines with grain sizes >100 nm up to 1 μm or larger in size. The fine-grained rims consist of distinct sulfide-rich and sulfide-poor regions that are irregular in character and appear to have been compacted together, probably during lithification of the meteorite (Fig. 7.5). These observations contrast with data for Adelaide, a highly primitive carbonaceous chondrite. Adelaide matrix is dominated by unusually FeO-rich (Fa_{36}–Fa_{84}), fine-grained (0.1–0.2 μm) olivine (Brearley 1991). However, an amorphous FeO-rich phase does occur interstitially to the crystalline phases, but is a minor component of the matrix.

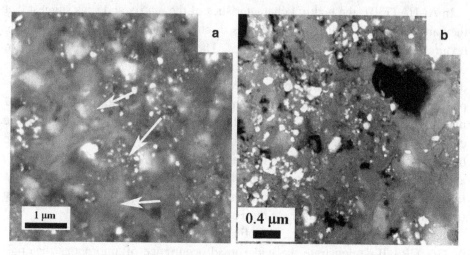

Figure 7.5 Electron microscope images showing the textural characteristics of fine-grained rims in the Y-791198 CM2 carbonaceous chondrite. (a) High-resolution field emission gun SEM (FEGSEM) BSE image of a typical region of a fine-grained rim showing that it consists of irregularly shaped silicate-rich domains (arrowed) with high abundances of very fine-grained iron, nickel sulfide grains. The sulfide-poor regions contain no sulfide grains, but exhibit higher atomic number constrast than the silicate material in the adjacent sulfide-rich regions. (b) DF-STEM image of a region of a fine-grained rim illustrating the complex structure of the rim. The rim consists of domains with variable abundances of sulfides and distinct atomic number contrast which appear to represent individual grains or clusters of grains that were accreted into the rims. From Chizmadia & Brearley (2008).

7.6.4 Summary of key observations

Some important conclusions can be drawn from the studies summarized above (see Table 7.1). First, the majority of pristine chondrites studied, with the exception of Adelaide, have matrices which contain a high proportion of amorphous silicate material. This characteristic is now recognized as a common feature of pristine chondrites. Iron oxide-rich olivines, the dominant silicate mineral in the matrices of petrologic Type 3.1 chondrites and above, are either rare or absent in petrologic Type 3.0 chondrites. The relative proportions of crystalline and amorphous materials in matrices is variable between the chondrite types. For example, Acfer 094 (unique) contains a relatively high proportion of crystalline silicates (50 vol%), whereas the CR3.0 chondrites QUE 99177 and MET 00426 and the CM2 Y791198 contain rare grains of olivine and pyroxene. Third, matrices consist of ranges in grain size from $\sim 1\,\mu m$ down into the nanometer size range ($\sim 10\,nm$). Fourth, matrices are a complex mixture of unequilibrated materials that formed under different conditions and/or locations and/or times and represent a snapshot of a range of processes that occurred within the solar nebula. Fifth, at least some matrices

Table 7.1 *Summary of key properties of matrix and fine-grained rims in pristine chondritic meteorites*

Meteorite	Type	Mineralogy	Textural characteristics
ALHA77307[1]	CO3.0	Si–Fe–Mg amorphous material, olivine, troilite, Fe,Ni metal, magnetite	Mainly submicron crystalline phases embedded in amorphous silicate. Distinct compositional domains
Acfer 094[2]	Unique Type 3	Si–Fe–Mg amorphous material, olivine, enstatite, pyrrhotite	Mainly submicron olivines and pyroxenes embedded in groundmass of amorphous silicate
QUE 99177[3]	CR3.0	Si–Fe–Mg amorphous material, pyrrhotite, pentlandite	Heterogeneously distributed nm-sized sulfides embedded within very abundant amorphous silicate material. Distinct compositional domains
MET 00426[3]	CR3.0	Si–Fe–Mg amorphous material, pyrrhotite, pentlandite	Heterogeneously distributed nm-sized sulfides embedded within very abundant amorphous silicate
Y-791198[4]	CM2	Si–Fe–Mg amorphous material, pyrrhotite, pentlandite, olivine, phyllosilicates	Distinct micron-sized sulfide-rich and sulfide poor domains, with sulfide embedded within amorphous silicate material
Adelaide[5]	Unique Type 3	Iron-bearing olivine, amorphous silicate, enstatite, pentlandite, magnetite	Well-shaped and irregular olivine grains, with interstitial amorphous material

[1] Brearley (1993), [2] Greshake (1997), [3] Abreu (2007), [4] Chizmadia & Brearley (2008), [5] Brearley (1991).

(e.g. CM2 Y-791198, CO3.0 ALHA77307) consist of distinct aggregates of grains that can be considered to be units with well-defined mineralogical and compositional characteristics. Finally, mixing of these diverse components was extremely thorough and occurred on a submicron scale. All these observations place different constraints on the processes of dust coagulation in chondrite matrices, that are discussed below.

7.7 What do chondrite matrices tell us about the grain size of nebular dust?

A key question in addressing the issue of dust coagulation concerns the grain size of dust. This is closely linked to the formation mechanisms of the dust as discussed above. It is important to note that larger grains reaching sizes of up to $10\,\mu m$ (e.g. Fig. 7.6a,b) do occur in the matrices of primitive chondrites. However, these grains typically represent <1 vol% of primitive matrix and hence are a very minor component.

Crystalline silicate grains. Based on the data presented above, defining the grain size of crystalline silicate grains is comparatively straightforward. Olivines and pyroxenes lie in the range 100 nm to a few microns at most. Most grains are in the submicron range (e.g. Figure 7.4e, Figure 7.6a). The abundance of such grains varies from chondrite to chondrite from extremely low in the CR and CM chondrites, to 20–30% in meteorites such as ALHA77307 and Acfer 094. There is a general view that most of the well-crystallized forsteritic and LIME (low-iron, manganese-enriched) olivine grains are primary nebular condensates (e.g. Klock *et al.* 1989; Scott & Krot 2005) which have preserved their original grain shapes (subrounded or faceted) and dimensions. Most crystalline olivines occur as isolated grains in matrices, embedded within amorphous material (Fig. 7.6b,c). Based on the very low occurrence of micron-sized and larger olivine grains we conclude that these grains did not undergo any significant degree of coagulation prior to their final accretion into chondrite parent bodies (see later discussion), unless olivine-grain aggregates were too fragile to survive for an extended period of time in the solar nebula.

Amorphous silicate grains. Constraining the grain size of amorphous silicate materials is complex, for two reasons. First, because of their amorphous character, these materials show no diffraction contrast in the transmission electron microscope. This makes it difficult to delineate distinct grains and requires the use of textural and/or compositional criteria to distinguish between grains. Textural criteria are often ambiguous and the analytical capability of distinguishing amorphous grains on compositional grounds at the submicron scale has only recently become available. Second, fine-grained amorphous materials are susceptible to compaction and lithification that could have modified their grain sizes. We can make the following inferences about grain size from the data that are available. A conservative estimate of the minimum grain size of amorphous silicate grains is < 0.4 μm and < 0.1 μm is probably more realistic, but it could be significantly less. This size constraint is based on TEM observations of fine-grained rims from just two meteorites, Y-791198 and ALHA77307, which show that the rims consist of distinct interlocking regions. These regions have been interpreted as representing distinct units

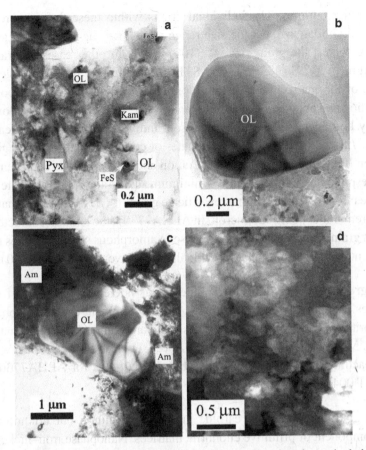

Figure 7.6 Transmission electron microscope images showing the typical characteristics and grain sizes of crystalline and amorphous phases in fine-grained rims in primitive chondritic meteorites. (a) TEM image of a region of a fine-grained rim in ALHA77307 consisting of a complex mixture of submicron olivine (OL), pyroxene (Pyx), sulfide, and kamacite (Kam) grains embedded in an amorphous silicate matrix. (b) Magnesium-rich olivine (OL) grain in a fine-grained rim in the CM2 carbonaceous chondrite Y-791198, also surrounded by amorphous (Am) or nanocrystalline silicate material and sulfides. (c) typical Mg-rich olivine (OL) grain embedded in very fine-grained amorphous or nanocrystalline silicate material in a fine-grained rim in the ALHA77307 (CO3.0) chondrite. Note that the grain is completely crystalline and contains no evidence of defects such as dislocations. (d) DF-STEM image of a region of a fine-grained rim in ALHA77307 showing fluffy aggregates of amorphous material. Figures (a) and (c) are from Brearley (1993) and Figure (b) is from Chizmadia & Brearley (2008).

or aggregates of grains that accreted to form the fine-grained rims (e.g. Brearley 1993; Chizmadia & Brearley 2008). In ALHA77307, TEM and scanning transmission electron microscope (STEM) imaging (Figure 7.6d) shows that the matrix and fine-grained rims consist of a complex mixture of fluffy aggregates of grains with irregular outlines. The individual grains within these fluffy aggregates are typically of similar size, but the size range varies from aggregate to aggregate. In some aggregates, the grain size is < 100 nm, but in other cases can range from 0.1–0.5 μm. In Y-791198, SEM and STEM imaging reveals the complex mosaic structure of the matrix, with distinct domains distinguishable based both on the composition and abundance of sulfide grains (Fig. 7.5). Some of these domains are texturally homogeneous and may represent individual amorphous silicate grains (e.g. Fig. 7.5), but they could be aggregates of compositionally similar grains that have been lithified during aqueous alteration and now appear to represent a single grain. Despite uncertainties as to the minimum size of amorphous silicate grains in chondrite matrices, it is clear that few silicate grains (< 5 %) are larger than 1 μm in size, the remainder being submicron. In ALHA77307, the textural characteristics of the finest grain fraction are similar to those in amorphous silicate smokes produced experimentally, suggesting that individual grains have grain sizes < 100 nm.

Sulfide grains. The situation for sulfide grains is clearer. In the most primitive chondrites, iron and iron nickel sulfides are the dominant crystalline phases and, in meteorites such as Y-791198, are a significant component of the matrix (e.g. Fig. 7.5). Most grains are <100 nm in size (i.e. are true nanoparticles) with a few grains reaching up to ∼ 0.4 μm. This observation holds for ALHA77307, Acfer 094, MET 00426 and QUE 99177.

Carbide grains. Iron–nickel carbide grains are a minor, but ubiquitous, crystalline component of primitive chondrite matrices. Nanophase iron–nickel carbide grains occur in the matrices of several CM and CR chondrites (Brearley 2004; Abreu & Brearley 2006). These grains frequently occur in clusters of several grains and all have rims of magnetite, which is likely the result of parent-body aqueous alteration. All the carbide grains are associated with carbonaceous material, which indicates a genetic relationship between them. The most plausible origin for these grains is that they were originally nanophase metal grains that underwent carburization during Fischer–Tropsch catalysis reactions in the solar nebula (Brearley 2004; see also Chapter 5). This suggestion is supported by the fact that on many grains the carbonaceous material has clearly undergone partial graphitization, a process that could not have occurred at the low temperature of aqueous alteration experienced by CM and CR chondrites. Therefore precursor metal grains of the carbides had grain sizes of 10–50 nm.

Carbonaceous grains. Carbonaceous material in chondritic meteorites is highly complex and has diverse origins. Most studies of carbonaceous material have focused on the analysis of residues of insoluble organic material extracted from the meteorites by acid dissolution (e.g. Alexander *et al.* 2007). Recently, TEM and NanoSIMS (described in Appendix 2) have been used to locate carbonaceous material *in situ* in carbonaceous chondrite matrices, providing information on the grain sizes of this material. There are multiple different occurrences of carbonaceous material ranging from < 2 nm to micron-sized (e.g. Brearley 2004; Garvie & Buseck 2004; Busemann *et al.* 2006; Nakamura-Messenger *et al.* 2006).

Interstellar grains. Interstellar grains, such as silicon carbide, graphite, diamond, corundum, spinel, and hibonite, have been studied extensively (Zinner 1998, 2003; Bernatowicz *et al.* 2006; and see Chapter 2) and their grain sizes are well known (see Lodders & Amari 2005). The abundances of these grains vary among chondritic meteorites, but are typically present at the tens to a few hundred parts per million levels (see Lodders & Amari 2005). The grain sizes of interstellar grains are also variable, ranging from 2 nm for diamond to 0.1–20 μm for silicon carbide. Oxide grains, such as hibonite, spinel, and corundum, generally have size ranges from ~ 0.2 to 3 μm. The recent discovery of interstellar silicate grains in carbonaceous chondrites (Nguyen & Zinner 2004; Nguyen *et al.* 2007) has shown that these grains (crystalline olivines and pyroxenes and amorphous silicates with compositions consistent with olivine and pyroxene) are small (0.1–0.3 μm).

In summary, despite the caveats discussed above, we can state with confidence that the bulk (> 95%) of material in highly un-equilibrated chondrite matrices has individual grain sizes (as defined earlier) less than 1 μm. We can, with equal confidence, conclude that a significant proportion of this dust had grain sizes that were < 100 nm in size, although exactly how much remains uncertain without more detailed studies. Importantly, a major part of the primitive dust preserved in chondrites has grain sizes that overlap with the range that is detectable by the spectroscopic techniques discussed above.

7.8 Dust coagulation: how and when?

We now consider what the characteristics of nebular dust preserved in chondrites can tell us about dust coagulation. Specifically, we discuss two principal questions: (1) what processes may have contributed to dust coagulation, and (2) what stages of dust coagulation are represented in matrices, i.e. when did dust coagulation take place? As discussed below, the meteoritic evidence indicates that dust coagulation occurred in the nebular mid-plane throughout disk evolution. It is therefore possible that fine-grained dust in chondrite matrices represents a mixture of materials that coagulated at different times throughout nebular evolution. Alternatively, processing of dust in the nebular mid-plane could have been so effective that no record of early dust coagulation is preserved in chondrite matrices. Instead, the characteristics of fine-grained dust may be a reflection of processes in the nebular mid-plane that occurred shortly before the accretion of chondrite parent bodies.

The data discussed above show that the size of individual dust grains in chondrite matrices (and fine-grained rims) was in the nanometer- to submicron range. Further, data from the two chondrites, ALHA77307 and Y-791198, indicate that fine-grained rims and matrices consist of aggregates of these grains, most aggregates having distinct textural, compositional, and mineralogical characteristics (e.g. Figs. 7.4d and 7.5). What process(es) was (were) involved in the initial coalescence of nanometer to submicron grains into larger aggregates? The answer is certainly complex and is intimately linked to the processes that formed the individual grains in the first place.

The coarse-grained components of chondritic meteorites consist of components (CAIs, chondrules, metals, sulfides) that formed under different conditions, in different environments, and at different times within the solar nebula. The dust preserved in fine-grained rims and matrices is a complex and diverse mixture of materials including interstellar grains (inorganic and organic) that have survived processing in the protoplanetary disk, equilibrium and disequilibrium condensates formed in the solar nebula, fragments of chondrules, and organic materials formed at low temperatures by gas–solid catalysis reactions. All these materials are intimately mixed on the submicron scale, forming the highly un-equilibrated assemblage of dust that accreted into asteroidal parent bodies. This mixture of materials represents the final product of a complex sequence of dust-grain formation, coagulation, transport, and mixing. What can we say about the timing of dust coagulation, as represented by the matrices of chondritic meteorites? Based on a wide range of chronological data for chondrites (see Chapters 1 and 9), the processes of dust formation and dust coagulation in the solar nebula are constrained to have occurred within the first 5–6 Myr of Solar System evolution. Additional data place further constraints on the time period of dust coagulation recorded by chondrites (e.g. Chapter 9).

The widespread occurrence of chondrules in chondritic meteorites is considered by cosmochemists as evidence that chondrule formation was an important process in the nebular mid-plane (see Chapter 8 and e.g. Cuzzi & Alexander 2006). Further, chondrule formation occurred by multiple events and was highly efficient. Although the compositions of chondrule precursors can be constrained within reasonable boundaries, the physical nature of chondrule precursor materials is poorly constrained (e.g. Jones et al. 2005). Most researchers regard chondrule precursors as being fluffy aggregates of fine-grained silicate grains (plus metal and sulfide) that were large enough to produce millimeter-sized melt spherules when subjected to transient high-temperature events (Jones et al. 2005). To produce low-density dust aggregates with sufficient mass, dust coagulation must have been a highly efficient process that occurred repeatedly and continuously. Although the initial chondrule precursors likely consisted of very fine-grained dust, as chondrule-forming events became more frequent, some proportion of chondrules underwent fragmentation as a result of collisions between individual chondrules. This fragmentation contributed coarser-grained fragments to the fine-grained dust precursors, which were recycled into the next generation(s) of chondrules (e.g. Rubin & Krot 1996; Jones *et al.* 2005). Such fragments, the so-called relict grains, that occur widely in chondrules, are recognized by their anomalous chemical or isotopic compositions. As the abundance of chondrules increased through time, it is likely that the fraction of chondrule fragments in the precursor dust also increased. However, the data presented above show that matrices and fine-grained rims in primitive chondrites contain a very low proportion of coarse-grained material, which is unequivocally derived from chondrules. This observation suggests that the evidence of dust coagulation involved in chondrule formation (i.e. the formation of "dustballs," the possible precursors to chondrules) is not well preserved in chondrite matrices.

In summary, it is self-evident that dust coagulation must have occurred throughout the period of nebular evolution when chondrule formation was taking place. However, it is not self-evident that the dust present in fine-grained rims and matrices represents the coagulation of material that might have been the precursors to chondrules. Instead, as discussed below, most dust preserved in matrices probably represents material that was processed or formed during chondrule-forming events and represents a late stage of dust formation and coagulation that occurred shortly before the accretion of asteroidal parent bodies.

7.9 Constraints on dust coagulation from amorphous silicates

Implicit in the discussion above is the view that high-temperature events caused melting of nebular dust to form chondrules. As well as melting, high-temperature events could also have caused evaporation and annealing of dust (see Chapter 8). We now discuss these additional processes based on evidence from chondrite matrices and suggest that the formation of chondrules played a significant role in the coagulation of dust. A key observation is that a significant component of primitive chondrite matrices consists of amorphous silicate material, rather than crystalline condensate phases as was previously thought (e.g. Scott *et al.* 1988). The origin of this amorphous material is of key importance, because it provides constraints on when the process of dust coagulation occurred.

There are several potential sources of amorphous materials in chondrite matrices, including interstellar materials, the products of disequilibrium condensation in the nebula, or shock processing of crystalline materials by nebular shock waves.

Amorphous interstellar silicate grains do occur in chondrite matrices (e.g. Stroud 2005), but the abundance of these materials is very low (ppm), based on isotopic constraints. Shock amorphization by nebular shock waves is certainly a possibility, but if such a process occurred it did not preserve the original shapes of crystalline grains and therefore seems relatively unlikely. Brearley (1993) argued that the most likely mechanism for the formation of amorphous silicates was disequilibrium condensation of evaporated dust in the nebula. A plausible scenario for such a process is during the high-temperature events which formed chondrules (e.g. Brearley 1993; Wasson & Trigo-Rodríguez 2004; Nuth *et al.* 2005).

During such events nebular dust will melt to form chondrules, but a component of the more volatile-rich material will also be evaporated. The behavior of this evaporated dust following a high-temperature event depends on the cooling rate of the dust, which is a function of the optical density of the dust in the nebular mid-plane. Constraints from chondrules indicate chondrule cooling rates on the order of $10\,K\,hr^{-1}$ to $1000\,K\,hr^{-1}$ (see Chapter 8). Assuming the dust cooled at the same rate as chondrules, at the higher cooling rates homogeneous nucleation of crystalline silicate phases is unlikely and the formation of amorphous disordered materials is favored. Experimental studies show that condensation under high cooling rates forms porous, fluffy aggregates of amorphous nanometer-scale grains with non-stoichiometric compositions (Day & Donn 1978; Stephens & Kothari 1978; Nuth & Donn 1983). These observations indicate that some coalescence of the smallest grains occurs during disequilibrium condensation. Cooling rates in these experiments are higher than those experienced by evaporated dust in the solar nebula, but the general principle that amorphous materials will undergo some coagulation during condensation is likely. These fluffy, open-structured aggregates of amorphous grains undergo significant compaction to produce the low-porosity structures that are present in chondrite matrices. In this model, a significant component of the dust present in chondrite matrices and fine-grained rims represents a residue of material evaporated during chondrule formation. It may therefore represent a highly processed material that is the result of continued and cyclic episodes of chondrule formation. This conclusion has two important implications for the astrophysical observations discussed earlier. First, it indicates that dust coagulation not only occurs by aggregation of pre-existing monomers at low temperatures, but also during the recondensation process itself at high tempertures, when individual monomers may aggregate together to form fluffy aggregates. Hence dynamic processes within the disk not only modify the chemistry of solid materials, but also the physical characteristics of the grains, producing millimeter-sized objects (i.e. chondrules), as well as driving an alternative grain coagulation process. Second, recondensation is also a mechanism by which a population of smaller grains are produced continually in the nebula. Transport of these grains to the surface of the protoplanetary disk may occur if they are coupled to the gas as a result of turbulence in the disk.

In the CM chondrite Y-791198, the matrix consists of distinct domains with different abundances of sulfide nanoparticles (Chizmadia & Brearley 2008). Sulfide-poor domains contain essentially no sulfide grains, whereas sulfide-bearing domains contain elevated contents of sulfide nanoparticles. The close textural relationship between sulfides and amorphous silicates (i.e. the former appear to be embedded in the latter) implies some sort of genetic relationship. The most plausible interpretations are that (a) the grains formed contemporaneously during disequilibrium condensation, or that (b) the condensation process formed an amorphous silicate phase that contains elevated concentrations of sulfur. During subsequent annealing, the sulfide nanoparticles crystallized from the amorphous material. Unfortunately, data to evaluate these two mechanisms are not available, but it is known that, during disequilibrium condensation, nucleation of simple binary compounds, such as sulfides and oxides, is energetically more favorable than the formation of complex structures such as silicates (Stephens & Kothari 1978). The formation of a loosely consolidated smoke of coagulated amorphous silicate grains and nanophase sulfides could plausibly explain the observed textures.

In summary, although there are uncertainties in the discussion above, arguments can be made that coagulation of dust occurred to a significant degree during disequilibrium condensation processes related to chondrule formation. In at least one primitive chondrite, structures are preserved which are consistent with grain aggregates comprising coagulated nano-phase silicate and sulfide grains. Further studies of chondritic meteorites and an improved understanding of the effects of lithification on open porous smoke structures are necessary to understand fully the origin of such textures, however.

7.10 When did dust coagulation occur?

Dust coagulation was, presumably, a process that took place continually throughout a significant period of nebular evolution. However, a fundamental question concerns exactly what period of nebular evolution is sampled by chondrite matrices and fine-grained rims. Are we observing dust that coagulated over an extended period of nebular history or simply material that represents a snapshot of the late stages of dust coagulation as the nebula dissipated?

The presence of rims of fine-grained material (e.g. Figure 7.4a) on chondrules in many chondrites provides useful constraints on when coagulation of fine-grained dust into larger aggregates occurred. Several models for the origin of these rims have been presented in the literature. Most models support formation of the rims by accretionary processes in the solar nebula (e.g. Metzler *et al.* 1992), but there are also models that argue for rim formation on asteroidal parent bodies (e.g. Trigo-Rodríguez *et al.* 2006). We do not examine the relative merits of these different models here but focus only on nebular models.

There is a type of rim on some chondrules in ordinary and CR carbonaceous chondrites that provides evidence for dust coagulation and accretion onto chondrules while the chondrules were still hot. These rims show evidence of sintering of material and textures indicating that they formed before metal was lost from the chondrules (see Lauretta *et al.* 2006). These sintered rims are, however, comparatively rare and have not been studied in as much detail as the so-called fine-grained rims, which we focus on below. Fine-grained rims are extremely common in CM and CO chondrites, and also occur in CV, CR, and ordinary chondrites. The textural evidence from fine-grained rims indicates that coagulation of fluffy submicron-scale condensate grains into larger aggregates (tens of microns) did not occur immediately following chondrule formation. This conclusion is based on the observation that fine-grained rims are a heterogeneous mixture of materials with different compositions, formational histories, and thermal histories intimately mixed on a fine scale. These constituents clearly did not form in the same event, but represent materials formed at different locations and times in the protoplanetary disk. In particular, the presence of exotic materials, such as interstellar grains, intimately associated with materials of nebular origin, shows that mixing of dust must have occurred after chondrule-forming events began to wane in either frequency or intensity. The remarkable heterogeneity of the dust preserved in fine-grained rims implies that some period of time elapsed before the final coagulation of dust and accretion onto chondrules occurred.

This mineralogical heterogeneity seems to be inconsistent with the formation or processing of dust during a single chondrule-forming event. Although the exact spatial scale of individual chondrule-forming events remains uncertain, Cuzzi & Alexander (2006) have argued that these events were probably of the order of several hundreds of kilometers across. It seems likely that all the dust that was processed within such events would have undergone a similar thermal history and hence be homogeneous in composition and character. If coagulation of individual submicron grains produced by disequilibrium condensation occurred shortly following a specific chondrule-forming event then we would expect minimal variation in mineralogy and composition between individual dust grains. Clearly this is not the case, as discussed above.

Data for the bulk compositions of fine-grained rims in ALHA77307 (CO3.0) and Y-791198 (CM) support the view that rims accreted late in nebular history. Although on a micron to submicron scale, the rims in each meteorite have heterogeneous compositions, on the scale of 10 microns, individual fine-grained rims have remarkably similar bulk compositions (e.g. Metzler *et al.* 1992; Brearley 1993;

Brearley *et al.* 1995; Chizmadia & Brearley 2008). This implies that essentially all chondrules in both meteorites accreted dust from the same reservoir at a similar time, although the dust reservoirs sampled by rims in each meteorite was different. It is possible that nebular dust did not evolve in composition through the entire period of chondrule formation, which would invalidate this conclusion. However, this seems unlikely given that this dust must have undergone thermal processing (evaporation, recondensation, annealing, etc.) during chondrule-forming events and would have undergone progressive fractionation with time. Collectively, these data suggest that most of fine-grained rim materials were formed by coalescence of dust grains into larger aggregates that accreted after the major episodes of chondrule formation had finished. It therefore seems most probable that they represent the last stages of dust coagulation, shortly before asteroidal accretion occurred, rather than recording an extended history of nebular dust coagulation.

The timescales for accretion of dusty rims onto chondrules have been investigated by several workers (e.g. Kring 1988; Metzler *et al.* 1992) and range from minutes to tens of thousands of years. Although the mechanisms of aggregation of dust grains into larger aggregates are not fully understood, Cuzzi (2004) and Ormel *et al.* (2008) have developed models that take into account the major processes involved in the accretion process. The Cuzzi (2004) model examined dust accretion in a weakly turbulent environment and estimated time periods for rim formation on chondrules of 10^2–10^3 years under a plausible range of nebular conditions (turbulence, dust/gas ratio etc.). These timescales are at least two to three orders of magnitude longer than typical chondrule cooling rates. This model therefore strengthens the view that accretion of fine-grained rims occurred after chondrule formation events, possibly after chondrules had cooled to ambient nebular conditions, supporting the mineralogical and textural observations described above. This allowed sufficient time for intimate mixing of components with very different thermal histories and implies that the last stage of coagulation of fine-grained dust occurred shortly before final accretion of asteroidal parent bodies. However, this conclusion is at odds with the fundamentals of dust coagulation, which require that if dust particles come into contact they will stick together, e.g. Dominik *et al.* (2007). If most grains formed in the same environment, i.e. associated with chondrule-forming events, coagulation would be expected to occur, precluding mixing of dust from different environments.

7.11 Astronomical versus meteoritic constraints

As discussed above, the data from astronomical observations and from chondritic meteorites probe very different regions of the protoplanetary disks. In addition, the grain size information obtained by the two approaches may not be equivalent. Care must therefore be taken in drawing information from these two different sources together to make inferences about dust coagulation processes. On the other hand, combining these two highly complementary data sets may also reveal important clues about the evolution of dust in protoplanetary disks. The bulk of the available astronomical data relates to grain size distributions in the *surface layers* of disks, as well as the mid-planes of the outer regions (>50 AU) – regions that are probably not sampled by chondritic meteorites. In the surfaces of disks, particularly in the inner disk, dust particles generally have grain size distributions and crystallinity different from interstellar dust (see also Chapters 6 and 8).

Nevertheless, the typical grain size observed in the disk surface is only an order of magnitude larger than interstellar grains (a few microns, rather than ~ 0.1 µm). The reason that grain growth is not more dramatic is likely a combination of settling to the mid-plane of larger grains, as well as a replenishment of smaller grains through fragmentation, vertical mixing, and nucleation from the gas phase in the innermost part of the disk. Conversely, evidence for the presence of millimeter- to centimeter-sized grains in the outer disk mid-planes is strong.

A complex picture emerges from these studies. First, radiative transfer modeling combined with numerous observations show that systemic temperatures are low (200–300 K) at radii of a few astronomical units from solar-type stars (e.g. D'Alessio *et al.* 2005; Pontoppidan *et al.* 2007a; Schegerer *et al.* 2008). However, the data from chondrites show unequivocally that temperatures sufficiently high to melt and/or evaporate silicate dust were commonly attained at the mid-plane. Such processes clearly had a major influence on dust coagulation, because they provide an additional dynamic environment in which grain growth can occur by collisional coagulation. Unfortunately, the only region of disks that has not yet been probed by astronomical techniques is the inner-disk (1–50 AU) mid-plane, including the region believed to be the source material of chondritic meteorites. However, relating the observations from chondrites to astronomical observations will finally become a reality with ALMA, which will observe dust in the inner-disk mid-plane. If chondrites are a true sample of the nebular mid-plane in the inner disk, then these processes resulted in an essentially bimodal grain size distribution with coarse-grained chondrules and CAIs (a few hundred microns to centimeters) and fine-grained rims and matrices (micron to nanometer-sized grains). Interestingly, although detailed grain size distributions are not yet available for this fine-grained material, the grain size range appears to be intermediate between that of interstellar dust and dust at the surfaces of protoplanetary disks. If significant amounts of dust from the disk surface are transported to the nebular mid-plane, the grain size distribution has fundamentally changed, presumably by thermal processing during, for example, chondrule-forming events as discussed above (see also Chapter 8).

The fine-grained dust fraction in chondrites does not appear to represent material that survived settling to the mid-plane from the disk surface. Instead, this material probably represents the coagulation of dust that formed by evaporation and rapid condensation during chondrule-forming events. It therefore provides no direct link to astronomical observations of dust coagulation at the disk surface, unless these grains communicated with the disk surface through vertical mixing.

Processing of dust by high-temperature events is preserved in chondrites and indicates a complex history of dust evolution at the nebular mid-plane, but the key question remains as to how important such processes were within the inner Solar System and in extrasolar protoplanetary disks in general. Was all dust within the inner 4–5 AU of the protoplanetary disk processed through such chondrule-forming events or was chondrule formation only a relatively localized phenomena, both temporally and spatially, hence having limited implications for global processes at the mid-plane? Chondrule formation does not appear to have been temporally limited. Estimates of average inner-disk lifetimes for low-mass stars are of the order of 3 Myr, but with a large dispersion of 1–10 Myr (Haisch *et al.* 2001; Meyer *et al.* 2007; and see Chapter 9 for a detailed review). Chronological data for both long- and short-lived radionuclides for chondrules indicates that most chondrules formed between ∼ 1 and ∼ 3 Myr after CAI formation (Kita *et al.* 2005). There is even evidence that chondrule formation commenced shortly after CAI formation (Bizzarro *et al.* 2004). Collectively, the data indicate that chondrule formation was an important process for most of the lifetime of the protoplanetary disk.

The spatial extent of chondrule formation as a function of radial distance from the proto-Sun is poorly constrained. Although asteroid spectroscopy has provided key information about the different spectral types of asteroids (e.g. Gradie & Tedesco 1982) within the Asteroid Belt, links between individual meteorite types and asteroids are not well established. It does appear, however, that outer-belt asteroids generally have low albedos (i.e. are spectrally dark) and are generally considered to be organic- and phyllosilicate-rich. If so, this suggests that although they may be chondritic in the broadest terms, they are probably not chondrule-rich like the ordinary, enstatite, and most of the carbonaceous chondrites. These data place general constraints on the outer limit of the disk region where chondrule formation can be considered to be an effective process. The inner part of the Asteroid Belt contains asteroids which are most consistent with achondrites (i.e. differentiated asteroids, Binzel & Xu 1993), in addition to the second-largest class of asteroids, the S-class, which may be the parent bodies of the ordinary chondrites. Although the association of S-class asteroids with the ordinary chondrites remains a subject of debate (Chapman 1996), it indicates that chondrule formation may have extended to the inner edge of the Asteroid Belt.

References

Abreu, N. M. 2007, Ph.D. thesis, University of New Mexico.

Abreu, N. M. & Brearley, A. J. 2006, *Meteoritics and Planetary Science*, **41**, 5372.

Alexander, C. M. O. D., Fogel, M., Yabuta, H., & Cody, G. D. 2007, *Geochimica et Cosmochimica Acta*, **71**, 4380.

Beckwith, S. V. W., Sargent, A. I., Chini, R. S., & Guesten, R. 1990, *Astronomical Journal*, **99**, 924.

Bernatowicz, T., Groat, T., & Daulton, T. 2006, in *Meteorites and the Early Solar System II*, ed. D. Lauretta & H. J. McSween, University of Arizona, 109–126.

Bernstein, M. P., Sandford, S. A., & Allamandola, L. J. 2005, *Astrophysical Journal Supplement Series* **161**, 53.

Bianchi, S., Gonçalves, J., Albrecht, M., *et al.* 2003, *Astronomy and Astrophysics*, **399**, L43.

Binzel, R. P. & Xu, S. 1993, *Science*, **260**, 186.

Bizzarro, M., Baker, J. A., & Haack, H. 2004, *Nature*, **431**, 275.

Blum, J. 2004, *Astromical Society of the Pacific Conference Series*, **309**, 369.

Blum, J. & Schräpler, R. 2004, *Physical Review Letters*, **93**, 115503.

Blum, J. & Wurm, G. 2008, *Annual Review of Astronomy and Astrophysics*, **46**, 21.

Bohlin, R. C., Savage, B. D., & Drake, J. F. 1978, *Astrophysical Journal*, **224**, 132.

Bouwman, J., Henning, T., Hillenbrand, L. A., *et al.* 2008, *Astrophysical Journal*, **683**, 479.

Bouwman, J., Meeus, G., de Koter, A., *et al.* 2001, *Astronomy and Astrophysics*, **375**, 950.

Brearley, A. J. 1991, *Lunar and Planetary Institute Conference Abstracts*, **22**, 133.

Brearley, A. J. 1993, *Geochimica et Cosmochimica Acta*, **57**, 1521.

Brearley, A. J., Bajt, S., & Sutton, S. 1995, *Geochimica et Cosmochimica Acta*, **59**, 4307.

Brearley, A. J. & Jones, R. 1998, *Reviews in Mineralogy*, **36**, 398.

Brearley, A. J. 2004, Lunar and Planetary Science Conference, League City, TX, March 15–19, Abstract 1896 (CDrom).

Burrows, C. J., Stapelfeldt, K. R., Watson, A. M., *et al.* 1996, *Astrophysical Journal*, **473**, 437.

Busemann, H., Young, A. F., Alexander, C. M. O. D., *et al.* 2006, *Science*, **312**, 727.

Chapman, C. R. 1996, *Meteoritics and Planetary Science*, **31**, 699.

Chizmadia, L. & Brearley, A. 2008, *Geochimica et Cosmochimica Acta*, **71**, 602.

Cho, J. & Lazarian, A. 2007, *Astrophysical Journal*, **669**, 1085.

Chokshi, A., Tielens, A. G. G. M., & Hollenbach, D. 1993, *Astrophysical Journal*, **407**, 806.

Cuzzi, J. 2004, *Icarus*, **168**, 484.

Cuzzi, J. & Alexander, C. M. O. 2006, *Nature*, **441**, 483.

D'Alessio, P., Merín, B., Calvet, N., Hartmann, L., & Montesinos, B. 2005, *Revista Mexicana de Astronomia y Astrofísica*, **41**, 61.

Day, K. & Donn, B. 1978, *Astrophysical Journal*, **222**, 45.

Dominik, C., Blum, J., Cuzzi, J. N., & Wurm, G. 2007, in *Protostars and Planets V*, ed. B. Reipurth, D. Jewitt, & K. Keil, University of Arizona Press, 783–800.

Dominik, C. & Tielens, A. G. G. M. 1997, *Astrophysical Journal*, **480**, 647.

Draine, B. T. & Lee, H. M. 1984, *Astrophysical Journal*, **285**, 89.

Dullemond, C. P. & Dominik, C. 2005, *Astronomy and Astrophysics*, **434**, 971.

Dullemond, C. P., Henning, T., Visser, R., *et al.* 2007, *Astronomy and Astrophysics*, **473**, 457.

Floss, C. & Stadermann, F. J. 2008, Lunar and Planetary Science, XXXIX, Abstract 1280.

Furlan, E., Hartmann, L., Calvet, N., *et al.* 2006, *Astrophysical Journal Supplement Series*, **165**, 568.

Garvie, L. & Buseck, P. 2004, *Earth and Planetary Science Letters*, **224**, 431.

Geers, V. C., Augereau, J.-C., Pontoppidan, K. M., *et al.* 2006, *Astronomy and Astrophysics*, **459**, 545.

Geers, V. C., van Dishoeck, E. F., Visser, R., *et al.* 2007, *Astronomy and Astrophysics*, **476**, 279.

Gradie, J. C. & Tedesco, E. F. 1982, *Science*, **216**, 1405.

Greshake, A. 1997, *Geochimica et Cosmochimica Acta*, **61**, 437.

Grossman, J. & Brearley, A. 2005, *Meteoritics and Planetary Science*, **40**, 87.

Habart, E., Natta, A., & Krügel, E. 2004, *Astronomy and Astrophysics*, **427**, 179.

Haisch, Jr., K. E., Lada, E. A., & Lada, C. J. 2001, *Astrophysical Journal*, **553**, L153.

Henning, T., Dullemond, C. P., Wolf, S., & Dominik, C. 2006, *Dust Coagulation in Protoplanetary Disks*, Cambridge University Press.

Indebetouw, R., Mathis, J. S., Babler, B. L., *et al.* 2005, *Astrophysical Journal*, **619**, 931.

Jones, R., Grossman, J., & Rubin, A. 2005, *American Institute of Physics Conference Proceedings*, **341**, 251.

Kemper, F., Vriend, W. J., & Tielens, A. G. G. M. 2005, *Astrophysical Journal*, **633**, 534.

Kessler-Silacci, J., Augereau, J.-C., Dullemond, C. P., *et al.* 2006, *Astrophysical Journal*, **639**, 275.

Kessler-Silacci, J. E., Dullemond, C. P., Augereau, J.-C., *et al.* 2007, *Astrophysical Journal*, **659**, 680.

Kim, S.-H., Martin, P. G., & Hendry, P. D. 1994, *Astrophysical Journal*, **422**, 164.

Kita, N., Huss, G., Tachibana, S., *et al.* 2005, *American Institute of Physics Conference Proceedings*, **341**, 558.

Klock, W., Thomas, K., McKay, D., & Palme, H. 1989, *Nature*, **339**, 126.

Kramer, C., Richer, J., Mookerjea, B., Alves, J., & Lada, C. 2003, *Astronomy and Astrophysics*, **399**, 1073.

Kring, D. 1988, Ph.D. thesis, Harvard University.

Krist, J. E., Stapelfeldt, K. R., Ménard, F., Padgett, D. L., & Burrows, C. J. 2000, *Astrophysical Journal*, **538**, 793.

Krot, A. N., Keil, K., Goodrich, C., Scott, E., & Weisberg, M. 2003, in *Meteorites, Comets and Planets*, ed. A. M. Davis, Treatise on Geochemistry, vol. 1, Elsevier, 143–200.

Kudo, T., Tamura, M., Kitamura, Y., *et al.* 2008, *Astrophysical Journal*, **673**, L67.

Lauretta, D., Nagahara, H., & Alexander, C. 2006, in *Meteorites in the Early Solar System II*, ed. D. Lauretta & H. J. McSween, University of Arizona Press, 431–459.

Lodders, K. & Amari, S. 2005, *Chemie der Erde*, **65**, 93.

Lommen, D., Wright, C. M., Maddison, S. T., *et al.* 2007, *Astronomy and Astrophysics*, **462**, 211.

Mathis, J. S., Rumpl, W., & Nordsieck, K. H. 1977, *Astrophysical Journal*, **217**, 425.

McCabe, C., Duchêne, G., & Ghez, A. M. 2003, *Astrophysical Journal*, **588**, L113.

Metzler, K., Bischoff, A., & Stoffler, D. 1992, *Geochimica et Cosmochimica Acta*, **56**, 2873.

Meyer, M. R., Backman, D. E., Weinberger, A. J., & Wyatt, M. C. 2007, in *Protostars and Planets V*, ed. B. Reipurth, D. Jewitt, & K. Keil, University of Arizona Press, 573–588.

Mukai, T., Ishimoto, H., Kozasa, T., Blum, J., & Greenberg, J. M. 1992, *Astronomy and Astrophysics*, **262**, 315.

Nakamura-Messenger, K., Messenger, S., Keller, L. P., Clemett, S. J., & Zolensky, M. E. 2006, *Science*, **314**, 1439.

Natta, A., Testi, L., Neri, R., Shepherd, D. S., & Wilner, D. J. 2004, *Astronomy and Astrophysics*, **416**, 179.

Newton, J., Bischoff, A., Arden, J., *et al.* 1995, *Meteoritics*, **30**, 47.

Nguyen, A., Stadermann, F., Zinner, E., *et al.* 2007, *Astrophysical Journal*, **656**, 1223.

Nguyen, A. & Zinner, E. 2004, *Science*, **303**, 1496.

Nuth, J. A., III, & Donn, B. 1983, *Journal of Geophysical Research*, **88**, A847.

Nuth, J. A., III, Brearley, A. J., & Scott, E. R. D. 2005, in *American Institute of Physics Conference Proceedings*, **341**, 675.

Ormel, C., Cuzzi, J., & Tielens, A. G. 2008, *Astrophysical Journal*, **679**, 1588.

Ossenkopf, V. 1993, *Astronomy and Astrophysics*, **280**, 617.

Ossenkopf, V. & Henning, T. 1994, *Astronomy and Astrophysics*, **291**, 943.

Palme, H. & Fegley, B. J. 1990, *Earth and Planetary Science Letters*, **101**, 180.

Pascucci, I., Apai, D., Luhman, K., *et al.* 2008, ArXiv e-prints.

Perrin, M. D., Duchêne, G., Kalas, P., & Graham, J. R. 2006, *Astrophysical Journal*, **645**, 1272.

Pontoppidan, K. M., Dullemond, C. P., Blake, G. A., *et al.* 2007a, *Astrophysical Journal*, **656**, 980.

Pontoppidan, K. M., Stapelfeldt, K. R., Blake, G. A., van Dishoeck, E. F., & Dullemond, C. P. 2007b, *Astrophysical Journal*, **658**, L111.

Rodmann, J., Henning, T., Chandler, C. J., Mundy, L. G., & Wilner, D. J. 2006, *Astronomy and Astrophysics*, **446**, 211.

Rubin, A. & Krot, A. 1996, in *Chondrules and the Protoplanetary Disk*, ed. R. Jones, E. Scott & R. Hewins, Cambridge University Press, 173–180.

Schegerer, A. A., Wolf, S., Ratzka, T., & Leinert, C. 2008, *Astronomy and Astrophysics*, **478**, 779.

Scott, E., Barber, D., Alexander, C., Hutchison, R., & Peck, J. 1988, in *Meteorites and the Early Solar System*, ed. J. Kerridge & M. Matthews, University of Arizona Press, 718–745.

Scott, E. & Krot, A. 2005, *Astrophysical Journal*, **623**, 571.

Shirley, Y. L., Wu, J., Shane Bussmann, R., & Wootten, A. 2008, *Astronomical Society of the Pacific Conference Series*, **387**, 401.

Shu, F. H., Shang, H., & Lee, T. 1996, *Science*, **271**, 1545.

Stapelfeldt, K. R., Krist, J. E., Menard, F., *et al.* 1998, *Astrophysical Journal*, **502**, L65.

Stephens, J. & Kothari, B. 1978, *Moon and Planets*, **19**, 139.

Stroud, R. 2005, *American Institute of Physics Conference Proceedings*, **341**, 645.

Takeuchi, T. & Lin, D. N. C. 2002, *Astrophysical Journal*, **581**, 1344.

Tamura, M., Hough, J. H., Greaves, J. S., *et al.* 1999, *Astrophysical Journal*, **525**, 832.

Trigo-Rodríguez, J., Rubin, A. E., & Wasson, J. T. 2006, *Geochemica et Cosmochimica Acta*, **70**, 1271.

van Boekel, R., Min, M., Leinert, C., *et al.* 2004, *Nature*, **432**, 479.

van Boekel, R., Min, M., Waters, L. B. F. M., *et al.* 2005, *Astronomy and Astrophysics*, **437**, 189.

van Boekel, R., Waters, L. B. F. M., Dominik, C., *et al.* 2003, *Astronomy and Astrophysics*, **400**, L21.

Vuong, M. H., Montmerle, T., Grosso, N., *et al.* 2003, *Astronomy and Astrophysics*, **408**, 581.

Wasson, J. T. & Trigo-Rodríguez, J. M. 2004, *Lunar and Planetary Institute Conference Abstracts*, **35**, 2140.

Weidenschilling, S. J. 1977, *Monthly Notices of the Royal Astronomical Society*, **180**, 57.

Weingartner, J. C. & Draine, B. T. 2001, *Astrophysical Journal*, **548**, 296.

Weisberg, M., McCoy, T., & Krot, A. 2006, in *Meteorites and the Early Solar System II*, ed. D. Lauretta & H. J. McSween University of Arizona Press, 19–52.

Weisberg, M. & Prinz, M. 1998, *Meteoritics and Planetary Science*, **33**, 1087.

Weisberg, M., Prinz, M., Clayton, R., & Mayeda, T. 1993, *Geochimica et Cosmochimica Acta*, **57**, 1567.

Wolf, S., Padgett, D. L., & Stapelfeldt, K. R. 2003, *Astrophysical Journal*, **588**, 373.

Wurm, G. & Blum, J. 1998, *Icarus*, **132**, 125

Wurm, G., Paraskov, G., & Krauss, O. 2005, *Icarus*, **178**, 253.

Zinner, E. 1998, *Annual Reviews in Earth and Planetary Sciences*, **26**, 147.

Zinner, E. 2003, in *Meteorites, Comets and Planets*, ed. A. Davis, Treatise on Geochemistry, vol 1, Elsevier, 17–39.

8

Thermal processing in protoplanetary nebulae

Dániel Apai, Harold C. Connolly Jr., and Dante S. Lauretta

Abstract Crystalline silicates around other stars demonstrate that protoplanetary material is often heated or processed. Similarly, primitive Solar System materials (chondrule components, IDPs, Stardust samples, comet grains) provide multiple lines of evidence for repeated dramatic heating events that affected most or all the protoplanetary materials in the first few million years. The existence of such powerful heating events is not predicted or understood from planet-formation models, yet may have had important implications on the status and composition of planetary raw materials. Here we synthesize the astronomical and meteoritic evidence for such events and discuss proposed models. By matching astronomical analogs to events in the young Solar System we attempt to reconstruct a possible scenario for the thermal processing of protoplanetary materials consistent with all evidence. We also highlight details where the astronomical and cosmochemical views are difficult to reconcile and identify key directions for future research.

Crystalline silicates are tracers of high temperatures (>1000 K), yet they are often observed in cool outer regions of protoplanetary disks (< 350 K). Their origin is one of the puzzles that offer insights into the thermal evolution of protoplanetary disks. These crystals may have formed *in situ* in the cold disk in past heating events or may have been mixed outward from the hot inner disk; perhaps both mechanisms played a role. Similarly, once-molten silicate spherules – chondrules – and refractory inclusions delivered from the cold Asteroid Belt (~ 180 K) by primitive chondritic meteorites are products of high temperatures. In many chondritic meteorites chondrules account for over 80% of the total mass, revealing that most of the primordial material in the Asteroid Belt, and perhaps elsewhere, has been heated and compacted. The Stardust mission also returned crystalline silicates and at least one grain that closely resembles refractory inclusions in chondritic meteorites, demonstrating the presence of thermally processed material even at large orbital radii.

The evidence for past heating is thus ubiquitous, both in the Solar System and around young stars. Yet, current disk evolution and planet formation models do not naturally explain the presence of high-temperature material in the outer disk. The importance of thermally processed material in the cold disks is twofold. First, it challenges and tests disk physics, including disk evolution, mixing, and shock wave propagation. Second, thermal processing significantly alters the pre-planetary material and influences the bulk composition of emerging planets, as well as the organics that are delivered post-formation.

The origin of these high-temperature products, and most prominently that of chondrules, has been the subject of intense research for over 120 years. Several excellent recent reviews explore different facets of this complex question (e.g. Jones *et al.* 2000; Connolly *et al.* 2006; Lauretta *et al.* 2006; Scott 2007; Wooden *et al.* 2007). During the last few years, however, our understanding of the question has been transformed by rapid progress thanks to space missions to comets (Deep Impact and Stardust) and advanced telescopic observations of protoplanetary disks (e.g. Spitzer Space Telescope, Very Large Telescope Interferometer). These changes warrant revisiting the question and motivate developing a unified picture of thermal processing in protoplanetary disks.

This chapter aims to compare the evidence for thermal processing in the Solar System and around young stars and identify possible heating processes. We first discuss the physics of thermal annealing, melting, and evaporation, followed by remote sensing evidence for heated material around young stars, followed by the discussion of the Solar System evidence. We end with the discussion of the heating processes that may explain different lines of evidence in search of an overall picture. Finally, we identify key open problems and suggest possible tests before concluding the chapter.

8.1 Thermal processing: annealing and evaporation

8.1.1 Thermal annealing of silicate grains

Silicate grains in the interstellar medium are mostly found in amorphous form (e.g. Kemper *et al.* 2005; Chapter 5), i.e. while the neighboring Si–O tetrahedrons may be ordered, units within the grains are oriented quasi-randomly. This is an energetically disfavored state; in contrast, a crystal structure with regular placements of the silicon and oxygen atoms in locations of the minimum energy represents an energetically ideal state. However, moving to a more ordered configuration requires breaking bonds, bending or stretching the existing structures, and relocating atoms. In other words, the probability of the transition to regular structure depends on the energy available in the system. The characteristic timescale for converting an amorphous domain to a crystalline domain is given by $t = \nu^{-1}e^{D/kT_a}$, where ν is the typical vibrational frequency of silicates, D is the activation energy of repositioning atoms within the lattice structure, and T_a stands for the kinetic temperature of the silicates (e.g. Lenzuni *et al.* 1995). Dust annealing experiments suggest typical values of 1–4×10^{13} s^{-1} for ν, and $41\,000$ K for D/k (e.g. Nuth & Donn 1982; and Chapter 5 for details).

The time to crystallize – or thermally anneal – grains is strongly dependent on the temperature: at $T_a \sim 630$ K it will take ~ 5 Myr (the lifetime of the disk), while at $T_a \sim 800$ K the timescale becomes comparable to the orbital periods in the inner disks (5 yr), and at $T_a \sim 1200$ K it takes just a few seconds.

The regular crystal lattice of thermally annealed grains will preferentially allow stretching and bending modes that can resonate within the lattice structure, while other modes will be damped. This leads to a strongly peaked phonon spectrum, in contrast to the broad and continuous phonon spectrum of the amorphous grains. The structured phonon energy distribution of silicates is observable in the mid-infrared spectrum, where stretching–bending transitions typical to crystalline silicates appear (see, e.g. Chapter 6 and Fig. 8.1).

Observations of abundant crystals around cool giant stars suggest that alternative, low-temperature pathways may also lead to the crystallization of olivine glasses, perhaps some also operating in protoplanetary disks (Molster *et al.* 1999). Electron-irradiation experiments at room temperature demonstrated that olivine glasses have crystallized, probably through elemental diffusion induced by the ionizing radiation (Carrez *et al.* 2002; Kimura *et al.* 2008). In addition, exothermic chemical reactions – such as graphitization of a carbon-rich layer – may also provide energy locally to crystallize amorphous silicate grains, although further experiments are needed to demonstrate that this mechanism can produce detectable amounts of crystalline silicates under realistic astrophysical conditions (Kaito *et al.* 2007).

8.1.2 Condensation

The process of condensation of minerals in the early solar nebula has long been invoked to explain the chemistry and mineralogy of primitive chondritic meteorites (e.g. Cameron 1963). Their observed bulk compositions show volatile-element depletions that are clearly smooth functions of calculated condensation temperature in a gas of solar composition (Davis 2006). Despite this success in explaining the bulk composition of chondrites, the diverse mineralogy of these bodies is not reproduced well in the condensation sequence calculations. To date, there is no incontrovertible evidence for direct condensation of rocky meteoritic material in the

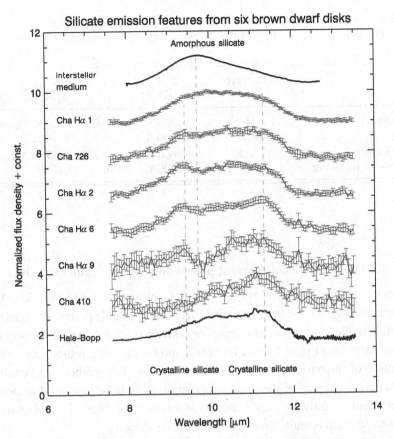

Figure 8.1 Mid-infrared spectra reveal the ubiquitous presence of crystalline silicates, even around cool brown dwarfs. From Apai *et al.* (2005).

solar nebula. Many meteorite inclusions and matrix grains have been hypothesized to be condensates, or to have equilibrated with vapor at high temperatures in the protoplanetary disk or nebula. In particular fluffy Type-A calcium–aluminum-rich inclusions (CIAs), spinel–pyroxene aggregates, and hibonite-bearing objects have been proposed as nebular condensates. Specific CAI components, such as minerals deposited in voids in Type-B CAIs, platinum-group element metal nuggets and relict hibonite and fassaite in igneous CAIs, are also candidates for nebular condensates. Chemical signatures such as non-fractionated rare-earth-element abundance patterns in CAIs and chondrules are cited as evidence for condensation. For the more volatile elements fine-grained components, such as FeO-rich olivine in matrix and inclusion rims, are proposed nebular condensates. Direct condensation of minerals is also one possible pathway for producing fine crystalline silicate dust.

Table 8.1 *Typical stellar and disk parameters for young stars. The columns show typical spectral types at ~1 Myr, effective temperatures, luminosity, disk masses, and mass accretion rates.*

Type	Spectral type	T_{eff} [K]	L [$L_\odot$]	M_{disk} [M_J]	$\dot{M}$ [$M_\odot$/yr]
Brown dwarf	>M6	<2900 K	<0.03	<6	$<10^{-11}$
Cool star	M0–M5	~3200 K	0.2	30	10^{-9}
Young Sun-like star	F, G, K	~3800 K	1.3	70	10^{-7}
Intermediate-mass star	A/B	>7000 K	40	300	—

8.2 Observations of thermal processing in protoplanetary disks

The shortest-wavelength stretching–bending transitions of Si–O bonds lie at 10 μm, forming a prominent band in the infrared spectrum of silicate grains. The band shape is sensitive to the lattice structure, grain or domain size, and chemical composition of the silicates (see Fig. 8.1 and Chapters 5 and 6), allowing remote sensing characterization of the basic properties of silicate grains. The methods and challenges of the spectral analysis of the dust composition have been discussed in details in Chapter 6 and we only emphasize here that infrared spectroscopy probes only the micron-sized grain population at the disk surface directly.

Because the lattice structure and the chemical composition of the silicate grains are strongly influenced by heating, infrared spectra can be used to identify evidence for the past thermal processing of small grains in protoplanetary disks. In the following, we will discuss the available observational evidence and its possible interpretations, before placing it in the context of thermal processing in the Solar System.

The 10 μm window has long been used to identify silicate emission features and characterize amorphous dust grains (e.g. Rydgren *et al.* 1976; Hanner *et al.* 1995), but only recently did instrumental sensitivities enable the systematic studies of sharp and faint features characteristic of crystalline grains. In 1995 the European Space Agency's Infrared Space Observatory (ISO) opened the window to sensitive, high-resolution infrared dust spectroscopy, with large ground-based telescopes at dry and high-altitude sites (Chilean Andes and Mauna Kea) reaching similar sensitivities by 2002 at lower spectral resolution and limited wavelength range. The Spitzer Space Telescope, launched by NASA in 2003, has provided a uniquely sensitive platform for infrared spectroscopy at a broad range of wavelengths (5–40 μm), albeit at a spectral resolution lower than that of the ISO.

8.2.1 Crystalline silicates in disks: the missing parameter

The earliest detailed studies of silicate dust in protoplanetary disks targeted those brightest in the mid-infrared, where high quality spectra could be obtained even by severely flux-limited observations. Cohen & Witteborn (1985) reported the earliest detection of crystalline silicate emission from the environment of young stars and interpreted it as evidence for dust having been transformed from its pristine state in the interstellar medium to the material known to be contained in the comets and perhaps primitive meteorites. Interestingly, this observation and explanation pre-dated the evidence that young stars are surrounded by disks and not by spherical envelopes.

The characteristic radius where the mid-infrared-emitting crystalline silicate species are located may be estimated knowing that the emitting area increases with radius ($\sim r^2$), while the disk surface temperature (T_s) decreases with radius ($T_s(r) \sim r^{-1/2}$ or $T_s(r) \sim r^{-3/4}$, see Chapter 3) and with it decreases the disk luminosity. The trade-off between the outward increasing surface area and the decreasing surface brightness defines the radius that dominates the mid-infrared emission. Conclusions drawn on the dust composition will, in effect, constrain dust properties at this characteristic radius. The characteristic radius is wavelength-dependent: longer wavelengths probe cooler disk regions and thus larger disk radii.

The temperature distribution is not only a function of radius, but also depends on the stellar luminosity, the disk geometry, and may depend on the accretion rate (see Table 8.1 and Section 3.3): for example, at a given radius irradiated flared disks will be warmer than flat disks. Naturally, hotter stars will heat their disks to higher temperatures at a given radius: thus, mid-infrared spectroscopy probes different radii in different disks.

Sensitive observations enabled comparative surveys of the brightest disks around intermediate-mass and Sun-like stars. These observations revealed a wealth of different silicate emission features, many of them with sharp peaks of crystalline silicates. The motivation of these surveys was to use crystallinity as a proxy for the evolution of the dust component as well as for the overall disk. Indirectly, these are thought to be linked to the formation of rocky planets.

Surprisingly, the early ground-based surveys could not find any correlation between the presence of crystalline silicates and the key stellar and disk parameters (e.g. van Boekel *et al.* 2003; Honda *et al.* 2006). An inherent difficulty of recognizing such correlations in the spectra of intermediate-mass stars is the ubiquitous presence of the sharp polycyclic aromatic hydrocarbon (PAH) peaks, often contaminating the crystalline silicate peaks.

More sensitive space-based studies with Spitzer helped secure identification of crystalline silicate peaks in PAH-free spectra of Sun-like stars, resolving this problem (e.g. Sargent *et al.* 2006; Bouwman *et al.* 2008; Watson *et al.* 2009). These studies, however, could all but reinforce the lack of strong correlations with age, spectral type, stellar temperature, and accretion signatures (Bouy *et al.* 2008). In a rigorous statistical study Pascucci *et al.* (2008) showed that the spectral shape of the silicate emission spectra of 44 co-eval single and binary Sun-like stars do not correlate with multiplicity or with the disk flaring angle, disproving earlier claims for the opposite.

The lack of any robust correlation between the presence of crystalline silicates and any fundamental system parameters in these studies suggests that crystallinity is not directly related to the disk evolution, but it is a consequence of a yet unidentified process or event, which may occur at different evolutionary stages in different disks (see also Sargent *et al.* 2009). Such processes may include disk instabilities, formation of giant planets, shock waves, equilibrium vs. non-equilibrium condensation, destruction of differentiated planetesimals, and changes in the efficiency of turbulent mixing. In the search for the key processes setting crystallinity we discuss its dependence on the disk radius, the presence of crystals around cool stars and around stars with very high accretion rates, and consider constraints from dust compositional studies.

8.2.2 Crystallinity – the radial dependence

Rapid developments in the infrared instrumentation also offer a way to constrain the location and spatial extent of the crystalline silicate peaks in protoplanetary disks. This is an important constraint on the radial distribution and mixing of the crystalline material. In Spitzer/Infrared Spectrograph spectra of disks around seven young Sun-like stars Bouwman *et al.* (2008) found a correlation between the presence of crystalline silicates in the inner (~ 1 AU) and in the outer disks (5–15 AU), as probed at short (10 μm) and long (20–30 μm) wavelengths. Using a linear combination of temperature-weighted opacities and the assumption that the disk continuum emission can be modeled as a polynomial, they found that the inner disk is enriched in enstatite relative to forsterite, while the cooler outer disk has more forsterite than enstatite. Such a relative abundance pattern is in contrast with one-dimensional chemical equilibrium accretion disk models, but is not inconsistent with the results of two-dimensional calculations (e.g. Gail 2004; Wehrstedt & Gail 2008).

For bright disks around Sun-like and intermediate-mass stars it is also possible to acquire spatially resolved mid-infrared spectra using long-baseline infrared interferometry. This technique reaches milli-arcsecond resolution and enables the spectral comparison of inner and outer disks. Studying three disks with the very large telescope interferometer (VLTI), van Boekel *et al.* (2005) found much higher crystalline silicate abundances in the inner disks (< 2 AU) than in their corresponding outer disks (> 2 AU), relative to the small amorphous grain content (see Fig. 8.2). In a similar study, Ratzka *et al.* (2007) showed that the crystalline peaks are much

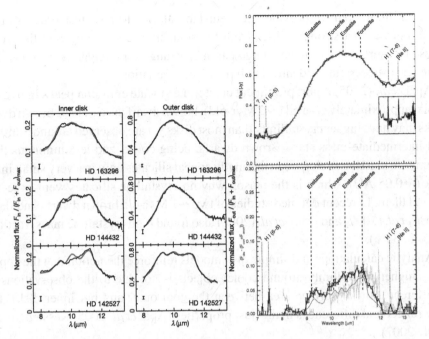

Figure 8.2 Left: mid-infrared spectra of the inner and the outer regions of three disks around intermediate-mass stars (van Boekel *et al.* 2005). Right: spectrum of the outer disk (upper panel) and the inner disk (lower panel) of the young Sun-like star TW Hya (Ratzka *et al.* 2007). In all four sources the crystalline silicate peaks are much more pronounced in the inner disks.

more prominent in the correlated spectrum of the 8 Myr-old star TW Hya than in its unresolved (total) spectrum (see Fig. 8.2). These observations suggest that crystals are more abundant in inner disks and suggest limited mixing within inner and outer disks. In contrast, in the disk around a brown dwarf candidate studied by Bouy *et al.* (2008) long-wavelength crystalline silicate features are much more pronounced than the 10 μm features. Simple radiative-transfer-based dust compositional analysis suggests a crystallinity higher in outer disks than in inner disks.

Note, however, that because the observations probe the amorphous/crystalline silicate mass fraction, they are also consistent with a radius-independent amount of crystalline silicates, if the inner disks are depleted in submicron-sized amorphous silicate grains.

8.2.3 High-temperature products around cool stars

The superior sensitivity of the Spitzer Space Telescope also opened a new window to the study of silicate emission features around cool stars – too faint to be studied by other instruments. With typical luminosities of ~1% and masses of ~5% of the Sun, these sources provide excellent laboratories to test dust processing in environments that strongly differ from the solar nebula or young Sun-like stars. These measurements provide a large parameter range for comparative studies and have been successful in identifying trends and correlations.

Apai *et al.* (2005) report prominent crystalline silicate emission peaks in five out of six, approximately co-eval (~ 2 Myr-old), brown dwarf disks. These brown dwarf disks may have higher crystallinity than most disks yet are observed around Sun-like and intermediate-mass stars. Brown dwarfs, being cooler and less luminous than their more massive analogs, can thermally anneal silicates only in a very small inner disk (< 0.05 AU), which is the reason why no crystalline silicates were expected around them. In recent detailed studies of two ~ 1 Myr-old brown dwarf candidates Merín *et al.* (2007) and Bouy *et al.* (2008) also found high crystalline mass fractions (15% to 33%).

Analytic calculations and simple disk models reinforce the notion that the apparently higher crystallinity around cooler objects is not due to the observations of a different ratio of surface area between the cool outer and hot inner disks, but they reflect real differences in the dust properties (Apai *et al.* 2005; Kessler-Silacci *et al.* 2007).

The abundance of silicate crystals in these disks provides strong constraints on thermal processing: whatever processes are responsible for the presence of crystals around Sun-like stars must also be capable of very efficiently producing crystals around cool stars and in very low-mass disks.

While no quantitative models exist yet for dust evolution and mixing in disks around very low-mass stars or brown dwarfs, a few qualitative statements can be made. Because the distance of the mid-infrared emitting disk region from the hot inner disk where silicates can be annealed is much smaller for a brown dwarf (~ 0.1 AU) than for a Sun-like star (~ 0.8–1 AU), radial mixing and outward diffusion of the crystals may be much more efficient. Disks around cool stars and brown dwarfs were also found to be flatter than disks around Sun-like stars owing to the overall larger grains and more efficient dust settling (Apai *et al.* 2008) and this difference may expose crystalline-rich material close to the disk mid-plane (see also Watson *et al.* 2009). Owing to their lower disk masses ($\sim 5\%$) and ~ 15 times shorter orbital periods these disks would also be ideal sites for evaluating the efficiency of shocks as an energy source for crystallization.

8.2.4 Runaway disk accretion and the Fu Ori disks

During the last century over a dozen young stars have been observed to undergo major outbursts, enhancing their luminosity by about hundred-fold. Observations show that the luminosity decreases slowly, in the course of several decades (e.g.

Hartmann & Kenyon 1996). Although the exact nature of these outbursts is not understood, it is thought that the extra luminosity is provided by runaway accretion events with mass accretion rates $\sim 10^{-4}$ $M_{\odot}$ yr^{-1} (Zhu *et al.* 2007). The rapidity of the outbursts and the amount of energy released suggest that the accretion disks became unstable and delivered much of their mass to the star. Possibly, all FU Orionis stars (Fu Ori-type stars or Fuors) are oddities – perhaps they are unstable hierarchical triples with massive disks (Reipurth & Aspin 2004) or have very unusual disk or envelope geometries. But more likely they are typical young stars that undergo outbursts characteristic to *all* young Sun-like stars. This scenario is supported by detailed, high-resolution studies that find no peculiar circumstellar features that could trigger outbursts (e.g. Quanz *et al.* 2007; Kóspál *et al.* 2008) and by detailed radiative transfer models that reveal the presence of a very hot inner disk (Zhu *et al.* 2007; Greene *et al.* 2008), consistent with the basic predictions of disk instability models (e.g. Bell & Lin 1994; Armitage *et al.* 2001).

If such outbursts are typical ingredients of disk evolution, the observed incidence rate implies that these are episodic and represent the dominant way of transporting mass through the disks. During such enhanced accretion events the disk heating is dominated by accretion heating, probably reversing the vertical temperature gradient. Mid-plane temperatures reach or exceed 1000 K within 1 AU (e.g. Zhu *et al.* 2007), leading to the evaporation of volatiles, and the annealing or melting of silicates throughout the inner disk (for a review, see Bell *et al.* 2000).

The FU Ori outbursts may have repeatedly heated the solar nebula, but the typical duration of these events ($\sim 50 - 100$ yr) make them unlikely sources for chondrule or CAI formation. However, free-floating refractory grains or pebbles, such as CAIs or chondrules, may have survived these episodes and may now carry clues on these events. That CAIs have been repeatedly heated (Connolly *et al.* 2006, and references therein) – some after an epoch of alteration (Beckett *et al.* 2000) – appears to be consistent with such episodes of global disk heating.

Surprisingly, mid-infrared spectroscopy with Spitzer, ISO, and the interferometer VLTI did not detect spectral features from crystalline silicate emission (Green *et al.* 2006; e.g. Quanz *et al.* 2006, 2007) in the outer or the inner disks of Fuors. This finding is rather puzzling: because of the very high accretion rates, Fuor disks are hot and should possess copious amounts of thermally processed dust. A possible, but unsatisfying, explanation may be that fresh, infalling material from the protostellar envelope would hide the material already processed in the disk. The systematic lack of high-temperature products in Fuor disks is unexpected and makes the role of the outbursts in the thermal processing of the protoplanetary dust difficult to understand.

8.2.5 Transition disks

A small subset of young protoplanetary disks exhibit inner disks that are strongly depleted or entirely devoid of fine dust, while they are surrounded by massive cold, outer disks. The typical sizes of these inner-disk holes range from a few to tens of astronomical units. While some of the large inner holes may be dynamically opened by unseen stellar or substellar companions (e.g. Ireland & Kraus 2008), others may be caught in the rapid inside-out clearing of the disk (e.g. Strom *et al.* 1989; Najita *et al.* 2007). If so, they may be the Rosetta stone for understanding how protoplanetary disks disperse, perhaps the strongest boundary condition for planet formation (see Chapter 9 for details).

If transition disks are indeed the last stage of the dust disk evolution, then dust in them should be similar to the fine dust preserved from the young Solar System, e.g. in chondritic matrices or perhaps in some interplanetary dust particles (IDPs). These Solar System grains are often rich in magnesium and poor in FeO, consistent with an origin as equilibrium condensation products. Surprisingly, at least five transition disks studied in detail (e.g. Sargent *et al.* 2006; Espaillat *et al.* 2007a,b) display very little or no crystalline silicate emission peaks in the integrated Spitzer spectra; overall, their 10 μm silicate emission features resemble that of the pristine interstellar dust. Some of the transitions disks, however, show evidence for crystalline silicates at longer wavelengths, originating at larger radii or at lower disk heights. Mid-infrared measurements by the VLTI reveal crystalline silicate grains at small radii in at least one of the disks that display amorphous silicate features in integrated light (Ratzka *et al.* 2007), demonstrating the radial inhomogeneity of the crystals (see Fig. 8.2).

These observations, taken together, show that at least the disk surface of many transition disks is rich in or dominated by fine amorphous silicate dust. Silicate crystals can be present in the inner or the outer disks, but probably only in small to modest abundances (< 5 wt%); these crystals can be concentrated in the inner or in the outer disks. If the inner hole is due to inside-out disk dispersal, then this happens at an early evolutionary stage when the dust disk is dominantly composed of interstellar-medium-like unprocessed grains. While this emerging picture suggests a complex disk structure, the pristine dust in transition disks appears to be inconsistent with the crystalline-rich, reprocessed, and equilibrated dust from the end of the dust-rich epoch of the Solar System, as preserved in chondritic matrices or in IDPs. This pronounced difference in dust evolution suggests that the Solar System has not evolved through this phase, and transition disks may represent an alternate pathway of disk evolution or be altogether different objects, such as circumbinary accretion disks (e.g. Ireland & Kraus 2008).

8.2.6 Magnesium-rich crystalline silicates

Silicates with olivine composition $(Mg_x Fe_{(1-x)})_2 SiO_4$ are common in chondrites, comets, IDPs, and in protoplanetary disks. The Mg-rich end-member of the olivine family is forsterite, also often termed as Fo_{100}; the Fe-rich end-member is fayalite (Fo_0). The interstellar medium contains a similar concentration of the FeO- and MgO-rich silicates (see Chapter 2). Correspondingly, *amorphous* silicate grains frequently have similar magnesium and iron abundances in protoplanetary disks, in cometary dust, and in chondritic IDPs. In stark contrast, *crystalline* dust is almost always dominated by Mg-rich grains in protoplanetary disks (e.g. Malfait *et al.* 1998; Bouwman *et al.* 2008), comet tails (e.g. Crovisier *et al.* 1997; Wooden *et al.* 2004; Harker *et al.* 2005; Lisse *et al.* 2006), in the most primitive and least processed chondritic matrices, and IDPs (for a review, see Wooden *et al.* 2007).

The depletion in FeO may be understood in at least two ways. First, the crystalline grains may be equilibrium condensates from a hot solar nebular composition gas with iron sequestered to metals or sulfides (see e.g. Chapter 4). In this case the condensed grains either had to condense slowly to form crystal domains, or had been reheated and thermally annealed at a later epoch. The second, alternative explanation is that ferromagnesian amorphous silicate grains were thermally annealed in a reducing environment, e.g. in the presence of carbon. Heating such precursors leads to the formation of metallic spheroids embedded between the forsterite crystals, as the initial FeO component is reduced (see e.g. Fig. 8.3 and Connolly *et al.* 1994; Jones & Danielson 1997; Leroux *et al.* 2003; Davoisne *et al.* 2006). Because carbon is ubiquitously present in primitive Solar System materials, this pathway offers a natural explanation to the observed FeO-poor silicate crystals. It is yet to be determined whether low-temperature crystallization processes, discussed in Section 8.1.1, would also lead to FeO depletion.

In view of the alternative pathways, the FeO depletion of crystalline silicates cannot be interpreted as an unambiguous marker of silicate recondensation nor as evidence for the presence of second-generation dust; rather, it is supporting evidence for the fact that these grains have been heated to high temperatures.

8.3 Thermal processing in the Solar System: chondrites

Much of the primitive planetary material in the Solar System, which includes primitive meteorites, IDPs, and samples returned from Stardust, have been processed repeatedly within the protoplanetary disk by energetic, high-temperature events during the course of the first few million years of its existence. The strongest evidence for such processing comes from primitive meteorites known as chondrites. Chondrites are typically referred to as defining $t = 0$ for the Solar System at

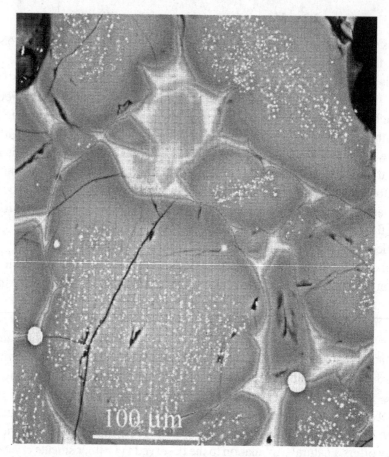

Figure 8.3 Heating dusty olivine grains in a reducing environment forms forsterite grains with small iron-rich blebs (<1 μm) toward the middle of the olivine grains and metallic globules (1–50 μm) in the glass matrix at the boundaries of the olivine grains. From Leroux *et al.* (2003).

4567.2 ± 0.6 Ma (Wadhwa *et al.* 2007 and references therein), which is actually the age of a group of inclusions within chondrites known as calcium–aluminum-rich inclusions (CAIs). The word primitive refers to the fact that the bulk compositions of all chondrites, within a factor of two, are solar in composition for all but the most volatile elements (Weisberg *et al.* 2006). This fact indicates that chondrites have not been through a planetary melting or differentiation process in their parent body, indicating that they have recorded the materials that were present and the processes that operated within the disk before or during planet formation.

Chondrites are classified into five major groups based on properties such as bulk composition, petrography, and oxygen isotopic ratios (Brearley & Jones 1998). They are further subdivided into several clans within these major types

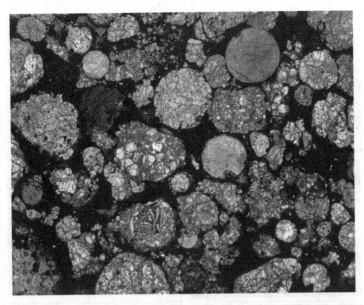

Figure 8.4 Chondrules dominate the volume and mass of primitive chondrites, such as in this un-equilibrated ordinary chondrite, Semarkona (LL3.0). Reprinted from *A Color Atlas of Meteorites in Thin Section* (D. S. Lauretta and M. Killgore, Golden Retriever Publications, 2005).

(Weisberg *et al.* 2006; Chapter 1). The key point is that subdivision of chondrites is based on differences in the basic chemical, mineralogical, and textural properties that are unique to each class. Chondrites also record difference in redox conditions before and during their accretion providing important clues to disk environments and properties such as oxygen, carbon, and sulfur abundances in these regions.

The anatomy of a chondrite can be broken down into refractory inclusions, chondrules, fragments of these objects, and fine-grained matrix (dust and organic molecules) that binds the rocks together. Of specific interest are the millimeter- to centimeter-sized silicate spheres known as chondrules and specific types of refractory inclusions, all of which are igneous rocks – rocks that have been melted, slowly cooled, and crystallized (see Figs. 8.4 and 8.5). The abundance of these objects varies within and among chondrites and thus the amount of material processed to high temperatures in the disk varied within and across different regions and stages of disk evolution.

The first-order observation is that these igneous rocks were heated to temperatures up to ~2200 K for seconds to minutes and then cooled between tens to hundreds of K hr^{-1} at pressures that stabilized silicate liquids. The energetic mechanism(s) that melted (or partially melted) these objects operated multiple times over millions of years. The intensity of these events was highly variable. The major assumption

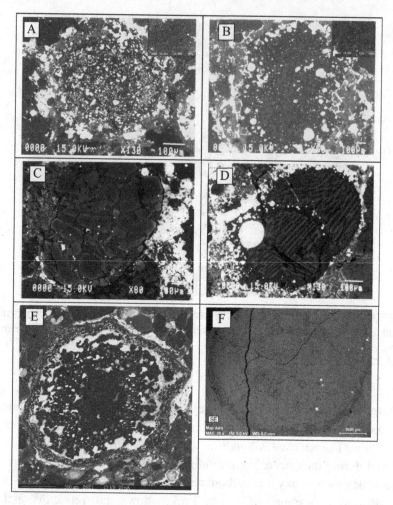

Figure 8.5 Different textural types of chondrules and a Type-B CAI shown in backscatter electron microscope images; samples from A through D suffered increasing levels of heating: (A) agglomeratic chondrules (scalebar 100 μm); (B) Type-IA (scalebar 100 μm); (C) Type-IIA (scalebar 100 μm); (D) compound barred chondrule set (scalebar 100 μm); (E) layered chondrules with fine grains (scalebar 500 μm); (F) a Type-B CAI (scalebar 1mm).

inherent in the community is that these igneous objects were free-floating wanderers within the mid-plane of the disk and that they were originally dust balls. The production of chondritic igneous objects, such as chondrules, is assumed to be a part of the formation of most planetesimals, suggesting that large fractions of dust in the mid-plane were melted before asteroidal bodies were formed. Thus, dust does not simply accrete to form planetesimals. Dust accretes to form millimeter-to centimeter-sized dust balls that then get melted, and remelted or heated multiple times, and are then accreted along with surviving or freshly injected dust to form planetesimals. From the point of view of astrophysics, these objects would not be predicted to exist if they did not exist. There is nothing intuitive in astronomy which states that before planetesimals are formed, igneous rocks are formed (Connolly *et al.* 2006). It remains to be proven, however, whether or not planet formation requires chondrules and igneous CAIs before the onset of large-scale planetesimal accretion. If it does, then the chemical differences in asteroids and planets are a direct reflection of transient heating.

8.3.1 Refractory inclusions

Refractory inclusions are a class of chondritic components that derive their classification in part from the fact that they are composed of some of the most refractory materials found in primitive planetary samples (Grossman 1972; MacPherson *et al.* 1988; Connolly 2005; Beckett *et al.* 2006). In meteoritics, refractory refers to the temperature at which materials condense or evaporate within the gaseous environment in the protoplanetary disk. Thus, refractory inclusions are composed of minerals that are among the first predicted to condense from a gas of solar or enhanced solar composition (Ebel 2006). Their abundance in chondrites can potentially range up to 15 vol%, although most appear to be only a few vol% (Grossman *et al.* 1988; Russell *et al.* 1998; Ebel *et al.* 2008; Hezel *et al.* 2008).

Most of these objects have experienced little (amoeboid olivine aggregates, AOAs) to potentially no melting (fine-grained or fluffy Type-A CAIs) after the condensation of their precursors. Those that did experience significant melting and crystallization are referred to either as the centimeter-sized Type-B CAIs or the millimeter-sized Type-C CAIs (see Fig. 8.5). Although debated in the community, it is possible that another type of inclusion, compact Type-A is also igneous, increasing the overall abundance of melted CAIs (Connolly *et al.* 2006). The thermal history of these objects has been reviewed elsewhere (Jones *et al.* 2000; Connolly & Desch 2004; Connolly *et al.* 2006); here we will review the critical variables appropriate to the goals of this book.

Shortly after the condensation of the first minerals to form within the Solar System, thermal processing of refractory inclusions began. The overall processing of these materials into igneous rocks was short-lived, perhaps only 100 000 years (Bizzarro *et al.* 2007). The critical constraints on the thermal histories of these objects is essentially limited to Type-B CAIs (Connolly & Desch 2004), which have been generated by the experimental reproduction of these objects in the laboratory. Type-B CAIs are composed mostly of melilite, anorthite, aluminous spinel, and a titanium-rich pyroxene known in the parlance as fassaite (for a summary of minerals common to the Solar System see Appendix 1). The general consensus is that initially, the precursor to Type-B CAIs was composed of anhydrous minerals, mostly silicates, which formed centimeter-sized dust balls that were melted. These objects experienced peak melting temperatures of $\sim$1700 K, which is constrained by the behavior of the mineral melilite upon crystallization (Stolper 1982; Stolper & Paque 1986; Beckett *et al.* 2006). The duration of the heating must be less than a few tens of hours to maintain observed minor-element abundances in spinels (Connolly & Burnett 2003). Many of these objects were likely heated multiple times (Connolly *et al.* 2006), potentially after an epoch of alteration within the disk as free-floating wanderers, requiring that the heating mechanism was repeatedly experienced. The rate at which they radiated their energy away, or the cooling rate they experienced, was $0.5-50$ K hr^{-1}, with $\sim$2 K hr^{-1} being the best analog to natural materials (Stolper & Paque 1986; Connolly *et al.* 2006).

The pre- and post-formation thermal regimes for Type-B CAIs has not been constrained rigorously. As a general consensus within meteoritics, in order to maintain the fractionated chemistry of individual silicate phases in Type-B CAIs reheated after crystallization, these CAIs are thought not to have been heated to temperatures near their solidus for long periods of time. However, the exact temperature/time histories have not been defined. Unlike chondrules, which we discuss below, Type-B CAIs could have experienced preheating temperatures much higher than those experienced by most chondrules, but, like the post-crystallization regime, the preheating thermal history of Type-B CAIs and other refractory inclusions are essentially quantitatively unconstrained.

8.3.2 Chondrules

Chondrules display a wide variety of compositions and textural types (see Fig. 8.5). The volume fraction occupied by chondrules in chondritic meteorites ranges from 85% (ordinary chondrites) down to 0% (CI chondrites). Without question the most abundant type of chondrule is dominated by ferromagnesian (Fe, Mg-rich) silicates (Figs. 8.4 and 8.5; Lauretta *et al.* 2006). Ferromagnesian chondrules are primarily composed of olivine, pyroxene (minor Ca-rich pyroxene), glass, spinels, Fe, Ni-rich metal, FeS, and other minor phases ($<$ 1 vol%). These are subdivided into FeO-poor and FeO-rich, or Type-I and Type-II chondrules (Brearley & Jones 1998; Jones *et al.* 2000, and references therein; Lauretta *et al.* 2006, and references therein). The division is delineated by the Mg-number [or Mg# defined as $100 \times$ Mg/(Mg + Fe)] of the olivine and low-Ca pyroxene: $>$ 90 is FeO-poor, and $<$ 90 is FeO-rich, which roughly translates in a bulk difference of 10 wt% FeO: $<$ 10 wt% = FeO-poor and $>$10 wt% = FeO-rich.

Most research that has placed constraints on the thermal history of chondrules has been focused on ferromagnesian chondrules, with experimental reproduction of these objects providing the most powerful constraints (Hewins *et al.* 2005). Initially, chondrule precursors were an assembly of anhydrous minerals that formed millimeter-sized dust balls. These were then melted. The peak melting temperature experienced by chondrules is defined by the peak melting temperature experienced by one textural type of chondrules, barred olivine (Fig. 8.5) chondrules, ~ 2200 K, for minutes to seconds (Lofgren & Lanier 1990; Radomsky & Hewins 1990; Connolly & Hewins 1995). Longer times require lower temperatures and vice versa. The bulk composition of a chondrule could be significantly altered by heating if evaporation were an important process. The fact that these chondrules are not refractory-rich residues that derived their composition from less refractory materials constrains their overall heating duration to minutes or seconds.

Transient heating of individual chondrules occurred multiple times. The presence of easily identified relict grains within chondrules supports the hypothesis that chondrules went through an epoch of abrasion, and fragments of previous generations of chondrules were incorporated into later-formed chondrules. Furthermore, the observation of igneous rims around some chondrules provides additional, independent evidence that chondrules experienced multiple heating events varying in intensity.

The textures of chondrules combined with the fractionated elemental abundances within their phases place strong constraints on the cooling rates (Hewins *et al.* 2005). It is the porphyritic olivine chondrules (e.g. Fig. 8.5C) that define the rate at which energy is radiated from the molten spheres. Porphyritic chondrules compose up to 85 vol% of all chondrule textural types (Gooding & Keil 1981; Connolly & Desch 2004) the majority of which cooled between 5 and 100 $K hr^{-1}$, although rates up to 1000 $K hr^{-1}$ may have occurred for some chondrules. The presence of a glassy mesostasis within chondrules indicates that they were quenched or frozen from a moderately high temperature, either slightly above or below their solidus temperatures ($\sim 1200-1400$ K) and that they did not continuously cool down to lower temperatures and completely crystallize.

Rough constraints on the pre-melting conditions experienced by Fe,Mg-rich chondrules exist. These are largely based on the presence of moderately volatile elements such as Na and S within chondrules. Depending on the overall pressure, the presence of these elements, especially S, constrains pre-melting conditions to between ~ 650 and 1200 K for less than a few hours (Connolly *et al.* 2006). Furthermore, carbonaceous material was included in some chondrule melts and played an important role in controlling the oxidation state of the resulting mineralogy (Connolly *et al.* 1994) requiring low ambient temperatures ($\sim 580-780$ K) before melting. It is generally accepted that chondrules have been heated rapidly, rather than gradually, from relatively cool ambient temperatures to peak temperatures (from subsolidus to peak melting temperatures in seconds).

The post-crystallization regime for Fe,Mg-rich chondrules is essentially quantitatively unconstrained. However, the presence of glassy mesostasis and fractionated element abundances within chondrule phases indicates that most chondrules definitively did not experience prolonged heating after they were cooled and quenched.

Chondrules provide additional evidence relevant to dust processing within the disk. At least 5% of chondrules are compound objects, which form when two chondrules collide while at least one is still molten (Wasson _et al._ 1995). This abundance constrains the concentration of solid material, relative to gas, to a value at least 45 times over the canonical solar value (Ciesla 2006). Chondrules often have fine-grained rims of mineral dust surrounding them. Petrologic evidence shows that accretion of these fine-grained dust rims began while chondrules were still molten (Lauretta & Buseck 2003). These rims can be trapped between plastically deformed or compound chondrules, indicating that the chondrules were still hot when they collided. Chondrule rims are also the host of presolar material, suggesting that much of this material accreted after chondrules cooled back down to ambient nebular conditions (Huss & Lewis 1995; Alexander _et al._ 1998). The presence of these rims indicates that a significant quantity of micron-sized dust was present in chondrule formation regions and survived the transient heating events. In light of these constraints, we review the various models proposed for producing the transient heating events responsible for chondrule formation.

The second most abundant class of chondrules is the Al-rich chondrules (given many names over the last two decades), which are compositionally much more diverse than Fe,Mg-rich chondrules. Some of these may be intermediate between CAIs and FeO-poor chondrules (Connolly & Desch 2004; Tronche _et al._ 2007). Very few constraints exist on their thermal histories.

Other components of chondrites such as AOAs (>90[+] vol% forsterites), their cousins, and agglomeratic chondrules have all experienced some level of melting, albeit very minor in most cases (Weisberg _et al._ 2003; Zanda _et al._ 2005). The reasons why AOAs were not melted extensively like chondrules and Type-B CAIs is unknown and no quantitative constraints on their thermal histories exist.

8.3.3 Chondritic matrix

Refractory inclusions and chondrules are embedded in fine-grained (0.1 to 1 μm) silicate-rich dust that was present in the proto-solar nebula at the time of the accretion of the chondritic parent bodies ($\sim$2 Myr). This compacted dust, known as matrix, may offer the best comparison to the micron-sized dust grains studied in protoplanetary disks around other stars (see Chapters 6, 7, and Section 8.2). The matrix carries highly heterogeneous dust from different locations in the proto-solar nebula as well as very small amounts (< 100 ppm) of presolar grains (for a recent review, see Chapter 7 and Scott 2007). The fine-grained matrix is very sensitive to parent-body alterations, but some of the least-altered matrices in relatively pristine carbonaceous chondrites offer constraints on their formation and thermal history. These matrices are dominated by Mg-rich crystalline silicates and Fe-rich amorphous silicates, resembling the silicate dust composition derived for protoplanetary disks (see Section 8.2.6). The fact that the crystals are FeO-poor suggests condensation under equilibrium conditions rather than annealing. The crystalline grains show ortho- and clinopyroxene domains that intergrew, revealing peak temperatures > 1300 K and typical cooling rates of $\sim$1300 K hr^{-1}, broadly similar to the thermal processing of chondrules. The very small amount (< 100 ppm) of presolar grains that survived reprocessing in the nebula underlines the efficiency of the widespread heating events that transformed almost all dust in the inner Solar System.

8.3.4 Interplanetary dust particles

Porous, anhydrous IDPs are likely the best samples of nebular dust available for study. They have broadly chondritic compositions, high C concentrations (>10 wt%) and porosities exceeding 30%. These particles are largely composed of 200–400 nm grains, but crystals up to several microns are also common. The smallest particles form micron-sized aggregates that are combined with similar or even larger mineral grains (Rietmeijer *et al.* 2002). The most abundant crystalline silicates are nearly pure enstatite and forsterite, with lesser amounts of ferrous (Mg# < 50) olivines and pyroxenes present (Rietmeijer 1997; Bradley 2003). Enstatite crystals have distinctive platelet and whisker morphologies (Bradley *et al.* 1983). Amorphous silicates occur as 100–500 nm spheroids made of glass with embedded metals and sulfides (GEMS), which likely form by irradiation of crystals (Bradley & Dai 2004). Pyrrhotite is abundant and refractory minerals, like those found in CAIs in chondrites, also occur. The carbonaceous material consists of poorly graphitized carbon, aliphatic organics, and PAHs. This organic material is concentrated in discrete inclusions and also acts as a glue to cement the particles together (Flynn *et al.* 2003).

The mineralogy, structure, and chemistry of IDPs provide some constraints on nebular processes that modified the protoplanetary dust (Nuth *et al.* 2005). The occurrence of enstatite whiskers and platelets provides strong evidence for an origin as condensates. These crystals also contain structural intergrowths that are characteristic of rapid cooling at $\sim$ 1000 K hr^{-1} from 1300 K. The amorphous GEMS grains must have avoided significant thermal annealing in order to preserve their glassy texture. Annealing to crystallinity of amorphous magnesium silicates that comprise most GEMS requires temperatures in excess of 1100 K. Rarer amorphous GEMS require even higher temperatures ($\sim$1400 K). However, given the small size of these grains, heating to this temperature is more likely to result in complete evaporation and recondensation of the amorphous GEMS. The carbonaceous phases present in other IDPs have been used to imply low-temperature formation processes such as Fischer–Tropsch synthesis ($\sim$ 800 K, see Chapter 5 for details) or carbonization and graphitization ($\sim$ 590 K).

8.4 Heating mechanisms

As discussed above, a large majority of primitive planetary materials have been processed by transient heating events. These materials represent the best analogs of the dust that are observed in protoplanetary disks. Figure 8.6 illustrates the plausible range of key parameters of the heating events (peak temperatures and cooling rate) for both Solar System material and for crystalline silicates. Figure 8.7 shows the disk regions with evidence for thermal processing in disks of the Sun and other stars. Transient heating events are central in establishing the chemistry, mineralogy, and grain sizes of these samples. The majority of solid material in the inner Solar System experienced transient high-temperature events prior to accretion into planetesimals. However, what mechanism or mechanisms produced the transient heating is still unknown. Over the last 100 years a plethora of potential mechanisms have been hypothesized. We review the more widely accepted hypotheses that also relate to the goals of this book. Astronomical observations designed to characterize the nature and source of transient heating events in protoplanetary disks are desperately needed to resolve this long-standing controversy.

8.4.1 Shock waves

Flash heating by a shock front is the leading explanation of chondrule formation and it is the most quantitative model to date (Ciesla & Hood 2002; Desch & Connolly 2002; Desch *et al.* 2005). The interactions with the particles cause the gas to achieve higher temperatures and pressures both upstream and downstream of the shock than would be reached otherwise. Chondrules are kept warm by the reservoir of hot shocked gas. The cooling rate of this reservoir depends on how fast dust grains and chondrules radiate away the energy contained within the gas. The calculated cooling rates of the particles agree with the cooling rates inferred for chondrules. These models may also explain the formation of crystalline grains in the cooler outer disks (Harker & Desch 2002).

A variant on the shock wave model is the surface melting and ablation of small planetesimals in nebula shock waves (Genge 2000). Bodies between $\sim$1 mm and

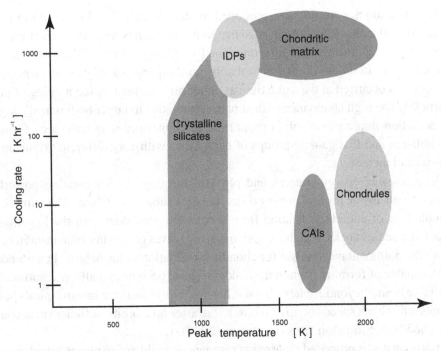

Figure 8.6 Illustration of the typical peak temperatures and cooling rates for high-temperature products. For details on the individual group of objects, see the corresponding sections in the text.

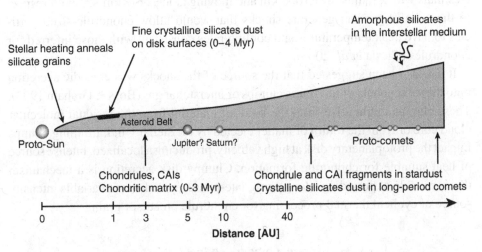

Figure 8.7 Disk regions with evidence for thermal processing. For comparison, the key structures expected for the young Solar System are drawn. Note that for small micron-sized grains the figure shows the radius where they are detected or can be traced back, but due to the intense mixing in the disks these grains likely originate from much broader regions. For details on the individual group of objects, see the corresponding sections in the text.

500 m in diameter would have produced molten droplets by ablation during gas drag in nebular shocks. It is suggested that most chondrules formed via ablation will have broadly chondritic compositions. The formation of chondrules by ablation of planetesimals in shock waves was probably most important at a late stage in nebular history and occurred at the same time as chondrules formed by the melting of dust particles. The high abundance of dust particles relative to larger bodies at all stages of accretion implies that only a proportion of chondrules may have been formed by ablation and that genetic groups of chondrules with very different origins may co-exist in meteorites.

The possible energy sources and physical mechanisms for creating powerful shock fronts inside protoplanetary disks remain a subject of debate. The numerical simulations of chondrule heating from shocks are consistent with the hypothesis that large-scale shocks, e.g. those due to density waves or gravitational instabilities, were the dominant mechanism for chondrule formation in the nebula. In a gaseous disk capable of forming Jupiter, the disk must have been marginally gravitationally unstable at and beyond Jupiter's orbit. This instability can drive inward spiral shock fronts with shock speeds of up to $10 \, \text{km s}^{-1}$ at asteroidal orbits, sufficient to account for chondrule formation (Boss & Durisen 2005).

Early on it was proposed that Jovian resonances could excite planetesimal eccentricities enough to cause collisional disruption and melting of dust by bow-shocks in the nebular gas (Weidenschilling *et al.* 1998). Follow-on work showed that planetesimals with a radius of ~ 1000 km and moving at least $\sim 8 \, \text{km s}^{-1}$ with respect to the nebular gas can generate shocks that would allow chondrule-sized particles to have peak temperatures and cooling rates consistent with those inferred for chondrules (Ciesla *et al.* 2004).

It has also been suggested that the source of the shocks was episodic accretion onto the solar nebula of low-mass clumps of interstellar gas (Boss & Graham 1993). These clumps might arise from the infall of material from the residual molecular cloud core or from the return of matter ejected by the stellar wind. Infalling clumps impact the protoplanetary disk at high velocity, producing a localized, intense source of heat suitable for chondrule formation. Clumpy disk accretion is a mechanism that is capable of providing a long-lived, late-phase, episodic, and variable intensity source of cyclical thermal processing for chondrule precursor grains.

8.4.2 X-Ray flares

Observations of the Orion nebular cluster by the Chandra X-ray observatory reveal the presence of frequent (one every ~ 6 days), large (up to 0.5 AU), and highly energetic stellar flares (with inferred magnetic fields as large as 3000 G) (Favata *et al.* 2005). The duration of the flares varies from less than an hour to almost three days. Such powerful flares likely have important effects on the solar nebular disk material (Feigelson & Montmerle 1999). Magneto-hydrodynamic simulations that follow an X-ray flare event generate shock waves in the upper region of circumstellar disks. Shock speed and gas density vary over a wide range with some shock waves suitable for chondrule formation (Nakamoto *et al.* 2005). In late stages of solar nebular evolution, the surface density is expected to decrease and dust grains are expected to settle gravitationally to the mid-plane. If the local dust-to-gas mass ratio near the mid-plane increases above cosmic abundances by factors of 10^3, current sheets reach high enough temperatures to melt millimeter-sized dust grains and may provide the mechanism to form chondrules (Joung *et al.* 2004).

8.4.3 X-Wind

The X-wind model is another chondrule-formation mechanism that is tied to the early, active Sun (Shu *et al.* 1996, 1997, 2001). The X-wind is a theoretical construct in which coronal winds from a central star and the inner edge of its accompanying accretion disk join together to form magnetized jets of gas capable of launching material above the disk at velocities of hundreds of kilometers per second. In this model, chondrule and CAI precursors are transported through the protoplanetary disk to the X-wind region. There, they are irradiated by energetic flares and/or optical photons from the Sun, then flung outward to land back into the disk. In the X-wind model, grains are processed by radiation from the Sun as they are launched above the disk in bipolar outflows. These grains would have lost their volatiles upon being heated, then decoupled from the gas to rain back onto the solar nebula. As these grains fell onto the solar nebula in the comet-formation region, there would be mixing between two components: the solar composition materials that were already present and the more refractory crystalline grains. Thus, asteroids and comets that grew from these materials would be depleted in volatile elements, and those depletions would correlate with the amount of crystalline materials they contain. This prediction is contrary to the chondrule–matrix compositional complementarity that is characteristic of chondritic meteorites.

A major prediction of the X-wind model is that comets will contain CAIs. This prediction was seemingly vindicated when Stardust returned samples from Comet Wild-2 that contain refractory mineral grains indistinguishable from chondritic CAIs. The concept of X-wind production of CAIs, that are subsequently flung out into the protoplanetary nebula, does not face the same chemical objections as does the genesis of chondrules: CAIs do not have any complementarity with their host rocks. Hence, a strong argument may be made that they were formed in one place and then scattered amongst the chondrite-forming region. However, recently, CH chondrite-like fragments have also been discovered in Stardust samples: these must have been mixed outward from the inner disk, and the same mechanisms may be sufficient to explain the presence of the CAIs in Stardust (Nakamura *et al.* 2008; Schmitz *et al.* 2008; Weisberg & Connolly 2008).

8.4.4 Lightning

It has also been suggested that chondrule origin by lightning in the solar nebula is consistent with many features of chondrules (e.g. Pilipp *et al.* 1992, 1998; Desch & Cuzzi 2000). Electric fields strong enough to trigger breakdown easily could have existed over regions large enough ($\gg$100 km) to generate powerful discharges of electrical energy (10^{16} erg), assuming a lightning bolt width of <10 electron mean free paths. The discharges would have been sufficiently energetic to have formed the chondrules.

8.4.5 Impacts

A few models exist to produce chondrules by volcanism on planetesimals (Hutchison & Graham 1975; Kennedy *et al.* 1992) or by collisions between partially molten planetesimals (Hutchison *et al.* 1988). These mechanisms imply formation of large planetesimals that experienced partial melting and differentiation prior to chondrule formation. Such processes leave distinct geochemical and isotopic signatures, providing a direct test of these hypotheses. In fact, there are igneous fragments in primitive ordinary chondrites that contain chemical signatures consistent with partial melting of chondritic material (Hutchison & Graham 1975; Graham *et al.* 1976; Hutchison *et al.* 1988; Hutcheon & Hutchison 1989; Kennedy *et al.* 1992). Recent studies show that a small fraction of Type-I chondrules in CV chondrites may have inherited some precursor components, mainly forsteritic olivine and metal, from earlier generations of differentiated planetesimals (Libourel & Krot 2007; Chaussidon *et al.* 2008). However, such igneous fragments are relatively rare and they have never been shown to be the primary source of chondrule precursors. Furthermore, analysis of chondrule bulk compositions are not consistent with the major- or trace-element fractionations that are characteristic of partial melting and differentiation within some type of planetary body.

There is another way to generate chondrules via impacts: recondensation of chondritic material vaporized in large planetesimal impacts. One enigmatic group of chondrites may have formed in this manner. Lead–lead ages of chondrules in the metal-rich CB (Bencubbin-like) carbonaceous chondrites Gujba (4562.7 ± 0.5 Myr) and Hammadah al Hamra 237 (4562.8 ± 0.9 Myr) suggest that they formed during a single-stage, highly energetic event (Krot *et al.* 2005). Both the relatively young ages and the single-stage formation of the CB chondrules are inconsistent with formation during a nebular shock wave. Chondrules and metal grains in the CB chondrites may have formed from a vapor-melt plume produced by a giant impact between planetary embryos after dust in the protoplanetary disk had largely dissipated.

8.4.6 Outward transport

The Stardust samples collected from Comet 81P/Wild 2 contain grains that are chemically, mineralogically, and isotopically similar to the refractory inclusions and chondrules in primitive chondrites (Zolensky *et al.* 2006; Nakamura *et al.* 2008; Schmitz *et al.* 2008). This observation indicates that either CAI and chondrule formation was widespread in the very earliest stages of Solar System formation or that large-scale mixing occurred in the solar nebula, carrying materials from the hot inner regions to cooler environments far from the Sun. Either way, we can infer from these data that at least one cometary body may be more similar to chondrites than previously suspected (Gounelle *et al.* 2006; Weisberg & Connolly 2008) and that thermal processing of materials to high temperatures may have been a key process in forming different types of planetary bodies.

The ubiquity of high-temperature products in the outer Solar System and in the cold regions of protoplanetary disks is difficult to reconcile with the idea of a slowly evolving, initially amorphous disk. Recent simulations suggest a more tubulent picture: during the collapse of the molecular-cloud core a large fraction of the submicron dust mass lands at small disk radii and is heated to high temperatures (>1000 K, Bockelée-Morvan *et al.* 2002). Some of these heated grains are subsequently mixed outward ($10-20$ AU), leading to a very high fraction of thermally processed dust throughout the disk, also influenced by the angular momentum of the natal cloud core (Dullemond *et al.* 2006). With continuing accretion more unprocessed material falls onto the disk, possibly lowering the relative fraction of high-temperature products.

Recent results from a new two-dimensional model show that this outward transport of high-temperature materials around the mid-plane of protoplanetary disks is also efficient for grains as large as CAIs (Ciesla 2007). That outward transport occurs most efficiently around the mid-plane of the disk suggests that the fraction of crystalline grains in a disk will correlate with the amount of settling that occurs, which increases as solids grow. These correlations are indeed observed because those disks that are inferred to have large crystallinity fractions based on the observed spectra are best fitted by models in which the solid subdisk is flat, rather than flared (e.g Apai *et al.* 2005; Watson *et al.* 2009). These flows could also preserve 0.1- to 1-cm CAIs in the solar nebula for the millions of years between their formation and their incorporation into meteorite parent bodies. An important consequence of these models is that the thermally processed grains remained in contact with the nebular gas throughout their transport, allowing volatiles to condense and dust to accrete on their surfaces in cooler environments.

8.5 How would Solar System formation look to an outside observer

From the view of meteoritics, we can make certain predictions of what an "ideal" observer could see in the protoplanetary disk during the first few million years. We can apply this predictive model to other disks. Clearly, dust was present before CAIs were formed. This dust would be refractory and be similar, if not identical, to minerals in fluffy Type-A CAIs. We would expect to see this type of mineral dust during the very earliest stages of the disk formation and through at least the first 100 000 years of the disk; some of these CAI dust grains may have been still preserved in the matrix of primitve chondites (Nittler *et al.* 1994; Choi *et al.* 1999; Scott 2007). Such minerals as olivine might be observed spectrally, but the majority of material should be such minerals as aluminous spinels, melilite, fassaite, anorthite, and potentially minor amounts of hibonite, perovskite, etc. It is possible that CAI precursors initially condensed as liquids (Yoneda & Grossman 1995), however, once melting started we would observe crystalline material. The region where we observe the refractory dust is likely limited to the inner disk, although we cannot rule out other areas or that we would observe this dust globally.

This hot phase of the forming planetary system could have occurred during or immediately after the protostellar birth ($< 10^5$ yr) and the system would still be deeply embedded in a massive dusty envelope, similarly to protostars observed in the Taurus star-forming region. Owing to the high dust extinction inner disks are very difficult to observe at this stage. If the CAI formation occurs at 1 Myr or beyond, the characteristic infrared spectrum of CAI components may be observable (Posch *et al.* 2007). The intensely accreting turbulent disk will carry most of the refractory CAI grains into the star and the surviving population, if any, may show a characteristic size distribution influenced by the efficiency of gas-to-dust coupling.

Soon after this initial stage of CAI dust we would begin to observe a decrease in the abundance of refractory dust, perhaps as a result of dust processing and CAI formation. The region(s) of CAI formation might be 10–1000 km in scale and we should observe an increase in luminosity of this region (or regions) for several hours with decreases occurring over several days (Cuzzi & Alexander 2006). Additional refractory material might condense during this time and we should initially note a decrease in refractory dust as it is processed into CAIs, then perhaps a slight increase in the abundances.

By approximately 1–3 Myr, the accretion rate and disk mass would have decayed substantially, leading to a less turbulent, cooler, and lower-mass disk, possibly with enhanced dust-to-gas ratio; dust grains would settle toward the disk mid-plane. The observed materials would also change, from refractory to the most common planetary material dust, such as olivine and pyroxene. These would probably be MgO-rich, consistent with existing observations of crystalline dust in protoplanetary disks of this age (for a review see Wooden *et al.* 2007). By this stage, very little to no pristine grains (small, amorphous, FeO-rich silicates) would be present in the disk. This lack of unprocessed dust is in contrast with the best astronomical analogs, the 1–4 Myr-old T Tauri disks, perhaps suggesting that these may not be as good young Solar System analogs as widely believed.

We know with confidence that the Solar System is much richer in less-refractory material than it is in refractory materials. In the subsequent stage a fraction of the refractory dust was processed into chondrules. The number and relative mass of chondrules is far greater than CAIs, although the size of the regions where they are processed is about equal to that of CAIs. Therefore we would see an increase in the local luminosity at the sites of the heating events, with an appreciable fraction of the dust being sequestered into chondrules. If shock waves driven by a massive planet were present, the planet would deform the disk structure and possibly open an observable gap.

Large-scale mixing would be evident from the earliest, hot phases of the nebula throughout the first million years: the nebular gas would be thoroughly mixed. The viscous evolution of the disk and large-scale convective flows would carry millimeter-sized or even larger grains from the inner disk to the outer edges of the disk. By the time chondrule formation concluded most of the fine dust would have been lost from the region corresponding to today's Asteroid Belt: this timescale (2–3 Myr) is very similar to the half-life of dusty disks observed around young stars, suggesting that dust loss is global. Over a similar, but poorly constrained period the gas would have been removed from the disk (5–10 Myr), ending the window for giant planet formation. No evidence exists for nebular thermal processing after the chondrule formation concluded, consistent with the observed lack of fine dust and gas around young stars. Beyond 5–10 Myr parent-body processes and collisional dust recycling would dominate the disk evolution and the planet-formation processes.

8.6 Promising future experiments

Piecing together the observations of thermal processing in protoplanetary disks reveals several promising avenues for future studies. We identify the following topics as those most needed to understand the thermal processing of protoplanetary materials:

- Large, systematic infrared spectroscopic studies of protoplanetary disks at wavelengths between 7 and 60 μm to test correlations between crystallinity and independent disk parameters
- Photometric surveys to search for and constrain the frequency of transient heating events in protoplanetary disks
- Compositional studies of the youngest observable protostars: is there evidence for CAI-like material?
- Temporal variability of the crystalline silicate features
- Search for the best analogs for the astronomically observed crystalline dust: mid-infrared laboratory spectroscopy of chondritic matrices is needed
- More quantitative and detailed models for heating by lightning, X-rays, ablation, and X-wind scenarios that provide testable predictions
- Turbulent mixing and dust disk evolution models for a range of stellar and disk masses: correlations are much easier to observe and more difficult to fit than simulations restricted to the minimum-mass solar nebula
- What fraction of the primitive (silicate) dust left over from the early Solar System can be unprocessed, i.e. ISM-like? Is there evidence for a crystallinity–age correlation in the chondritic matrices?
- Sample returns from additional comets, outer main belt and Trojan asteroids, and Kuiper Belt objects representing distinct regions of the early Solar System.

References

Alexander, C. M. O., Russell, S. S., Arden, J. W., *et al.* 1998, *Meteoritics and Planetary Science*, **33**, 603.

Apai, D., Luhman, K., & Liu, M. C. 2008, *Astronomical Society of the Pacific Conference Series*, **384**, 383.

Apai, D., Pascucci, I., Bouwman, J., *et al.* 2005, *Science*, **310**, 834.

Armitage, P. J., Livio, M., & Pringle, J. E. 2001, *Monthly Notices of the Royal Astronomical Society*, **324**, 705.

Beckett, J. R., Connolly, H. C., & Ebel, D. S. 2006, in *Meteorites and the Early Solar System II*, ed. D. S. Lauretta & H. Y. McSween, University of Arizona Press, 399–429.

Beckett, J. R., Simon, S. B., & Stolper, E. 2000, *Geochimica et Cosmochimica Acta*, **64**, 2519.

Bell, K. R., Cassen, P. M., Wasson, J. T., & Woolum, D. S. 2000, in *Protostars and Planets IV*, ed. V. Mannings, A. Boss, & S. S. Russell, University of Arizona Press, 897–926.

Bell, K. R. & Lin, D. N. C. 1994, *Astrophysical Journal*, **427**, 987.

Bizzarro, M., Ulfbeck, D., Trinquier, A., *et al.* 2007, *Science*, **316**, 1178.

Bockelée-Morvan, D., Gautier, D., Hersant, F., Huré, J.-M., & Robert, F. 2002, *Astronomy and Astrophysics*, **384**, 1107.

Boss, A. P. & Durisen, R. H. 2005, *Astrophysical Journal*, **621**, L137.

Boss, A. P. & Graham, J. A. 1993, *Icarus*, **106**, 168.

Bouwman, J., Henning, T., Hillenbrand, L. A., *et al.* 2008, *Astrophysical Journal*, **683**, 479.

Bouy, H., Huelamo, N., Pinte, C., *et al.* 2008, *Astronomy & Astrophysics*, **486**, 877.

Bradley, J. 2003, in *Astromineralogy*, ed. T. K. Henning, Lecture Notes in Physics, vol. 609, Springer 217–235.

Bradley, J. P., Brownlee, D. E., & Veblen, D. R. 1983, *Nature*, **301**, 473.

Bradley, J. P. & Dai, Z. R. 2004, *Astrophysical Journal*, **617**, 650.

Brearley, A. J. & Jones, R. 1998, *Reviews in Mineralogy*, **36**, 3.

Cameron, A. G. W. 1963, *Icarus*, **1**, 339.

Carrez, P., Demyk, K., Leroux, H., *et al.* 2002, *Meteoritics and Planetary Science*, **37**, 1615.

Chaussidon, M., Libourel, G., & Krot, A. N. 2008, *Geochimica et Cosmochimica Acta*, **72**, 1924.

Choi, B.-G., Wasserburg, G. J., & Huss, G. R. 1999, *Astrophysical Journal*, **522**, L133.

Ciesla, F. J. 2006, *Meteoritics and Planetary Science*, **41**, 1271.

Ciesla, F. J. 2007, *Science*, **318**, 613.

Ciesla, F. J. & Hood, L. L. 2002, *Icarus*, **158**, 281.

Ciesla, F. J., Hood, L. L., & Weidenschilling, S. J. 2004, *Meteoritics and Planetary Science*, **39**, 1809.

Cohen, M. & Witteborn, F. C. 1985, *Astrophysical Journal*, **294**, 345.

Connolly, H. C. & Burnett, D. S. 2003, *Geochimica et Cosmochimica Acta*, **67**, 4429.

Connolly, H. C., Hewins, R. H., Ash, R. D., *et al.* 1994, *Nature*, **371**, 136.

Connolly, H. C., Jr. 2005, *Astronomical Society of the Pacific Conference Series*, **341**, 215.

Connolly, H. C., Jr. & Desch, S. J. 2004, *Chemie der Erde / Geochemistry*, **64**, 95.

Connolly, H. C., Jr., Desch, S. J., Ash, R. D., & Jones, R. H. 2006, in *Meteorites and the Early Solar System II*, ed. D. S. Lauretta and H. Y. McSween, University of Arizona Press, 383–397.

Connolly, H. C. Jr. & Hewins, R. H. 1995, *Geochimica et Cosmochimica Acta*, **59**, 3231.

Crovisier, J., Leech, K., Bockelée-Morvan, D., *et al.* 1997, *Science*, **275**, 1904.

Cuzzi, J. N. & Alexander, C. M. O. 2006, *Nature*, **441**, 483.

Davis, A. M. 2006, in *Meteorites and the Early Solar System II*, ed. D. S. Lauretta & H. Y. McSween, University of Arizona Press, 295–307.

Davoisne, C., Djouadi, Z., Leroux, H., *et al.* 2006, *Astronomy and Astrophysics*, **448**, L1.

Desch, S. J., Ciesla, F. J., Hood, L. L., & Nakamoto, T. 2005, *Astronomical Society of the Pacific Conference Series*, **341**, 849.

Desch, S. J. & Connolly, H. C. Jr. 2002, *Meteoritics and Planetary Science*, **37**, 183.

Desch, S. J. & Cuzzi, J. N. 2000, *Icarus*, **143**, 87.

Dullemond, C. P., Apai, D., & Walch, S. 2006, *Astrophysical Journal*, **640**, L67.

Ebel, D. S. 2006, in *Meteorites and the Early Solar System II*, ed. D. S. Lauretta & H. Y. McSween, University of Arizona Press, 253–277.

Ebel, D. S., Brunner, C. E., & Weisberg, M. K. 2008, in *Lunar and Planetary Science Conference Abstracts*, **39**, 2121.

Espaillat, C., Calvet, N., D'Alessio, P., *et al.* 2007a, *Astrophysical Journal*, **664**, L111.

Espaillat, C., Calvet, N., D'Alessio, P., *et al.* 2007b, *Astrophysical Journal*, **670**, L135.

Favata, F., Flaccomio, E., Reale, F., *et al.* 2005, *Astrophysical Journal Supplement*, **160**, 469.

Feigelson, E. D. & Montmerle, T. 1999, *Annual Review of Astronomy and Astrophysics*, **37**, 363.

Flynn, G. J., Keller, L. P., Feser, M., Wirick, S., & Jacobsen, C. 2003, *Geochimica et Cosmochimica Acta*, **67**, 4791.

Gail, H.-P. 2004, *Astronomy and Astrophysics*, **413**, 571.

Genge, M. J. 2000, *Meteoritics and Planetary Science*, **35**, 1143.

Gooding, J. L. & Keil, K. 1981, *Meteoritics,* **16**, 17.

Gounelle, M., Spurný, P., & Bland, P. A. 2006, *Meteoritics and Planetary Science,* **41**, 135.

Graham, A. L., Easton, A. J., Hutchison, R., & Jerome, D. Y. 1976, *Geochimica et Cosmochimica Acta,* **40**, 529.

Green, J. D., Hartmann, L., Calvet, N., *et al.* 2006, *Astrophysical Journal,* **648**, 1099.

Greene, T. P., Aspin, C., & Reipurth, B. 2008, *Astronomical Journal,* **135**, 1421.

Grossman, J. N., Rubin, A. E., Nagahara, H., & King, E. A. 1988, *Meteorites and the Early Solar System,* ed. J. F. Kerridge & M. Shapley Matthews, University of Arizona Press, 619–659.

Grossman, L. 1972, *Geochimica et Cosmochimica Acta,* **36**, 597.

Hanner, M. S., Brooke, T. Y., & Tokunaga, A. T. 1995, *Astrophysical Journal,* **438**, 250.

Harker, D. E. & Desch, S. J. 2002, *Astrophysical Journal,* **565**, L109.

Harker, D. E., Woodward, C. E., & Wooden, D. H. 2005, *Science,* **310**, 278.

Hartmann, L. & Kenyon, S. J. 1996, *Annual Review of Astronomy and Astrophysics,* **34**, 207.

Hewins, R. H., Connolly, Lofgren, G. E., Jr., & Libourel, G. 2005, *Astronomical Society of the Pacific Conference Series,* **341**, 286.

Hezel, D. C., Russell, S. S., Ross, A. J., & Kearsley, A. T. 2008, *Meteoritics and Planetary Science,* **43**, 1879.

Honda, M., Kataza, H., Okamoto, Y. K., *et al.* 2006, *Astrophysical Journal,* **646**, 1024.

Huss, G. R. & Lewis, R. S. 1995, *Geochimica et Cosmochimica Acta,* **59**, 115.

Hutcheon, I. D. & Hutchison, R. 1989, *Nature,* **337**, 238.

Hutchison, R., Alexander, C. M. O., & Barber, D. J. 1988, *Royal Society of London Philosophical Transactions Series A,* **325**, 445.

Hutchison, R. & Graham, A. L. 1975, *Nature,* **255**, 471.

Ireland, M. J. & Kraus, A. L. 2008, *Astrophysical Journal,* **678**, L59.

Jones, R. H. & Danielson, L. R. 1997, *Meteoritics and Planetary Science,* **32**, 753.

Jones, R. H., Lee, T., Connolly, Jr., H. C., Love, S. G., & Shang, H. 2000, *Protostars and Planets IV,* ed. V. Mannings, A. Boss, & S. S. Russell, University of Arizona Press, 927.

Joung, M. K. R., Mac Low, M.-M., & Ebel, D. S. 2004, *Astrophysical Journal,* **606**, 532.

Kaito, C., Miyazaki, Y., Kumamoto, A., & Kimura, Y. 2007, *Astrophysical Journal,* **666**, L57.

Kemper, F., Vriend, W. J., & Tielens, A. G. G. M. 2005, *Astrophysical Journal,* **633**, 534.

Kennedy, A. K., Hutchison, R., Hutcheon, I. D., & Agrell, S. O. 1992, *Earth and Planetary Science Letters,* **113**, 191.

Kessler-Silacci, J. E., Dullemond, C. P., Augereau, J.-C., *et al.* 2007, *Astrophysical Journal,* **659**, 680.

Kimura, Y., Miyazaki, Y., Kumamoto, A., Saito, M., & Kaito, C. 2008, *Astrophysical Journal,* **680**, L89.

Kóspál, Á., Ábrahám, P., Apai, D., *et al.* 2008, *Monthly Notices of the Royal Astronomical Society,* **383**, 1015.

Krot, A. N., Amelin, Y., Cassen, P., & Meibom, A. 2005, *Nature,* **436**, 989.

Lauretta, D. S. & Buseck, P. R. 2003, *Meteoritics and Planetary Science,* **38**, 59.

Lauretta, D. S., Nagahara, H., & Alexander, C. M. O. 2006, *Meteorites and the Early Solar System II,* ed. D. S. Lauretta & H. Y. McSween, University of Arizona Press, 431–459.

Lenzuni, P., Gail, H.-P., & Henning, T. 1995, *Astrophysical Journal,* **447**, 848.

Leroux, H., Libourel, G., Lemelle, L., & Guyot, F. 2003, *Meteoritics and Planetary Science,* **38**, 81.

Libourel, G. & Krot, A. N. 2007, *Earth and Planetary Science Letters*, **254**, 1.

Lisse, C. M., VanCleve, J., Adams, A. C., *et al.* 2006, *Science*, **313**, 635.

Lofgren, G. & Lanier, A. B. 1990, *Geochimica et Cosmochimica Acta*, **54**, 3537.

MacPherson, G. J., Wark, D. A., & Armstrong, J. T. 1988, *Meteorites and the Early Solar System*, ed. J. F. Kerridge & M. Shapley Matthews, University of Arizona Press, 746–807.

Malfait, K., Waelkens, C., Waters, L. B. F. M., *et al.* 1998, *Astronomy and Astrophysics*, **332**, L25.

Merín, B., Augereau, J.-C., van Dishoeck, E. F., *et al.* 2007, *Astrophysical Journal*, **661**, 361.

Molster, F. J., Yamamura, I., Waters, L. B. F. M., *et al.* 1999, *Nature*, **401**, 563.

Najita, J. R., Strom, S. E., & Muzerolle, J. 2007, *Monthly Notices of the Royal Astronomical Society*, **378**, 369.

Nakamoto, T., Kita, N. T., & Tachiban, S. 2005, *Antarctic Meteorite Research*, **18**, 253.

Nakamura, T., Noguchi, T., Tsuchiyama, A., *et al.* 2008, *Lunar and Planetary Science Conference Abstracts*, **39**, 1695.

Nittler, L. R., Alexander, C. M. O., Gao, X., Walker, R. M., & Zinner, E. K. 1994, *Nature*, **370**, 443.

Nuth, J. A., III., Brearley, A. J., & Scott, E. R. D. 2005, in *Astronomical Society of the Pacific Conference Series*, **341**, 675.

Nuth, J. A., III & Donn, B. 1982, *Astrophysical Journal*, **257**, L103.

Pascucci, I., Apai, D., Hardegree-Ullman, E. E., *et al.* 2008, *Astrophysical Journal*, **673**, 477.

Pilipp, W., Hartquist, T. W., & Morfill, G. E. 1992, *Astrophysical Journal*, **387**, 364.

Pilipp, W., Hartquist, T. W., Morfill, G. E., & Levy, E. H. 1998, *Astronomy and Astrophysics*, **331**, 121.

Posch, T., Mutschke, H., Trieloff, M., & Henning, T. 2007, *Astrophysical Journal*, **656**, 615.

Quanz, S. P., Apai, D., & Henning, T. 2007, *Astrophysical Journal*, **656**, 287.

Quanz, S. P., Henning, T., Bouwman, J., Ratzka, T., & Leinert, C. 2006, *Astrophysical Journal*, **648**, 472.

Radomsky, P. M. & Hewins, R. H. 1990, *Geochimica et Cosmochimica Acta*, **54**, 3475.

Ratzka, T., Leinert, C., Henning, T., *et al.* 2007, *Astronomy and Astrophysics*, **471**, 173.

Reipurth, B. & Aspin, C. 2004, *Astrophysical Journal*, **608**, L65.

Rietmeijer, F. J. M. 1997, *Lunar and Planetary Science Conference Abstracts*, **28**, 1173.

Rietmeijer, F. J. M., Hallenbeck, S. L., Nuth, J. A., III, & Karner, J. M. 2002, *Icarus*, **156**, 269.

Russell, S. S., Huss, G. R., Fahey, A. J., *et al.* 1998, *Geochimica et Cosmochimica Acta*, **62**, 689.

Rydgren, A. E., Strom, S. E., & Strom, K. M. 1976, *Astrophysical Journal Supplement*, **30**, 307.

Sargent, B., Forrest, W. J., D'Alessio, P., *et al.* 2006, *Astrophysical Journal*, **645**, 395.

Sargent, B. A., Forrest, W. J., Tayrien, C., *et al.* 2009, *Astrophysical Journal*, **690**, 1193.

Schmitz, S., Brenker, F. E., Vincze, L., *et al.* 2008, *Lunar and Planetary Science Conference Abstracts*, **39**, 1137.

Scott, E. R. D. 2007, *Annual Review of Earth and Planetary Sciences*, **35**, 577.

Shu, F. H., Shang, H., Glassgold, A. E., & Lee, T. 1997, *Science*, **277**, 1475.

Shu, F. H., Shang, H., Gounelle, M., Glassgold, A. E., & Lee, T. 2001, *Astrophysical Journal*, **548**, 1029.

Shu, F. H., Shang, H., & Lee, T. 1996, *Science*, **271**, 1545.

Stolper, E. 1982, *Geochimica et Cosmochimica Acta*, **46**, 2159.
Stolper, E. & Paque, J. M. 1986, *Geochimica et Cosmochimica Acta*, **50**, 1785.
Strom, K. M., Strom, S. E., Edwards, S., Cabrit, S., & Skrutskie, M. F. 1989, *Astronomical Journal*, **97**, 1451.
Tronche, E. J., Hewins, R. H., & MacPherson, G. J. 2007, *Geochimica et Cosmochimica Acta*, **71**, 3361.
van Boekel, R., Waters, L. B. F. M., Dominik, C., *et al.* 2003, *Astronomy and Astrophysics*, **400**, L21.
van Boekel, R., Min, M., Waters, L. B. F. M., *et al.* 2005, *Astronomy and Astrophysics*, **437**, 189.
Wadhwa, M., Amelin, Y., Davis, A. M., *et al.* 2007, in *Protostars and Planets V*, ed. B. Reipurth, D. Jewitt, & K. Keil, University of Arizona Press, 835–848.
Wasson, J. T., Krot, A. N., Min, S. L., & Rubin, A. E. 1995, *Geochimica et Cosmochimica Acta*, **59**, 1847.
Watson, D. M., Leisenring, J. M., Furlan, E., *et al.* 2009, *Astrophysical Journal Supplement*, **180**, 84.
Wehrstedt, M. & Gail, H.-P. 2008, ArXiv e-prints, 0804.3377.
Weidenschilling, S. J., Marzari, F., & Hood, L. L. 1998, *Science*, **279**, 681.
Weisberg, M. K. & Connolly, H. C. 2008, *Lunar and Planetary Science Conference Abstracts*, **39**, 1981.
Weisberg, M. K., Connolly, Jr., H. C., & Ebel, D. S. 2003, *Lunar and Planetary Science Conference Abstracts*, **34**, 1513.
Weisberg, M. K., McCoy, T. J., & Krot, A. N. 2006, in *Meteorites and the Early Solar System II*, ed. D. S. Lauretta and H. Y. McSween, University of Arizona Press, 19–52.
Wooden, D. H., Desch, S., Harker, D., Gail, H.-P., & Keller, L. 2007, in *Protostars and Planets V*, ed. B. Reipurth, D. Jewitt, & K. Keil, University of Arizona Press, 815–833.
Wooden, D. H., Woodward, C. E., & Harker, D. E. 2004, *Astrophysical Journal*, **612**, L77.
Yoneda, S. & Grossman, L. 1995, *Geochimica et Cosmochimica Acta*, **59**, 3413.
Zanda, B., Hewins, R. H., Bourot-Denise, M., Bland, P. A., & Albarède, F. 2005, *Meteoritics & Planetary Science, Supplement*, **40**, 5216.
Zhu, Z., Hartmann, L., Calvet, N., *et al.* 2007, *Astrophysical Journal*, **669**, 483.
Zolensky, M. E., Zega, T. J., Yano, H., *et al.* 2006, *Science*, **314**, 1735.

9

The clearing of protoplanetary disks and of the proto-solar nebula

Ilaria Pascucci and Shogo Tachibana

Abstract Circumstellar disks are a natural outcome of the star-formation process and the sites where planets form. Gas, mainly hydrogen and helium, accounts for about 99% of the disk's initial mass while dust, in the form of submicron-sized grains, only for about 1%. In the process of forming planets circumstellar disks disperse: submicron dust grains collide and stick together to form larger aggregates; gas accretes onto the star, onto the cores of giant and icy planets, and evaporates from the disk surface. A key question in planet formation is the timescale and physical mechanism for the clearing of protoplanetary disks. How rapidly gas and dust disperse determines what type of planets can form.

In this chapter we compare the evolution of protoplanetary disks to that of the proto-solar nebula. We start by summarizing the observational constraints on the lifetime of protoplanetary disks and discuss four major disk-dispersal mechanisms. Then, we seek constraints on the clearing of gas and dust in the proto-solar nebula from the properties of meteorites, asteroids, and planets. Finally, we try to anchor the evolution of protoplanetary disks to the Solar System chronology and discuss what observations and experiments are needed to understand how common is the history of the Solar System.

9.1 The observed lifetime of protoplanetary disks

Observations at different wavelengths trace different disk regions (see e.g. Chapter 3). Therefore, determining when disks disperse requires multi-wavelength observations of disks around stars of different ages. The ages of young stars (younger than $\sim 100\,$Myr) are typically estimated by comparing their positions in the Hertzsprung–Russell diagram to predictions from pre-main-sequence evolutionary tracks. Systematic and random errors in these estimates have to do both with the accuracy of observationally determined quantities as well as the uncertainties in theoretical evolutionary tracks. A conservative estimate is that ages of individual stars are only accurate within factors of ~ 2–3 (Hillenbrand 2008). Relative ages of star-forming regions, associations, and clusters are better determined and can be used to explore evolutionary timescales for protoplanetary disks.

So far most of the literature has concentrated on the evolution of protoplanetary disks in nearby, low-density, low-mass star-forming regions like Taurus. However, recent studies on ^{60}Fe–^{60}Ni isotope systems in meteorites suggest that the Solar System formed nearby massive stars ($\geq 10 M_\odot$) likely in a dense cluster environment (Tachibana & Huss 2003; Mostefaoui *et al.* 2005; Tachibana *et al.* 2006). This finding motivated recent studies to explore how the lifetime of protoplanetary disks is affected by the presence of nearby high-mass stars. Surveys of protoplanetary disks in high-mass star-forming regions are summarized at the end of Section 9.1.1, while theoretical expectations are presented in Section 9.2. As discussed in Section 9.1.2, detections of spectral lines from molecular gas remain sparse even from disks in nearby low-mass star-forming regions; hence little is known about the evolution of gaseous disks in more distant dense clusters.

9.1.1 Constraints on the dispersal of the dust component

Optically thick dust disks around young stars present a characteristic spectral energy distribution (SED) with significant excess emission relative to the photospheric flux from near-infrared to millimeter wavelengths (see Fig. 9.1). Such broad SEDs are well reproduced by disk models in which gas and small dust grains are well mixed and distributed from a few stellar radii out to hundreds of astronomical units (see also Chapter 3). As grains grow and settle to the disk mid-plane the overall shape of the SED is expected to change with excess emission at short wavelengths vanishing first (Fig. 6 from Dullemond & Dominik 2005). Disks with little near-infrared emission but large mid- and far-infrared emission, often called transition disks, are believed to be caught in the phase of dispersing their inner dust regions (Section 9.1.3). Recent observations with the Spitzer Space Telescope have also found many disks with reduced emission at all infrared wavelengths in comparison to primordial disks suggesting that transitional disks may not be the only pathway to an evolved disk (Section 9.1.3). Finally, debris disks, in which dust is replenished by collisions of planetesimals, have little excess emission starting at wavelengths longer than $\sim 10 \mu m$. This emission can be modeled well with dust grains confined to narrow belts (e.g. Wyatt 2008). Figure 9.1 shows examples of SEDs for a primodial, a transition, and a debris disk.

The above discussion provides the basis for using the infrared excess relative to the photospheric flux as a tool to detect primordial dust disks and determine the timescale over which they disperse. We should note that emission at infrared

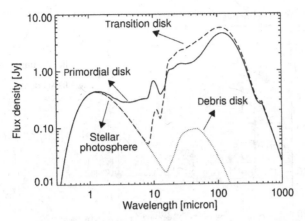

Figure 9.1 Examples of spectral energy distributions from young Sun-like stars with circumstellar dust disks. Optically thick dust disks (solid line) have excess emission relative to the stellar photosphere over a broad wavelength range, from near-infrared to millimeter wavelengths. Transition disks (dashed line) lack near-infrared excess emission, but have large mid- and far-infrared emission. Debris disks (dotted line) have small excess emission starting at wavelengths typically longer than 10 µm. Primordial and transition disks often show a prominent 10 µm silicate emission feature from warm dust grains in the disk atmosphere.

wavelengths is sensitive to the presence of small dust grains, not larger than a few microns in size. Therefore infrared observations effectively trace the dispersal of small dust grains in disks.

Already in the early 1990s it was recognized that more than 50% of the solar-type pre-main-sequence stars (T Tauri stars) have massive dust disks that disperse relatively fast (see e.g. Strom *et al.* 1993). Haisch *et al.* (2001b) expanded upon these earlier works by surveying hundreds of stars in six young clusters with ages between ∼ 0.3 and 30 Myr at J, H, K, and L bands (1.25, 1.65, 2.2, and 3.4 µm). They showed that the disk frequency is ≥ 80% in clusters younger than ∼ 1 Myr, in agreement with the high disk frequency in low-density star-forming regions like Taurus. They also find that the disk frequency sharply decreases with cluster age suggesting a disk lifetime (the time for all stars to lose their disks) of 6 Myr (see the dot–dashed line in Fig. 9.2).

The sensitivity of the Infrared Array Camera (IRAC) camera on board the Spitzer Space Telescope (Fazio *et al.* 2004) recently allowed to characterize in detail the decrease in disk frequency with stellar age and trace dust slightly cooler than that observed in the L-band, out to about 1 AU from T Tauri stars. Figure 9.2 shows the fraction of T Tauri stars (mostly K and M stars) with infrared excess at IRAC wavelengths (3.6, 4.5, 5.8, and 8 µm, full circles). In addition to the data (and references) presented in Hernández *et al.* (2008) we have included the disk statistics

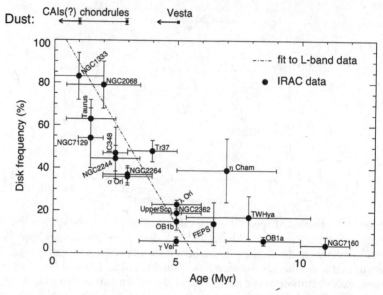

Figure 9.2 Fraction of stars with excess emission at IRAC wavelengths (between 3.6 and 8 μm) as a function of the age of the stellar group. In addition to the data presented in Hernández *et al.* (2008) and references therein, we have included the disk frequencies in the TW Hya association (Weinberger *et al.* 2004), and from the FEPS sample of Sun-like stars (Silverstone *et al.* 2006). The dot–dashed line is the least-squares fit to the L-band data from Haisch *et al.* (2001b). Above the plot we show a comparison to the formation timescale of CAIs, chondrules, and the asteroid Vesta in the Solar System. As we discuss in Section 9.4 there is evidence that CAIs formed early, in the first Myr of disk evolution.

in the TW Hya association from Weinberger *et al.* (2004), and the Formation and Evolution of Planetary Systems (FEPS) Spitzer Legacy survey sample from Silverstone *et al.* (2006). Such disk surveys generally include all stars, single stars as well as stars with one or more stellar companions. We shall see later that the disk fraction of single stars like the Sun is higher at young ages when binaries closer than < 20 AU are removed from the samples. Another remark is that the disk fractions in Fig. 9.2 include all the disk-bearing stars, those having optically thick emission from an almost primordial disk, as well as those with more evolved disks that may be closer to debris disks. As noted by Megeath *et al.* (2005) and Gautier *et al.* (2008) the η Chameleontis association is particularly rich in optically thick disks transitioning into optically thin ones. If transition disks preferentially evolve from massive disks (Section 9.1.3) then their higher frequency in η Chameleontis may be linked to a deficit of stellar companions at separation ≥ 30 AU (Brandeker *et al.* 2006) that could truncate the outer disk. Figure 9.2 shows that the decrease in disk fraction is less steep than that predicted from the L-band observations of Haisch *et al.* (2001b). This is in agreement with the theoretical expectation that disks in low-mass star-forming regions dissipate from inside out (Section 9.2). The IRAC observations in Fig. 9.2 show that the disk fraction is only a few percent at ~ 10 Myr suggesting that most young solar analogs clear out small dust grains inside ~ 1 AU over this timescale. Care should be taken in determining the dispersal of primordial dust from the decrease in disk frequency alone since both primordial and debris disks may be present in the age range ~ 5–10 Myr. A better approach might be the one proposed by Cieza *et al.* (2007) where both the slope of the excess emission in combination with the wavelength at which the infrared excess begins are taken into account. The difficulty in applying this approach to many stars in different clusters is that it requires observations sensitive to the stellar photosphere at several wavelengths.

Mid-infrared spectroscopy of solid-state features probes the population of dust grains in the disk atmosphere. Silicate emission features at ~ 10 and ~ 20 μm reveal that many young disks have grains 10 times larger than those observed in the interstellar medium, and a non-negligible (from a few to several percentage in mass fractions) contribution from crystalline grains (see Chapters 7 and 8). These observations demonstrate that substantial grain processing occurs very early, in the first million years of disk evolution. Unfortunately, so far no clear trend has been found between the extent of grain processing and the age of the star/disk system. Instead, solid-state features appear to be very diverse even in disks around stars of similar spectral type and age (Natta *et al.* 2007; Pascucci *et al.* 2008; Chapter 8).

The results described above likely apply not only to disks around single stars but also to disks in medium- and wide-separation binaries (separation ≥ 20 AU). Early investigations of T Tauri stars found no significant difference in the frequency of near- and mid-infrared excess emission between single and binary star systems (see, e.g. Mathieu *et al.* 2000). In addition, Pascucci *et al.* (2008) show that these disks also evolve in a similar way in the first few million years. Since two-thirds of the Sun-like stars in the solar neighborhood are members of multiple star systems and medium- and wide-separation binaries constitute more than half of these systems (Duquennoy & Mayor 1991), these results suggest that most Sun-like stars disperse their primordial dust within ~ 1 AU in less than ~ 10 Myr. However, the disk fraction of close-in binaries (separation < 20 AU) is found to be dramatically lower than that in medium- and wide-separation binaries: Kraus *et al.* (2008) reports that only 25% of the close-in binaries have a disk in Taurus in contrast to 75% of the wide-separation binaries and single stars. This suggests that most close-in solar-type stars remove their disks fast, in the first 1–2 Myr. Because about 20% of the Sun-like stars have a separation $< 20\,\text{AU}$, the disk fraction of single stars at young ages should be higher by $\sim 15\%$ than that presented in Fig. 9.2 (Kraus *et al.* 2008).

What is the evolution of the disk material outside 1 AU where giant planets may form? The combined sample of about 800 disks from the five star-forming regions observed by the "Cores 2 Disks" Spitzer Legacy Program (Evans *et al.* 2008) shows that all disks with infrared excess at IRAC bands also have excess emission at 24 μm with the Multiband Imaging Photometer for Spitzer (MIPS) camera (Merín *et al.* 2008). This indicates that the disk dispersal timescale out to $\sim 5\,\text{AU}$ around Sun-like stars is similar to that at $\sim 1\,\text{AU}$. It is worth noting that Spitzer surveys of star-forming regions and associations at 24 μm with the MIPS camera (Rieke *et al.* 2004) are typically flux-limited, hence cannot provide independent statistics on the disk frequency. An exception is the survey of 314 Sun-like stars carried out by the FEPS (Meyer *et al.* 2006). They find excess fractions of 5/30 and of 9/48 for stars 3–10 Myr and 10–30 Myr old, respectively (Meyer *et al.* 2008; Carpenter *et al.* 2009). Of the 14 sources bearing disks at 24 μm, only five have optically thick disks and only one of them is older than 10 Myr (Silverstone *et al.* 2006). This finding suggests that primordial dust out to $\sim 5\,\text{AU}$ also dissipates in less than 10 Myr, corroborating earlier indications from the Infrared Astronomical Satellite (IRAS) and Infrared Space Observatory (ISO) at 25 and 60 μm (Meyer & Beckwith 2000; Spangler *et al.* 2001).

Finally, the cold disk material outside $\sim 10\,\text{AU}$ is probed at submillimeter and millimeter wavelengths. The combination of spatially resolved disk emission and the slope of the SED at these wavelengths can be used to infer the size distribution of grains in the outer disk (Testi *et al.* 2003). This technique has shown that several disks around low- and intermediate-mass stars formed millimeter- and up to centimeter-sized particles already in the first million years of their evolution (see Chapter 7). However, so far only a dozen protoplanetary disks could be spatially resolved at millimeter wavelengths. Spatially unresolved observations have been obtained on large samples of sources and have been used to trace the evolution of dust disk mass. Beckwith *et al.* (1990) and more recently Andrews & Williams (2007) surveyed the Taurus Aurigae star-forming region and report that: (a) there is little evolution in the disk mass over the few million year age range of the Taurus sources; (b) less than 10% of the objects have no inner-disk signatures but millimeter detections suggesting that both infrared and millimeter disk emission disappears on a similar timescale. Carpenter *et al.* (2005) further showed that 10–30 Myr-old Sun-like stars have less massive disks than few million-year-old stars in Taurus. This finding suggests that significant evolution occurs in the circumstellar dust properties around Sun-like stars by the ages of 10–30 Myr. In summary, the current data indicate that most Sun-like stars clear out the dust in their inner and outer disk by $\sim 10\,\text{Myr}$.

A dispersal timescale of $\sim 10\,\mathrm{Myr}$ seems also to be characteristic of most low-mass stars even in clusters that have massive stars. Sicilia-Aguilar *et al.* (2005, 2006) surveyed two clusters in the Cep OB2 Association, Tr 37 and NGC 7160, that have or had a massive O star. They find that the fractions of low-mass stars with disks are similar to those in clusters with no massive stars (see also Fig. 9.2). This suggests that massive stars have only a limited effect on the dispersal of disks and hence on the formation of planets around most stars in a cluster. A similar result is found by Balog *et al.* (2007) in the NGC 2244 cluster that contains 7 O stars. Although there is a deficit of disks within 0.5 pc from the closest O stars, Balog *et al.* (2007) show that outside that radius, where most stars in the cluster are located, the disk fraction is comparable to the fraction of disks in clusters without a massive star. In conclusion, while massive stars can affect disk evolution in their immediate vicinity, most disks in the cluster remain unaffected.

It is important to point out that although $\sim 10\,\mathrm{Myr}$ is the timescale for most Sun-like stars to disperse their disks observations indicate ranges of an order of magnitude (~ 1 to $\sim 10\,\mathrm{Myr}$) for the clearing timescale of any individual object. There are many examples of young non-accreting T Tauri stars without circumstellar disks in star-forming regions that are a only few million years-old (Padgett *et al.* 2006; Cieza *et al.* 2007). There are also a few cases of long-lived optically thick dust disks around accreting stars as old as $\sim 10\,\mathrm{Myr}$ or older (TW Hya, Calvet *et al.* 2002; St 34, Hartmann *et al.* 2005; White & Hillenbrand 2005; and PDS 66, Cortes *et al.* 2009).

All the discussion above concentrated on young solar analogs for a direct comparison with the Solar System. It is interesting to note that the observed disk lifetime is strongly dependent on the stellar mass. Disks around intermediate-mass stars ($M > 1.5\,\mathrm{M_\odot}$) dissipate in much less than $10\,\mathrm{Myr}$ while disks around stars with masses similar to the Sun or smaller persist for longer times (Haisch *et al.* 2001a; Sterzik *et al.* 2004; Hernández *et al.* 2005; Carpenter *et al.* 2006; Lada *et al.* 2006; Currie *et al.* 2007; Hernández *et al.* 2007; Riaz & Gizis 2008; Merín *et al.* 2008).

9.1.2 Constraints on the dispersal of the gas component

The initial mass and lifetime of gas in circumstellar disks affect both the formation of giant planets as well as the formation of terrestrial planets. According to the widely accepted scenario of giant-planet formation, rocky cores need to reach several $\mathrm{M_\oplus}$ before being able to accumulate a substantial amount of gas from the protoplanetary disk. Current models require from a few to 10 million years to form Jupiter-like planets at 5 AU (see e.g. Lissauer & Stevenson 2007), meaning that primordial gas should persist for as long in the disk. Whether terrestrial planets can form and whether their orbits will be circular as those of Earth and Venus strongly depend on the amount of residual gas at their formation time (see e.g. Kominami & Ida 2004). In spite of its relevance to planet formation, little is known about the time over which circumstellar gas clears out.

Excess emission in the optical/ultraviolet and the broadening of the Hα line are often observed toward young low-mass stars and indicate that hot (5000–10 000 K) gas is being accreted from the disk onto the star (see e.g. Calvet *et al.* 2000). Hartigan *et al.* (1995) found that there is a one-to-one correspondence between accretion signatures and near-infrared excess emission. Similarly to the excess emission from dust, accretion rates are found to decline with stellar age (see Fig. 9.3). However, the large uncertainties in mass accretion rates (~ 0.5 dex), the large spread in accretion rates at any age (~ 2 dex), and the few estimates for sources older than ~ 5 Myr do not allow the quantification of the rate of decay (see also Hartmann *et al.* 1998). A few stars as old as 10 Myr are found still to be accreting at a rate of $\leq 10^{-9}\,M_\odot\,\mathrm{yr}^{-1}$,

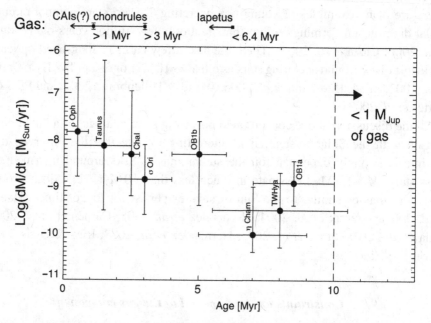

Figure 9.3 Mass accretion rates versus the age of the stellar group. The age error bars represent typical uncertainties, while the accretion rate error bars are the maximum and minimum values measured in each region. In addition to the data presented in Calvet *et al.* (2005), we have included the mass accretion rates from ρ Ophiuchi (Natta *et al.* 2006), and from σ Orionis (Gatti *et al.* 2008). Above the plot we show a comparison to the formation timescale of CAIs, chondrules, and the moon Iapetus in the Solar System (see Sections 9.3.1 and 9.3.2).

about 10 times lower than younger stars (Muzerolle *et al.* 2000; Lawson *et al.* 2004). These observations demonstrate that the phase of active gas accretion ends by 10 Myr for most stars but cannot constrain the evolution of the nebular gas in the cooler outer regions where planets form. An inner hole could develop in the disk (see Section 9.2) leaving the outer gas disk decoupled. In this situation no gas is accreting onto the star but primordial gas is left in the planet-forming region. Therefore, understanding when primordial gas disks disperse requires the use of gas diagnostics tracing the inner as well as the outer disk.

Ro-vibrational transitions from carbon monoxide (CO) are often detected in the near-infrared spectra of sources with optically thick inner disks (see e.g. Brittain *et al.* 2007). Line profile models suggest that CO lines trace disk regions out to about 1 AU from young Sun-like stars (see e.g. Najita *et al.* 2007a). Thus, they could be good tracers of the gas dispersal at disk radii comparable to those probed by dust continuum emission at near-infrared wavelengths. Similarly molecular hydrogen (H_2) ro-vibrational transitions have the potential to constrain the gas dispersal timescale out to a few astronomical units from the central star. So far the few sparse H_2 detections only reveal a tendency for a reduced mass of hot (~ 1500 K) gas in systems older than about 5 Myr (Bary *et al.* 2003; Carmona *et al.* 2007; Ramsay Howat & Greaves 2007; Bary *et al.* 2008; Bitner *et al.* 2008).

Tracing the region where Earth and Jupiter formed requires observations of gas lines in the mid- and far-infrared. The ISO provided a first glimpse of the gas content at these disk radii. Thi *et al.* (2001b,a) reported pure rotational H_2 S(0) and S(1) line detections from a large number of pre-main-sequence stars and also from three main-sequence stars with debris disks. These detections translated into large reservoirs of gas, suggesting a gas dispersal timescale longer than the accretion timescale. However, subsequent ground- and space-based infrared spectroscopy (Richter *et al.* 2002; Sheret *et al.* 2003; Sako *et al.* 2005; Chen *et al.* 2007) and ultraviolet observations (Lecavelier des Etangs *et al.* 2001) cast doubt on whether the observed lines originated in disks. More recently Hollenbach *et al.* (2005) and Pascucci *et al.* (2006) used the high-resolution spectrograph on board the Spitzer Space Telescope to survey 16 young solar analogs with ages between five and a few hundred million years. All stars in the systems have passed the phase of active gas accretion, but most of them have dusty disks (very likely debris disks). They report no detections of infrared lines from molecular hydrogen or from other abundant atoms and ions. Models of the line flux upper limits show that none of the disks has enough mass to form planets similar to Jupiter or Saturn in our Solar System. This result suggests that giant planets form early, probably during the phase of active gas accretion. Interestingly, the gas surface density upper limits at 1 AU are found to be smaller than 0.01% of the minimum-mass solar nebula for most disks. If terrestrial planets form frequently and their orbits are circularized by gas, then this result suggests that the circularization should occur early (Pascucci *et al.* 2006). These results are corroborated by the stringent upper limits on the gas disk mass from far-ultraviolet H_2 observations of two ~ 12 Myr-old edge-on disks (Roberge *et al.* 2005, 2006). Further constraints on the timescale for the clearing of the circumstellar gas may come from detecting ionized neon in non-accreting disks. This infrared line was recently discovered toward many young accreting Sun-like stars with disks (Espaillat *et al.* 2007a; Herczeg *et al.* 2007; Lahuis *et al.* 2007; Pascucci *et al.* 2007; Pascucci & Sterzik 2009). Models of disks irradiated by stellar X-ray and/or ultraviolet emission predict that the presence of [Ne II] emission traces a small amount of gas in the region within ~ 10 AU from Sun-like stars (Glassgold *et al.* 2007; Gorti & Hollenbach 2008). If this prediction is correct then this line could be an excellent probe of dissipating gas at disk radii where terrestrial and giant planets may form.

The cold outer-disk regions (≥ 200 AU) are mainly traced using rotational lines at millimeter wavelengths. As summarized in Dutrey *et al.* (2007), there is a good correspondence between gas line detections at millimeter wavelengths and the presence of optically thick dust disks. Most disks detected in the CO rotational lines are found to be large (~ 200–1000 AU) and with gas in Keplerian rotation. Zuckerman *et al.* (1995) investigated the gas dispersal timescale using as tracer the ^{12}CO $(2–1)$ transition. They surveyed 16 stars, the majority of which have ages ≤ 10 Myr and are more massive than the Sun. The detection of cold CO gas only in the eight youngest sources led Zuckerman *et al.* (1995) to conclude that most circumstellar gas clears out in less than 10 Myr. However, there are at least two important issues that need to be considered: the possible condensation of CO onto grains (see e.g. Aikawa *et al.* 1996) and the photodissociation of CO molecules (see e.g. Kamp & Bertoldi 2000). Both processes reduce the CO gas-phase abundance relative to H and thereby raise the amount of gas mass in the disk that would go undetected using CO millimeter transitions. In a recent paper Kamp *et al.* (2007) point out that the HI line at 21 cm could be a good tracer of dissipating disks because during the transition from a protoplanetary to a debris disk most of the gas mass should be in atomic hydrogen. Although current radio telescopes do not have the sensitivity to detect this faint line, models suggest that the Square Kilometer Array will be able to detect the HI 21 cm line from protoplanetary disks in nearby star-forming regions.

In summary, current observations indicate that most circumstellar gas mass disperses quickly on a timescale similar to (or maybe even shorter than) the dust-clearing timescale. Significant progress in this field is expected to occur in the next years with the launch of the Herschel Space Observatory. In particular, the

Herschel Open Time Key Program "Gas in Protoplanetary Systems"[1] will obtain spectra of far-infrared atomic and molecular lines from over 240 disks in nearby star-forming regions with ages in the critical 1–30 Myr range over which gas clears out and planets form.

9.1.3 Transition and evolved disks

Transition objects were discovered at the end of the eighties as a subgroup of young disks displaying small near-infrared excess but large mid- and far-IR excesses (Strom *et al.* 1989; Skrutskie *et al.* 1990; see also Fig. 9.1). The dearth of near-infrared excess emission indicates an optically thin inner cavity within the dust disk, believed to mark the disappearance of the primordial massive disk (Calvet *et al.* 2002; D'Alessio *et al.* 2005). Different processes have been proposed to explain this inside-out clearing, including grain growth (see e.g. Ciesla 2007a), disk photoevaporation (see e.g. Alexander *et al.* 2006b), magneto-rotational instability (Chiang & Murray-Clay 2007) and, most interestingly, dynamical clearing by a low-mass stellar companion or by a giant planet (see e.g. Quillen *et al.* 2004). Transition disks are rare: less than about 10% of the disk-bearing stars with ages <10 Myr are transition objects (Sicilia-Aguilar *et al.* 2006; Brown *et al.* 2007; Hernández *et al.* 2007, Merín *et al.* 2008). If all disks evolve from an optically thick to such an optically thin structure, then the low detection rate of transition disks suggests that the transition occurs quickly, in less than 1 Myr. However, transition disks may not be the only evolutionary pathway. Spitzer observations are revealing many disks with weak excess emission at all infrared wavelengths in comparison to primordial disks. Grains in this class of evolved disks may have grown substantially over a large range of disk radii without resulting in the formation of an optically thin inner-disk region (e.g. Lada *et al.* 2006; Cieza *et al.* 2007; Hernández *et al.* 2007). It is found that the fraction of evolved disks increases on going from regions of 3 to 10 Myr (Hernández *et al.* 2007) but numbers at any age are strongly dependent on the way evolved disks are separated from primordial disks. Merín *et al.* (2008) speculate that the disk primordial mass sets the subsequent evolution. Evolved disks would be the next evolutionary stage of objects with low primordial disk masses in which grain growth and protoplanet formation are slow. On the contrary, transition disks would be those with massive primordial disks where the fast formation of planets in conjunction with photoevaporation can lead to a clear out of the inner-disk region.

The Infrared Spectrograph (IRS) provided crucial data to characterize the mid-infrared rise in the spectral energy distribution of transition disks, which bear

---

[1] http://www.laeff.inta.es/projects/herschel/index.php

information on the extension of the optically thin inner cavity. Models of spectral energy distributions estimate inner cavities ranging from a few astronomical units (see e.g. DM Tau, Calvet *et al.* 2005) to more than 50 AU (see e.g. UX Tau A, Espaillat *et al.* 2007b), containing little (a few percent of lunar masses) or no dust in submicron-sized grains. In a few cases interferometric millimeter images detected reduced emission around the star extending out to the inner disk radius inferred from SED modeling (Andrews & Williams 2007; Hughes *et al.* 2007; Brown *et al.* 2008). These observations prove that inner holes are indeed present but multi-wavelength millimeter images are required to pin down the hole sizes and rule out dust opacity effects.

In a large sample of 34 transition disks from the c2d team Merín *et al.* (2008) show that the inner cavities inferred from SED models scale roughly linearly with the mass of the central star and disk masses inferred from millimeter observations are very low. These trends are compatible with disk photoevaporation (Section 9.2). A slightly different result is reached by Najita *et al.* (2007b) who studied the ensemble of transition disks in Taurus and report that transition objects have accretion rates typically 10 times lower than non-transition disks of similar age but median disk masses about four times larger. These properties are predicted by different planet-formation models and led Najita *et al.* (2007b) to suggest that the formation of giant planets plays a role in explaining the origin of some transition objects. Detecting giant planets in transition disks with radial-velocity monitoring is in principle possible but complicated by the numerous large spots on the surface of young stars that can mimic the radial-velocity variations induced by a massive planet (Huélamo *et al.* 2008; Prato *et al.* 2008; Setiawan *et al.* 2008). Circumbinary disks may also appear as transition disks (see the cases of CoKu Tau/4, Ireland & Kraus 2008; and CS Cha, Guenther *et al.* 2007). However, it appears that many other known transition disks (e.g. GM Aur, UX Tau A, LkCa 15, and SR 21) do not have any binary companions with the semi-major axis needed to clear their disks suggesting that another process is at work (Kraus, private communication).

9.2 Disk dispersal processes

The observations summarized in the previous sections indicate that most disks disperse their primordial dust and gas on a timescale shorter than ~ 10 Myr. Part of this primordial material may be incorporated into planets. However, planet formation is not the major disk-dispersal mechanism. In the case of our Solar System the mass of the planets amounts to less than one tenth of the minimum-mass solar nebula, which is the minimum disk mass required to reproduce the solar chemical composition (0.013–0.036 M$_\odot$ Hayashi *et al.* 1985; Desch 2007). With hydrogen and helium largely depleted in planets (see e.g. Weidenschilling 1977), other mechanisms than planet formation must have dominated the dispersal of the solar nebula. In the following we briefly describe four additional processes and refer to the review by Hollenbach *et al.* (2000) and Dullemond *et al.* (2007) for more details.

Accretion. Viscous stresses and gravitational torques within the disk transport angular momentum to the outer regions allowing disk matter to flow inward and accrete onto the star. Because the source of viscosity is still not well understood (see also Chapter 4), it is common to describe the viscosity via a dimensionless parameter α (Shakura & Syunyaev 1973). Using this simplification, the viscous dispersal timescale, i.e. the time for the disk to disperse via accretion, becomes inversely proportional to α and increases linearly with the radial distance from the star (Hartmann *et al.* 1998). While the inner-disk material accretes onto the star, material further out moves in and replenishes the inner disk. Thus, the disk-dispersal timescale from accretion alone is set by the timescale to disperse the mass at the outer disk.

Stellar encounters. Since most stars form in clusters (see e.g. Lada & Lada 2003) it is important to evaluate the effect of stellar encounters on the survival of protoplanetary disks. In the most destructive case the disk and the perturbing star move on coplanar orbits and in the same direction: matter is removed from the disk as close in as one third of the periastron distance (Clarke & Pringle 1993). Even for this most destructive case and for conditions typical to dense clusters like the Trapezium, Hollenbach *et al.* (2000) find that stellar encounters can only appreciably reduce the lifetime of the outer disk regions ($>100\,\text{AU}$).

Disk and stellar winds. Disk winds form from the interaction between the rotating magnetic field with the inner disk. They transport out about one tenth of the gas accreting onto the star and carry away the disk angular momentum that is lost from the disk in the accretion process (Konigl & Pudritz 2000; Pudritz *et al.* 2007). Once the phase of gas accretion ends, a spherically symmetric stellar wind is driven by the magnetic activity in the stellar chromosphere. The mass loss in these winds is likely to be a thousand times higher than the present-day solar mass loss (Güdel 2007). A wind blowing over the disk surface can help to remove disk material, but it is unlikely to be the major disk dispersal mechanism. Elmegreen (1979) shows that a centralized stellar wind could not have blown away the solar nebula. Hollenbach *et al.* (2000) point out that powerful winds are only present during disk accretion and in addition they may be deflected by an ultraviolet-heated disk atmosphere and photoevaporative flow.

Photoevaporation. High-energy photons either from the central star or from nearby massive stars heat and ionize the disk surface. The main heating photons can come from the stellar chromosphere or from the shock at the base of the accretion and lie in the far-ultraviolet (FUV) (6–13.6 eV) and extreme ultraviolet (EUV)

(>13.6 eV) regimes, while X-rays (≥ 0.1 keV) seem to play a minor role (Alexander *et al.* 2004). Where the thermal speed of the gas becomes larger than the local escape speed from the gravitationally bound system, gas starts to flow out of the disk. This characteristic radius is called the gravitational radius (r_g, see e.g. Hollenbach *et al.* 1994).

(a) Photoevaporation by the central star. In the best-studied case the disk heating and ionization are dominated by stellar EUV photons: evaporation is found to occur from inside out (Hollenbach *et al.* 1994), mostly from a radius that is $\sim 0.15\,r_g$ (Liffman 2003; Font *et al.* 2004). This radius corresponds to about 1 AU from a Sun-like star and scales with the mass of the central star. The case of ionization dominated by FUV photons has not been fully explored, but preliminary studies suggest that the evaporation occurs in the outer disk (see Dullemond *et al.* 2007). Thus, the combined effect of stellar EUV and FUV is to squeeze the disk into intermediate radii.

(b) External photoevaporation. The proplyds in Orion are the best examples of disks photoevaporated by the ultraviolet photons from nearby massive stars (see e.g. O'dell 2001). In the case of external photoevaporation most of the disk mass is lost at the outer-disk radius. Hollenbach *et al.* (2000) show that for conditions typical to the Trapezium external photoevaporation can efficiently (in less than 10 Myr) remove disk mass as close as 10 AU from Sun-like stars. Recently, Fatuzzo & Adams (2008) constructed the EUV and FUV radiation field in an ensemble of clusters and determined the percentage of disks that are destroyed by external photoevaporation before planet formation can take place. They assumed that planet formation is compromised if the disk evaporation time is less than 10 Myr at a disk radius of 30 AU and consider the distribution of cluster sizes in the solar neighborhood. With these assumptions they find that 25% of the disk population loses some of its planet-forming potential due to FUV radiation from the background cluster, whereas only 7% of the disk population is compromised by EUV radiation. These calculations suggest that planet formation will proceed unperturbed in most disks in agreement with the observations of disk frequencies in clusters (Section 9.1.1).

A comparison of disk-dispersal timescales, such as that presented in Fig. 1 of Hollenbach *et al.* (2000), suggests that viscous spreading and photoevaporation are the major dispersal mechanisms. Models combining these two mechanisms were first developed by Clarke *et al.* (2001) and later refined by Alexander *et al.* (2006a,b). The evolution of an accreting and photoevaporating disk can be summarized as follows. In the first 10^{6-7} yr viscous evolution proceeds relatively unperturbed by photoevaporation. Once the viscous accretion inflow rates fall below the photoevaporation rates a gap opens up close to r_g and the inner disk rapidly ($\sim 10^5$ yr) drains onto the central star. At this point direct ionization of the disk inner edge (the flux is not anymore attenuated by the inner-disk atmosphere) disperses the outer disk in $\sim 10^5$ yr (Alexander *et al.* 2006b). This latest model is consistent with three observables: (i) the rapid dispersal of disks; (ii) the almost-simultaneous loss of the outer disk with the inner disk; (iii) the SEDs of transition disks with inner regions being completely devoid of dust. Note that the model assumes that the EUV flux is dominated by the stellar chromosphere (Alexander *et al.* 2005) and hence remains high even when accretion ends. If, however, the EUV flux of young stars is dominated by the accretion at the base of the shock, then photoevaporation would be reduced in time as gas accretion diminishes.

9.3 Our Solar System

The Sun's parent molecular cloud collapsed to form the infant Sun and a surrounding protoplanetary disk at ~ 4.57 Ga as suggested by the age of the oldest Solar System materials (Amelin *et al.* 2002, 2006) and the helioseismic age of the Sun (Bonanno *et al.* 2002). Dust particles in the proto-solar disk were annealed or evaporated into the disk gas due to heating in the active stage of the disk, resulting in changes of structures and the chemical and isotopic compositions of dust particles (see also Chapter 8). Dust particles accumulated to form planetesimals in the later stage of the disk evolution. Most meteorites are fragments of planetesimals, and preserve records of processes that occurred in the proto-solar disk or in the planetesimals. In the first part of this section, we attempt to extract information on the clearing of the proto-solar disk from meteoritic components (Section 9.3.1).

Planet formation followed the planetesimal and protoplanet formation in the final stage of the disk evolution (Chapter 10). Gas giants, Jupiter and Saturn, captured disk gas due to their large gravities, and other planets, including Earth, may also have some evidence of disk-gas capture. In the second part of this section, we will seek constraints on the timing of dust and gas dispersal in the proto-solar disk from planets (Section 9.3.2).

9.3.1 Constraints from the meteoritic components

Bulk isotopic compositions of chondrites. Isotopic compositions of bulk chondrites are essentially uniform within variations of 0.1–0.01% except for light elements such as H, C, N, and O and for presolar grains (see e.g. Lodders 2003; Palme & Jones 2003). Presolar grains have isotopic compositions significantly different from those of Solar System materials, suggesting that they were dust particles formed in circumstellar environments and incorporated into the proto-solar molecular cloud (Chapter 2 and see e.g. Nittler 2003; Zinner 2005). Presolar grains are thus considered to be the first dust components that formed in the proto-solar disk. The rarity of presolar grains in chondrites (several ppb for silicon nitride to ~ 200 ppm for silicates; Nittler 2003; Zinner 2005; Chapter 2) and the essentially homogeneous isotopic compositions of chondrites strongly suggest that isotopic homogenization occurred in the early stages of disk evolution. Because the isotopic homogenization of chondritic materials seems to have occurred in micron to submicron scale, the most plausible process for homogenization is vaporization of pre-existing dust particles in the proto-solar disk, at least in the inner region of the disk, where chondrites formed. This implies that hot disk gas (> 2000 K) containing most of the metallic elements was once present in the very early stages of proto-solar disk evolution.

CAIs. Calcium–aluminum-rich inclusions (CAIs), the oldest known solids in the Solar System (Fig. 9.4), are enriched in refractory elements such as Ca, Al, and Ti, and are mainly composed of spinel ($MgAl_2O_4$), melilite ($Ca_2Al_2SiO_7$–$Ca_2MgSi_2O_7$), perovskite ($CaTiO_3$), hibonite ($CaAl_{12}O_{19}$), calcic pyroxene ($CaMgSi_2O_6$) with enrichment of Al and Ti, anorthite ($CaAl_2Si_2O_8$), and forsterite (Mg_2SiO_4) – see Chapter 8. These minerals are predicted to condense from hot disk gas of solar chemical composition (Chapter 4; and see e.g. Grossman 1972; Wood & Hashimoto 1993; Yoneda & Grossman 1995; Lodders 2003), while some CAIs showing signatures of mass-dependent isotopic fractionation of Mg and Si may have experienced kinetic evaporation from molten state.

Variations of oxygen isotopic compositions are seen among minerals in CAIs (e.g. Clayton 2005 and references therein; Chapter 4). Melilites have ^{16}O-poor compositions compared to spinel and calcic pyroxene. Such a difference in oxygen isotopes cannot be explained by mass-dependent isotopic fractionation of thermally activated processes, such as evaporation, condensation, and diffusion, in a single reservoir. Instead, the difference requires an exchange of oxygen isotopes between ^{16}O-rich CAI materials and ^{16}O-poor disk gas, suggesting the presence of disk gas in the CAI-forming region (see Chapter 4 for possible processes to produce mass-independently fractionated components). For instance, Yurimoto *et al.* (1998) found that ^{16}O-rich melilite crystals within a CAI in Allende meteorite (CV chondrite) co-exist with ^{16}O-poor melilite, the textual relationship supporting the idea of isotopic exchange of oxygen between ^{16}O-rich melilites and ^{16}O-poor disk gas.

AOAs. Amoeboid olivine aggregates (AOAs) are irregular shaped aggregates of Mg-rich olivine grains, and are likely to be condensates from disk gas (see e.g. Scott & Krot 2003). The oxygen isotopic compositions of AOAs are ^{16}O-poor, indicating that these aggregates condensed from ^{16}O-poor disk gas or experienced extensive exchange of oxygen isotopes with disk gas. The ^{26}Al–^{26}Mg systems in AOAs show that the initial $^{26}Al/^{27}Al$ ratios for AOAs are in the range of $(3.2-2.7)\times10^{-5}$ (Itoh *et al.* 2004), which is smaller than the canonical initial

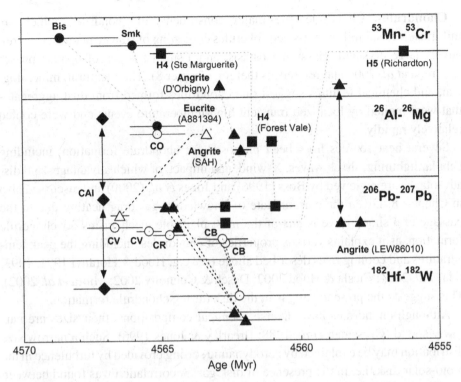

Figure 9.4 Absolute and relative ages of CAIs, chondrules, metamorphosed chondrites, and differentiated meteorites as obtained from Pb–Pb, ^{26}Al–^{26}Mg, ^{53}Mn–^{53}Cr, and ^{182}Hf–^{182}W systems. Vertical arrows connect the samples used to anchor the short-lived chronometers to the absolute ages derived from Pb–Pb systems. Dashed lines connect data from the same meteorite. The abbreviations Bis and Smk represent chondrules from Bishunpur and Semarkona chondrites. Revised from Figure 9 from Kita *et al.* (2005b). Data sources are references listed in Wadhwa *et al.* (2006) and in Kita *et al.* (2005b); for additional data points, see also Polnau and Lugmair (2001), Krot *et al.* (2005a), Markowski *et al.* (2007), Amelin (2008), and Connelly *et al.* (2008).

^{26}Al/^{27}Al ratio for CAIs (^{26}Al/^{27}Al $= 5.0 \times 10^{-5}$). If ^{26}Al was initially homogeneously distributed in the Solar System (or at least in the formation region of chondritic components) with the canonical value, the range of initial ^{26}Al/^{27}Al ratios of AOAs would correspond to a time interval that is 0.5–1 Myr after CAI formation (Itoh *et al.* 2004).

The lines of evidence from CAIs and AOAs suggest that disk gas was present in the earliest epoch of solid formation in the Solar System. However, it is not clear what stage of the proto-solar disk evolution corresponds to the earliest solid formation in the Solar System. This makes it difficult to compare the evolution of the proto-solar disk with that of protoplanetary disks (see Section 9.4).

Chondrules. Chondrules are major constituents of chondrites, which are millimeter- to submillimeter-sized spherules consisting of silicate phenocrysts (relatively large crystals), glassy mesostasis, and a small fraction of opaque phases (Lauretta *et al.* 2006 and references therein; Chapter 8). Their textural, mineralogical, and chemical features suggest that chondrules formed from dust aggregates that were melted by localized transient high-temperature events and were cooled relatively rapidly.

Several heat sources have been proposed for chondrule formation, including nebular lightning, shock waves, X-wind, and impact, of which advantages and disadvantages are reviewed by Boss (1996) and Jones *et al.* (2000) and discussed also in Chapter 8. Given the current state of knowledge, gas-drag heating due to the passage of a shock wave is one of the most plausible mechanisms for chondrule formation, as it explains various properties of chondrules, including the peak temperatures and cooling rates described above (see e.g. Hood & Horanyi 1991, 1993; Iida *et al.* 2001; Ciesla & Hood 2002; Desch & Connolly 2002; Miura *et al.* 2002). This suggests the presence of gas in the disk during chondrule formation.

Although chondrules have diverse chemical compositions, their sizes are narrowly sorted (Grossman *et al.* 1988; Brearley & Jones 1996). Such a narrow size distribution may be explained by aerodynamic sorting provided by turbulence in the proto-solar disk, i.e. in the presence of disk gas. A correlation was found between the volume of the chondrule and the volume of the accretionary fine-grained dust rim on chondrules in the Allende meteorite (Paque & Cuzzi 1997). These results could also be explained in terms of aerodynamic sorting in the gaseous disk.

Another important observation for chondrules is that little or no isotopic fractionations were found in major and/or volatile elements such as K (Alexander *et al.* 2000), Fe (Alexander & Wang 2001), Mg (Esat & Taylor 1990; Huss *et al.* 1996; Alexander *et al.* 1998), Si (Clayton *et al.* 1991), and S (Tachibana & Huss 2005). On the other hand, chondrules show volatility-related elemental fractionation (see e.g. Jones *et al.* 2000). Vacuum laboratory experiments for reproducing chemical characteristics of chondrules have shown that mass-dependent isotopic fractionation of elements is inevitable when chemical fractionation occurs owing to kinetic evaporation from chondrule melts, i.e. heavier isotope enrichment in a residual melt due to preferential evaporation of lighter isotopes (Yu *et al.* 2003). Experimental and theoretical studies of evaporation from chondrule melts have shown that isotopic fractionation can be decoupled from elemental fractionation if evaporation occurred in the presence of dust vapor (Cuzzi & Alexander 2006; Alexander *et al.* 2008), or evaporated gases recondensed during cooling (Nagahara & Ozawa 2000; Ozawa & Nagahara 2001). In either case, reactions between chondrule melts and high-temperature dust vapor are important for chondrule formation. The effective gas–melt reaction occurs in the presence of disk gas, which reduces the mean free path of gas species (e.g. Cuzzi & Alexander 2006), and thus the lack of isotopic fractionation within chondrules can be further evidence of the presence of disk gas in the time period of chondrule formation.

Ion-microprobe studies of ^{26}Al–^{26}Mg systems aiming to date common ferromagnesian chondrules in Semarkona (LL3.0), one of the least-metamorphosed ordinary chondrites, showed that the initial ^{26}Al/^{27}Al ratios for the chondrules are in the range $(0.87–0.47) \times 10^{-5}$, smaller than the canonical value for CAIs (Kita *et al.* 2000). Similar results were obtained from ferromagnesian chondrules in Bishunpur (LL3.1) and Krymka (LL3.1), both of which are also the least metamorphosed ordinary chondrites (Mostefaoui *et al.* 2002; Kita *et al.* 2005). Ferromagnesian chondrules from Y81020 (CO3.0), one of the least metamorphosed carbonaceous chondrites, also show excess ^{26}Mg, and the obtained range of initial ^{26}Al/^{27}Al ratios is comparable to that for chondrules from ordinary chondrites (Kunihiro *et al.* 2004; Kurahashi *et al.* 2008). These findings suggest that chondrules from both ordinary and carbonaceous chondrites formed contemporaneously (Fig. 9.4). The range of initial ^{26}Al/^{27}Al ratios of those chondrules indicates a time interval of 1–3 Myr after CAI formation. The data sets for chondrules with excess ^{26}Mg show that the age distribution of chondrules has a peak at 2 Myr. The observed ratios suggest that chondrule-forming events lasted at least for a few million years. They probably occurred nearly simultaneously within the formation regions of ordinary and carbonaceous chondrites. These may correspond to regions dominated by S-type and C-type asteroids in the present Solar System, unless migration of chondrules or planetesimals occurred in the proto-solar disk.

Such an age distribution of chondrules may correlate with the history of transient high-temperature events in the early Solar System. However, some chondrules contain relict grains derived from earlier generations of chondrules, which indicates that chondrules were heated repeatedly. Thus, initial ^{26}Al/^{27}Al ratios obtained from Al-rich minerals or mesostasis by *in situ* analyses using ion-microprobe analysis probably record the last heating event for individual chondrules, as Al-rich phases would have melted readily and their ^{26}Al–^{26}Mg systems would be reset by heating events. This suggests that the age distribution of chondrules determined by ^{26}Al–^{26}Mg chronometry might be different from the frequency distribution of high-temperature chondrule-forming events and that chondrule-forming events may have been more frequent in the earlier stages than at the peak of the ^{26}Al–^{26}Mg age distribution of chondrules.

Amelin *et al.* (2002) determined high-precision Pb–Pb ages of CAIs from Efremovka (CV3) and chondrules from Acfer 059 (CR) of 4.5672 ($\pm$0.0006) Ga and 4.5647 ($\pm$0.0006) Ga, respectively (Fig. 9.4). Amelin & Krot (2005) further reported absolute ages of chondrules from Allende, Gujba (CB), and Hammadah al Hamra 237 (CB) of 4.5667 ($\pm$ 0.0010) Ga, 4.5627 ($\pm$0.0005) Ga, and 4.5628

(± 0.0009) Ga, respectively (Fig. 9.4). Considering that chondrules from CB chondrites may have been produced by impact-related processes on the parent body and may not be products in the proto-solar disk (Rubin *et al.* 2003; Krot *et al.* 2005a), the difference between the Pb–Pb ages of CAIs and chondrules suggests that chondrule formation started contemporaneously with or shortly after CAI formation and lasted for 2 Myr. Petrographical evidence from CAI-bearing chondrules (Krot *et al.* 2005b) and chondrule-bearing CAIs (Itoh & Yurimoto 2003; Krot *et al.* 2005b) supports these dating results. The above conclusions appear to be consistent with the history of high-temperature chondrule-forming events estimated from the ^{26}Al–^{26}Mg chronometer, which in turn supports the proposed homogeneous distribution of ^{26}Al in the early Solar System.

We may thus put constraints from chondrules on the presence of dust and gas 2 Myr after the first solid formation. Again, it is not possible at present to correlate directly the chondrule formation period with the evolutionary stage of the protoplanetary disk, but high-temperature events capable of melting chondrule precursors could have occurred during the active stage of the protoplanetary disk (classical T Tauri stage).

There are a fraction of chondrules containing abundant noble gases with solar compositions in enstatite chondrites, the most reducing group of chondrites (Okazaki *et al.* 2001). High abundances of solar noble gases in enstatite-chondrite chondrules were interpreted as being a result of implantation of solar wind into chondrule precursor materials. This implantation into chondrule precursors should occur close to the Sun or in the disk after gas clearing. If the latter is the case then the dissipation of disk gas may have started at the time of formation of chondrules in enstatite chondrites. There are, unfortunately, no chronological data yet on gas-rich chondrules.

Metals and sulfides. Iron–nickel alloy is one of the ubiquitous components in chondrites. There are Fe–Ni metal grains with zoning of Ni in CH chondrites (Meibom *et al.* 2000). Because Ni is slightly less volatile than Fe, the first metal to condense from the gas of solar composition should have had a Ni-rich composition. The presence of zoned metal grains may suggest that relatively rapid condensation of Fe–Ni metal occurred from high-temperature disk gas in the proto-solar disk. However, it is not clear when and where those metallic grains formed.

Troilite (FeS) is also a common mineral in chondrites. Chondrules sometimes have rims of troilite, which may be recondensates of evaporated sulfur from chondrule precursors. The recondensation behavior of sulfur onto chondrules would be different from that of moderately volatile alkali elements described above. During cooling, sulfur would not re-enter the melt because the chondrule melt would have solidified by the time sulfur began to recondense. Instead, sulfur would recondense as sulfide veneers around chondrules (Zanda *et al.* 1995) or as opaque assemblages such as those found at chondrule boundaries, in chondrules rims, and in matrix (Lauretta *et al.* 2001; Lauretta & Buseck 2003). Although the presence of troilite rims on chondrules may be evidence for recondensation of sulfur in the presence of disk gas, condensation may have also occurred in vapor plumes formed by asteroidal impacts in the late stages of disk evolution (e.g. Krot *et al.* 2005a).

Asteroids and comets. Equilibrated chondrites and achondrites experienced thermal metamorphism and melting within their parental asteroids, respectively, and thus their ages represent the timing of thermal processes after the formation of asteroids. For instance, phosphates in Ste Marguerite, Forest Vale, and Richardton, which are equilibrated ordinary chondrites, show $^{207}Pb-^{206}Pb$ ages of 4.5627 ± 0.00006, 4.5609 ± 0.00007, and 4.5627 ± 0.00017 Ga (Fig. 9.4, Gopel *et al.* 1994; Amelin *et al.* 2005). Because their ages should record asteroidal thermal metamorphism, their parent bodies should have formed earlier than 4.563–4.561 Ga, implying that planetesimal formation may have occurred soon after or almost contemporaneously with chondrule formation. Asuka 881394 is a member of the basaltic eucrite meteorites, which are a group of HED meteorites considered to be derived from the asteroid 4 Vesta or related asteroids (vestoids), and its $^{207}Pb-^{206}Pb$ age is estimated to be 4.5665 ± 0.0003 Ga (Amelin *et al.* 2006). This suggests that a Vesta-sized asteroid (~ 500 km in diameter) may have formed in the very early stages of the disk evolution. However, $^{26}Al-^{26}Mg$, $^{53}Mn-^{53}Cr$, and $^{182}Hf-^{182}W$ ages of basaltic eucrites including Asuka 881394 are estimated to be ~ 4 Myr after CAIs (e.g. Wadhwa *et al.* 2006 and references therein; Fig. 9.4).

Although more chronological data sets are required to estimate the timing of thermal metamorphism and/or early crustal formation in asteroids, the current data sets indicate that the formation of asteroids occurred very rapidly (< 4 Myr after CAIs) in the early Solar System, implying that the timing of dust clearing was also < 4 Myr after CAIs. The CH and CB chondrites, of which chondrules may have been formed by asteroidal impacts $\sim 4-5$ Myr after CAIs (Krot *et al.* 2005a), are characterized by the lack of fine-grained matrix (e.g. Weisberg *et al.* 2006). This may imply that fine dust particles were absent in the disk when impact-related materials accumulated to form parent bodies of CH and CB chondrites.

Angrites are a rare group of basaltic achondrites formed by igneous crustal processes. Most angrites experienced little or no alteration and shock metamorphism after their early formation, and a fraction of angrites were known to have formed in the very early stages of disk evolution. Busemann *et al.* (2006) have shown that glassy components in the angrite D'Orbigny (4564.5 ± 0.0002 Ga, $^{207}Pb-^{206}Pb$ age, Amelin 2007), contain noble gases with solar-like compositions. Because D'Orbigny was probably a product of rapid cooling of magma near the surface of the angrite parent body and the elemental and isotopic patterns of noble gases in its glass cannot be those equilibrated to the disk gas, the solar-like noble gases in the glass should be inherited from materials accreted to form the angrite parent body (Busemann *et al.* 2006). Because the angrite parent body was not large enough to capture the nebular gas gravitationally, the presence of the solar noble-gas component in angrites suggests that effective solar-wind implantation due to dissipation of disk gas had already started before accretion of the angrite parent body.

As mentioned above, implantation of solar wind must have occurred in the absence of gas, i.e. after the dispersal of the nebular gas. The implantation of solar winds may also have occurred before or during the accretion of parent bodies of CI and CM chondrites, in which mineral grains containing solar flare tracks and associated spallation-produced Ne were observed (Goswami & Lal 1979; Goswami & MacDougall 1983; Hohenberg *et al.* 1990). Busemann *et al.* (2003) also showed possible retention of solar noble gas in enstatite chondrites. Although the formation ages of parent bodies of those chondrites have not yet been determined, this evidence implies the dissipation of disk gas before the accretion of chondrite parent bodies.

It should be noted that late irradiation of solar winds might also occur at the surface of asteroids or during the fragmentation and re-accretion of asteroids, which is a potential problem to constrain the timing of disk dispersal using irradiated solar wind components in meteorites. Nakamura (1999) reported that no solar noble gas was found in a fraction of a CM chondrite (Y791198) that preserved original petrologic features formed before accretion, and concluded that such a rock component having records of the pre-accretion history formed without irradiation of solar winds in the presence of disk gas. This implies that the observed signatures of solar wind implantation in CM chondrites shown above may be due to late irradiation after accretion of asteroids. Detailed chemical, mineralogical, and petrologic studies will thus be important to distinguish early irradiation signatures from those for late irradiation.

Recent analyses of cometary dust particles, obtained from Comet Wild 2 by the Stardust mission, showed that cometary dust grains have homogeneous isotopic compositions, which are identical to chondritic materials (McKeegan *et al.* 2006). This implies that isotopic homogenization due to vaporization occurred even in the comet-forming region or that disk materials were dynamically mixed. Dynamical mixing may have occurred due to hydrodynamics between gas and dust grains in the disk (Ciesla 2007b; Chapter 3), which also implies the presence of disk gas in the comet-forming region.

Both the analysis of Comet Wild 2 dust particles (Brownlee *et al.* 2006) and infrared spectroscopy of dust particles from comets Halley, Hale–Bopp, and Tempel 1 (Lisse *et al.* 2006) have shown that crystalline silicates are common constituents of comets. The presence of crystalline silicates in comets indicates that the crystallization of amorphous interstellar silicates occurred in the comet-forming region or crystalline materials were transported from the inner region of the disk (see also Chapters 5 and 8). In either case, the presence of disk gas in the comet-forming outer disk is required: the crystallization of amorphous silicates may have been caused by gas-drag heating due to shock waves (Harker & Desch 2002) and because dust transport from the inner disk may have occurred owing to hydrodynamics between gas and dust in the disk (Bockelée-Morvan *et al.* 2002; Dullemond *et al.* 2006; Ciesla 2007b; Chapter 8). However, there are unfortunately no chronological data present for such cometary grains.

9.3.2 Constraints from planets

Jovian planets and Neptunian planets. Jupiter, Saturn, Uranus, and Neptune have massive hydrogen-rich atmospheres that must have been accreted from the disk gas. They likely formed outside of the snowline, where larger amounts of solid materials (icy materials as well as dusty refractory components) allowed the formation of cores of several Earth-masses, heavy enough to trap disk gas. While current models can explain the *in situ* formation of a planet like Jupiter in a timescale similar to or shorter than the observed disk-dispersal timescale of 10 Myr (see e.g. Lissauer & Stevenson 2007), the formation of Uranus and Neptune remain problematic. Recently, Tsiganis *et al.* (2005) proposed that the formation regions of Uranus and Neptune may have been closer to the Sun than their present orbits to account for the eccentricities and inclinations of Solar System giant planets (the Nice model). The model suggests that Uranus and Neptune formed at 13.5–17 AU and 11–13 AU, respectively, and they migrated rapidly outward, exchanging their orbits. The migration of Neptune is also supported by the orbital distribution of Kuiper Belt objects (Malhotra 1993). The newly proposed mass distribution model of the proto-solar disk based on the Nice model (Desch 2007) showed that growth timescales of the cores of Jupiter, Saturn, Neptune, and Uranus could be 0.5, 1.5–2, 5.5–6, 9.5–10.5 Myr after formation of planetesimals, suggesting that the cores of all the Solar System's giant planets could form within a typical timescale for the gas dispersal of protoplanetary disks. However, it should be noted that the massive and truncated outer planetesimal disk that was assumed in the Nice model as a required initial condition that is not obvious to frequently occur. In addition, theoretical models themselves cannot be a direct measure of disk clearing in the proto-solar disk and it is important to independently constrain the gas dispersal timescale in the proto-solar disk.

Sample return missions for satellites, which formed in subsystems of gaseous and icy giants and experienced no significant geological processes, and subsequent chronological studies of returned samples are the best and most direct way to estimate the timing of giant-planet formation. However, such sample return missions are not easy from the viewpoint of current technology. A recent geophysical study of Iapetus, the most distant regular satellite of Saturn, suggests that the rotation state and shape of Iapetus and the presence of the equatorial ridge would have required heat from a short-lived radionuclide ^{26}Al (Castillo-Rogez *et al.* 2007). This further suggests the formation of Iapetus at ~ 3.4–5.4 Myr after the CAI formation (Castillo-Rogez *et al.* 2008), and could be one of the independent constraints on the formation timing of the Saturnian system. Another way to constrain the timing of giant-planet formation may be the determination of the age distribution of inner Solar System materials in future samples returned from short-period comets originating in the Kuiper Belt. As already discussed, the Stardust mission revealed that materials in the warm inner disk were transported into the cold outer disk (McKeegan *et al.* 2006) probably by the outward gas flow in the mid-plane of the disk (Ciesla 2007b). If the outward gas flow in the mid-plane was the main transport mechanism of inner-disk materials, the transport should have terminated after the formation of Jupiter; thus, the younger end of the age distribution of inner-disk materials preserved in cometary samples may constrain the duration of the material transport, i.e. the timing of the formation of Jupiter. Furthermore, if the formation age of the comet is also estimated possibly by determination of the timing of hydration or thermal metamorphism inside the comet, the timing of migration of Neptune to the present orbit might be constrained because the formation of Kuiper Belt comets should have pre-dated the Neptune migration.

Terrestrial planets. Runaway growth and the following oligarchic growth of planetesimals are proposed as the formation mechanism of Mars-sized protoplanets in the terrestrial-planet-forming region. Accretion of the protoplanets would form terrestrial planets, such as Earth and Venus (Chapter 10), and the accretion due to mutual orbital crossing would result in eccentricities of 0.1 for the last-formed planets, inconsistent with those for the present eccentricities of Earth and Venus (~ 0.03). As mentioned in Section 9.1.2, one of the mechanisms proposed to damp eccentricities is the drag force from gas in a mostly depleted gas disk (Kominami & Ida 2004). If this mechanism was responsible for circularizing the orbits of Earth and Venus, gas dispersal would already have started in the terrestrial-planet-forming region at the time of formation of the protoplanets, and would have continued during the final accretion stage of terrestrial planets. The ^{182}Hf–^{182}W systems of meteoritic, terrestrial, and lunar samples may potentially be useful chronometers for the giant impact event that formed the Moon and core formation in planetary bodies. Earlier studies showed that the core formation of terrestrial planets and the formation of the Earth–Moon system should have occurred within 30 Myr of the formation of the Solar System (Kleine *et al.* 2002; Yin *et al.* 2002; Jacobsen 2005; Kleine *et al.* 2005). However, Touboul *et al.* (2007) recently studied ^{182}Hf–^{182}W systems of lunar metals, which do not contain cosmogenic ^{182}W that is produced by neutron capture of ^{181}Ta, and inferred that the formation of the Moon occurred ~60 Myr after formation of the Solar System. This shows that there are still uncertainties in the ^{182}Hf–^{182}W chronometer, and thus the timing of the final accretion stage of terrestrial planets is poorly constrained.

Noble gases in the atmosphere of or in the interior of terrestrial planets may also constrain disk-gas dispersal. Although noble gases are rare in meteorites and in other planetary solids, there are several different components with distinct elemental and/or isotopic abundances, such as trapped solar and planetary components, *in situ* components formed by radioactive decay or by spallation by galactic cosmic rays, and an extrasolar component (e.g. Wieler *et al.* 2006). The trapped planetary component in meteorites may be explained by the elemental fractionation from the solar composition, where lighter gases are systematically depleted relative to heavier ones, while it may possibly also be an exotic presolar component (Huss & Alexander 1987). The trapped solar component in meteorites, which has elemental and isotopic patterns similar to that of the solar wind, is considered to be noble gases with the implanted solar wind. The Venusian atmosphere has a higher abundance of Ar compared to Earth's atmosphere. The elemental abundance pattern for Ar, Kr, and Xe is less fractionated compared to the planetary component, and seems to be similar to that of the solar component (Pepin 1991; Wieler 2002). These noble-gas signatures of the Venusian atmosphere may be attributed either to the gravitational capture of disk gas without significant elemental fractionation or accretion of volatile-rich comets that contain heavier noble gases as components adsorbed physically at low temperatures (Wieler 2002). The former case implies the presence of remnant disk gas in the Venus-forming region, while the latter case requires the presence of gas in the low-temperature region of the proto-solar disk.

Terrestrial oceanic island basalts that originated from the deep mantle contain Ne with the isotopic signature resembling a mixture of solar wind and solar energetic particles, indicating that planetesimals accreted into the Earth were irradiated by solar wind (see e.g. Trieloff *et al.* 2000, 2002). If this is the case, the column density of disk gas should have been low enough for solar wind to be implanted to planetesimals that formed the Earth. Note, however, that the Ne component in the Earth's deep mantle may also be explained by the mixing of the solar noble gas component and the atmospheric Ne, which supports the idea of the presence of remnant disk gas during the accretion stage of terrestrial planets (R. Okazaki, personal communication).

The abundances of noble gases in terrestrial planets are key to constraining the timing of the final stage of disk gas dispersal and its mechanism, but further investigations are surely required.

9.4 Discussion

The main difficulty in comparing the evolution of protoplanetary disks to that of the proto-solar nebula lies in anchoring two very different evolutionary frames. The evolution of protoplanetary disks is measured relative to the age of the star-forming regions and the associations that individual stars belong to. These ages are primarily determined from pre-main-sequence positions of stars in the Hertzsprung–Russell diagram compared with theoretical evolutionary tracks which likely under-predict low-mass stellar ages by 30–100% (Hillenbrand 2008b). The evolution of the proto-solar nebula is measured relative to the age of the CAIs, the oldest known solids in the Solar System. But at which disk evolutionary stage did CAIs form?

There is no observational evidence yet for the presence of CAIs in protoplanetary or more evolved debris disks. In the following we attempt to constrain their time of formation by comparing their formation environment to that of disks in different evolutionary stages. With this approach we are implicitly assuming that CAIs are a common outcome of the disk-evolution and planet-formation process, which may not be the case.

Refractory materials in chondrites (CAIs or CAI precursors) certainly formed in a gas-rich environment: they either condensed from hot disk gas (~ 2000 K) or experienced evaporation during high-temperature heating events in the disk (see also Chapter 8). Both scenarios point to CAIs forming in the early stages of disk evolution. Gas as hot as 2000 K is detected at the surface of protoplanetary disks a few million years old or younger and found to extend out to a few astromical units from Sun-like stars (Section 9.1.2). We do not know in detail the thermal structure of the inside of the disk, but models suggest that temperatures rapidly drop as the disk interior becomes optically thick. Temperatures above a thousand Kelvin should be present only well inside 1 AU from the star near to the disk mid-plane (e.g. Gorti & Hollenbach 2008). Another constraint for the region where CAIs formed comes from the pressure which should have been between 0.1–10 000 Pa according to the recent estimate by Grossman *et al.* (2008). For an ideal neutral hydrogen gas at 2000 K these pressures translate into volumetric gas densities of $\sim 4 \times 10^{12-17}$ H atoms per cm^3, typical for the mid-plane of protoplanetary disks. For comparison, disk models predict densities of 10^7 H atoms per cm^3 at only 0.2 AU above the disk mid-plane at a radial distance of 1 AU from Sun-like stars (see e.g. Glassgold *et al.* 2004). These arguments suggest that CAIs formed close to the dense disk mid-plane and probably within 1 AU from the Sun.

Millimeter observations of protoplanetary disks provide further evidence for early formation of the CAIs. As discussed in Section 9.1.1, these observations demonstrate that grains can grow up to the size of CAIs in the first million years of disk evolution. It is possible that some of these large grains found in the outer-disk regions are CAIs formed in the inner disk and transported outward. Dynamical mixing is also supported by the detection in comets of crystalline silicate grains that likely formed in the inner proto-solar nebula (Section 9.3.1; see also Chapter 8).

In summary, there is ample evidence that CAIs formed in the first million years of disk evolution, but how early is still a matter of debate. One popular line of thought is that CAIs formed very early in $\leq 10^5$ yr when the proto-Sun was rapidly accreting gas from its surroundings (Class 0 or I phase, e.g. Scott 2006). This is based on the narrow age spread of the CAIs as inferred from the ^{26}Al–^{26}Mg chronometer. However, such a narrow age distribution does not necessarily support the idea of the very early formation of CAIs. Thus, we will assume here that CAIs formed within the first one Myr of disk evolution and discuss the extreme possibility of CAIs forming essentially together with the Sun (0 Myr in our figures).

If CAIs formed in the first one Myr of disk evolution, then chondrules should have formed within the first 3 Myr (see Fig. 9.5 for a summary of the different

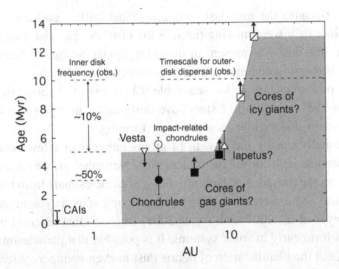

Figure 9.5 Summary of the timescales for the formation of chondrules, asteroids, and planets in the Solar System compared to the lifetime of disks around young stars. The Solar System chronology is based on the dating of the CAIs, which, we assume, formed within the first Myr of disk evolution. The inner-disk frequency is from infrared excess measurements of stars in different stellar groups (see Section 9.1.1). The timescale for the outer-disk dispersal is discussed in Sections 9.1.1 and 9.1.2. The Solar System chronology is summarized in Section 9.3. For the formation timescales of giant planets, we used those in Desch (2007) with the assumption that outer-disk planetesimals formed 2 Myr after CAIs.

Table 9.1 *Summary of the disk lifetimes and comparison with the formation timescale of chondrules, asteroids, and planets in the Solar System. Note that the Solar System chronology is based on the dating of the CAIs, the oldest known solids in the Solar System.*

Disk component	Lifetime (Myr)	Observable
		Protoplanetary disks
Micron-sized grains	5–10	Fraction of stars with IR excess
Millimeter grains	10–30	Millimeter fluxes – disk masses
Hot gas	10	Stellar accretion rates
Warm gas	$\simeq 10$	Detection of infrared gas lines
		Solar System
Fine dust	~ 2	Fine-grained dust rims on dated chondrules
Kilometer-sized bodies	<4	Dating of chondrites and differentiated meteorites
Gas	~ 2	No isotopic fractionation in chondrules

timescales). Because the fine-dust rims associated with chondrules were likely collected while they were moving through the dust and gas disk (see e.g. Ormel *et al.* 2008), fine dust was present in the solar nebula during the formation time of CAIs as well as of chondrules. This is consistent with the observed lifetime of fine dust in protoplanetary disks (see Table 9.1). Figure 9.2 shows that as many as $\sim 50\%$ of 3 Myr-old Sun-like stars have dusty disks, meaning that fine dust is available in many disks while chondrules are forming.

Observationally tracing the growth of grains larger than a few centimeters in size is not possible with current instrumentation. Nevertheless, we can obtain some information on the presence of larger bodies and planetesimals from the detection of debris dust. The existence of debris disks around stars as young as ~ 10 Myr (see e.g. Meyer *et al.* 2007 and Wyatt 2008) suggests that asteroid and Kuiper Belt analogs form early in some systems. It is possible that planetesimals formed even earlier, but the identification of debris dust in even younger systems is made difficult by the possible presence of primordial dust. The chronology of asteroids and terrestrial planets in our Solar System also supports the rapid formation of planetesimals and larger bodies. In spite of the uncertainties in using the Hf–W chronology (Section 9.3.2), it seems likely that asteroids began assembling within a few million years after the CAIs, perhaps contemporaneously with chondrules, and that the Moon–Earth system formed around 60 Myr after the CAIs.

To summarize, the evolution and dispersal of the dust in the proto-solar nebula seem to have followed a path similar to that of dust in most protoplanetary disks if CAIs formed at around 1 Myr in the disk evolutionary frame. If they formed very early in the first 100 000 yr, then the proto-solar nebula would have already assembled asteroids in the first few million years of its evolution. This does not seem to be the case for most Sun-like stars because at these ages they are surrounded by optically thick dust disks with the disk opacity dominated by small (submicron- to micron-sized) amorphous grains. However, it is true that planetesimals may form in the disk mid-plane and remain hidden by the large amounts of small primordial grains. Certainly, more effort should be directed in understanding whether CAIs form frequently in disks and in which phase of the disk evolution. A possible way to answer these questions could be via infrared spectroscopic identification of CAI components (see also Chapter 8). Posch *et al.* (2007) identified several prominent CAI-related bands at 9.3, 20.9, and 60 μm . There is also need for better determining when primordial disks disperse and evolve into the debris disks. Infrared excess measurements are not sufficient to identify the nature of the dust in the age range 5–10 Myr when there is evidence for both primordial and debris disks. As discussed in Section 9.1.1 a better approach may be to take into account the amount of the excess in combination with the wavelength at which the excess arises.

The evolution of the gas component is more difficult to assess both from the observations of disks and from the properties of Solar System bodies. We have discussed several indicators for the presence or absence of gas in the proto-solar disk but it remains difficult to quantify the amount of gas present. Chondrules are the youngest rocks in the Solar System that may have formed while gas was present: this fact sets only a lower limit to the gas dispersal of ~ 2 Myr after the CAIs. Chondrules with high abundances of noble gases may be the oldest rocks formed in the absence of nebular gas and their dating would be extremely valuable to set an upper limit for gas dispersal. This dispersal timescale would be relevant to the region where such chondrules formed; this unfortunately is yet unknown. Therefore, in addition to dating chondrules with a high abundance of noble gases, effort should be directed in understanding where they formed in the proto-solar disk. If these chondrules formed around 1 AU the gas surface density upper limit inferred from the solar implantation of noble gases may be relevant to the circularization of Earth's orbit. The debris disks observed so far point to an early circularization of the orbits of terrestrial planets if the major mechanism is gas drag (Section 9.1.2).

Studies of the gas content of protoplanetary disks with ages between 1 and 30 Myr are necessary to determine how rapidly the gas disperses and make a more direct comparison to the evolution and dispersal of dust in disks. As we discussed in Section 9.1.2, the dispersal of gaseous disks also provides an upper limit for the formation time of giant planets that can be compared to the time necessary to form Jupiter and Saturn in our Solar System. From a Solar System perspective it is interesting to expand on the constraints placed on the gas dispersal from the age determination of meteorites with implantation of solar wind, which provide us a time constraint on the gas dispersal in the Asteroid Belt. It is also important to have a time constraint from the moons of Jovian planets (see Castillo-Rogez *et al.* 2007, 2008 for an estimate of the formation of Iapetus), which likely applies to gas cooler than that traced by the formation time of chondrules.

Finally, it is important to remember that the Solar System probably formed in a cluster, near to massive stars, while most studies so far have focused on the evolution of disks in nearby low-mass star-forming regions. We feel that future observations should aim at a better characterization of the evolution of the inner disks in clusters for comparison with the evolution of the proto-solar nebula.

Acknowledgements I. Pascucci would like to thank J. Hernández and J. Muzerolle for their help in creating Figs. 9.2 and 9.3 and A. Krause, B. Merin, J. Castillo, and L. Hillenbrand for stimulating discussions. S. Tachibana would like to thank R. Okazaki and A. N. Krot for helpful discussions. The authors thank M. Trieloff and the anonymous referees for thoughtful comments that improved the manuscript.

References

Aikawa, Y., Miyama, S. M., Nakano, T., & Umebayashi, T. 1996, *Astrophysical Journal*, **467**, 684.

Alexander, C. M. O., Grossman, J. N., Ebel, D. S., & Ciesla, F. J. 2008, *Science*, **320**, 1617.

Alexander, C. M. O., Grossman, J. N., Wang, J., *et al.* 2000, *Meteoritics and Planetary Science*, **35**, 859.

Alexander, C. M. O., Mock, T., & Carlson, R. 1998, *Meteoritics and Planetary Science*, **33**, 9.

Alexander, C. M. O. & Wang, J. 2001, *Meteoritics and Planetary Science*, **36**, 419.

Alexander, R. D., Clarke, C. J., & Pringle, J. E. 2004, *Monthly Notices of the Royal Astronomical Society*, **354**, 71.

Alexander, R. D., Clarke, C. J., & Pringle, J. E. 2005, *Monthly Notices of the Royal Astronomical Society*, **358**, 283.

Alexander, R. D., Clarke, C. J., & Pringle, J. E. 2006a, *Monthly Notices of the Royal Astronomical Society*, **369**, 216.

Alexander, R. D., Clarke, C. J., & Pringle, J. E. 2006b, *Monthly Notices of the Royal Astronomical Society*, **369**, 229.

Amelin, Y. 2008, *Geochimica et Cosmochimica Acta*, **72**, 221.

Amelin, Y. 2007, *Lunar and Planetary Institute Conference Abstracts*, **38**, 1669.

Amelin, Y., Ghosh, A., & Rotenberg, E. 2005, *Geochimica et Cosmochimica Acta*, **69**, 505.

Amelin, Y. & Krot, A. N. 2005, *Lunar and Planetary Institute Conference Abstracts*, **36**, 1247.

Amelin, Y., Krot, A. N., Hutcheon, I. D., & Ulyanov, A. A. 2002, *Science*, **297**, 1678.

Amelin, Y., Wadhwa, M., & Lugmair, G. 2006, *Lunar and Planetary Institute Conference Abstracts*, **37**, 1970.

Andrews, S. M. & Williams, J. P. 2007, *Astrophysical Journal*, **659**, 705.

Balog, Z., Muzerolle, J., Rieke, G. H., *et al.* 2007, *Astrophysical Journal*, **660**, 1532.

Bary, J. S., Weintraub, D. A., & Kastner, J. H. 2003, *Astrophysical Journal*, **586**, 1136.

Bary, J. S., Weintraub, D. A., Shukla, S. J., Leisenring, J. M., & Kastner, J. H. 2008, *Astrophysical Journal*, **678**, 1088.

Beckwith, S. V. W., Sargent, A. I., Chini, R. S., & Guesten, R. 1990, *Astronomical Journal*, **99**, 924.

Bitner, M. A., Richter, M. J., Lacy, J. H., *et al.* 2008, *Astrophysical Journal*, **688**, 1326.

Bockelée-Morvan, D., Gautier, D., Hersant, F., Huré, J.-M., & Robert, F. 2002, *Astronomy and Astrophysics*, **384**, 1107.

Bonanno, A., Schlattl, H., & Paternò, L. 2002, *Astronomy and Astrophysics*, **390**, 1115.

Boss, A. P. 1996, in *Chondrules and the Protoplanetary Disk*, ed. R. Hewins, R. Jones, & E. Scott, Cambridge University Press, 257–263.

Brandeker, A., Jayawardhana, R., Khavari, P., Haisch, Jr., K. E., & Mardones, D. 2006, *Astrophysical Journal*, **652**, 1572.

Brearley, A. J. & Jones, R. H. 1996, *Reviews in Mineralogy and Geochemistry*, **36**, 3.1.

Brittain, S. D., Simon, T., Najita, J. R., & Rettig, T. W. 2007, *Astrophysical Journal*, **659**, 685.

Brown, J. M., Blake, G. A., Dullemond, C. P., *et al.* 2007, *Astrophysical Journal*, **664**, L107.

Brown, J. M., Blake, G. A., Qi, C., Dullemond, C. P., & Wilner, D. J. 2008, *Astrophysical Journal*, **675**, L109.

Brownlee, D., Tsou, P., Aléon, J., *et al.* 2006, *Science*, **314**, 1711.

Busemann, H., Baur, H., & Wieler, R. 2003, *Lunar and Planetary Institute Conference Abstracts*, **34**, 1665.

Busemann, H., Lorenzetti, S., & Eugster, O. 2006, *Geochimica et Cosmochimica Acta*, **70**, 5403.

Calvet, N., D'Alessio, P., Hartmann, L., *et al.* 2002, *Astrophysical Journal*, **568**, 1008.

Calvet, N., D'Alessio, P., Watson, D. M., *et al.* 2005, *Astrophysical Journal*, **630**, L185.

Calvet, N., Hartmann, L., & Strom, S. E. 2000, *Protostars and Planets IV*, ed. V. Mannings, A. P. Boss, & S. S. Russell, University of Arizona Press, 377–399.

Carmona, A., van den Ancker, M. E., Henning, T., *et al.* 2007, *Astronomy and Astrophysics*, **476**, 853.

Carpenter, J. M., Bouwman, J., Mamajek, E. E., *et al.* 2009, *Astrophysical Journal Supplement Series*, **181**, 197.

Carpenter, J. M., Mamajek, E. E., Hillenbrand, L. A., & Meyer, M. R. 2006, *Astrophysical Journal*, **651**, L49.

Carpenter, J. M., Wolf, S., Schreyer, K., Launhardt, R., & Henning, T. 2005, *Astronomical Journal*, **129**, 1049.

Castillo-Rogez, J. C., Matson, D. L., Sotin, C., *et al.* 2007, *Icarus*, **190**, 179.

Castillo-Rogez, J. C., Torrence, J., Man-Hoi, L., *et al.* *Icarus*, **190**, in press.

Chen, C. H., Li, A., Bohac, C., *et al.* 2007, *Astrophysical Journal*, **666**, 466.

Chiang, E. & Murray-Clay, R. 2007, *Nature Physics*, **3**, 604.

Ciesla, F. J. 2007a, *Astrophysical Journal*, **654**, L159.

Ciesla, F. J. 2007b, *Science*, **318**, 613.

Ciesla, F. J. & Hood, L. L. 2002, *Icarus*, **158**, 281.

Cieza, L., Padgett, D. L., Stapelfeldt, K. R., *et al.* 2007, *Astrophysical Journal*, **667**, 308.

Clarke, C. J., Gendrin, A., & Sotomayor, M. 2001, *Monthly Notices of the Royal Astronomical Society*, **328**, 485.

Clarke, C. J. & Pringle, J. E. 1993, *Monthly Notices of the Royal Astronomical Society*, **261**, 190.

Clayton, R. N. 2005, *Meteorites, Comets and Planets*, in Treatise on Geochemistry, vol. 1, ed. A. M. Davis, Elsevier, 129–145.

Clayton, R. N., Mayeda, T. K., Olsen, E. J., & Goswami, J. N. 1991, *Geochimica et Cosmochimica Acta*, **55**, 2317.

Connelly, J. N., Amelin, Y., Krot, A. N., & Bizzarro, M. 2008, *Astrophysical Journal*, **675**, L121.

Cortes, S. R., Meyer, M. R., Carpenter, J. M., *et al.* 2009, *Astrophysical Journal*, **697**, 1305.

Currie, T., Balog, Z., Kenyon, S. J., *et al.* 2007, *Astrophysical Journal*, **659**, 599.

Cuzzi, J. N. & Alexander, C. M. O. 2006, *Nature*, **441**, 483.

D'Alessio, P., Hartmann, L., Calvet, N., *et al.* 2005, *Astrophysical Journal*, **621**, 461.

Desch, S. J. 2007, *Astrophysical Journal*, **671**, 878.

Desch, S. J. & Connolly, Jr., H. C. 2002, *Meteoritics and Planetary Science*, **37**, 183.

Dullemond, C. P., Apai, D., & Walch, S. 2006, *Astrophysical Journal*, **640**, L67.

Dullemond, C. P. & Dominik, C. 2005, *Astronomy and Astrophysics*, **434**, 971.

Dullemond, C. P., Hollenbach, D., Kamp, I., & D'Alessio, P. 2007, in *Protostars and Planets V*, ed. B. Reipurth, D. Jewitt, & K. Keil, University of Arizona Press, 555–572.

Duquennoy, A. & Mayor, M. 1991, *Astronomy and Astrophysics*, **248**, 485.

Dutrey, A., Guilloteau, S., & Ho, P. 2007, in *Protostars and Planets V*, ed. B. Reipurth, D. Jewitt, & K. Keil, University of Arizona Press, 495–506.

Elmegreen, B. G. 1979, *Astronomy and Astrophysics*, **80**, 77.

Esat, T. M. & Taylor, S. R. 1990, *Lunar and Planetary Institute Conference Abstracts*, **21**, 333.

Espaillat, C., Calvet, N., D'Alessio, P., *et al.* 2007a, *Astrophysical Journal*, **664**, L111.

Espaillat, C., Calvet, N., D'Alessio, P., *et al.* 2007b, *Astrophysical Journal*, **670**, L135.

Evans, II, N. J., Dunham, M. M., Jørgensen, J. K., *et al.* 2009, *Astrophysical Journal Supplement Series*, **181**, 321.

Fatuzzo, M. & Adams, F. C. 2008, *Astrophysical Journal*, **675**, 1361.

Fazio, G. G., Hora, J. L., Allen, L. E., *et al.* 2004, *Astrophysical Journal Supplement Series*, **154**, 10.

Font, A. S., McCarthy, I. G., Johnstone, D., & Ballantyne, D. R. 2004, *Astrophysical Journal*, **607**, 890.

Gatti, T., Natta, A., Randich, S., and Testi, L. 2008, *Astronomy and Astrophysics*, **481**, 423.

Gautier, III, T. N., Rebull, L. M., Stapelfeldt, K. R., & Mainzer, A. 2008, *Astrophysical Journal*, **683**, 813.

Glassgold, A. E., Najita, J., & Igea, J. 2004, *Astrophysical Journal*, **615**, 972.

Glassgold, A. E., Najita, J. R., & Igea, J. 2007, *Astrophysical Journal*, **656**, 515.

Gopel, C., Manhes, G., & Allegre, C. J. 1994, *Earth and Planetary Science Letters*, **121**, 153.

Gorti, U. & Hollenbach, D. 2008, *Astrophysical Journal*, **683**, 287.

Goswami, J. N. & Lal, D. 1979, *Icarus*, **40**, 510.

Goswami, J. N. & MacDougall, J. D. 1983, *Journal of Geophysical Research*, **88**, 755.

Grossman, J. N., Rubin, A. E., Nagahara, H., & King, E. A. 1988, in *Meteorites and the Early Solar System*, ed. J. F. Kerridge, University of Arizona Press, 619–659.

Grossman, L. 1972, *Geochimica et Cosmochimica Acta*, **36**, 597.

Grossman, L., Simon, S. B., Rai, V. K., *et al.* 2008, *Geochimica et Cosmochimica Acta*, **72**, 3001.

Güdel, M. 2007, *Living Reviews in Solar Physics*, **4**, 3.

Guenther, E. W., Esposito, M., Mundt, R., *et al.* 2007, *Astronomy and Astrophysics*, **467**, 1147.

Haisch, Jr., K. E., Lada, E. A., & Lada, C. J. 2001a, *Astronomical Journal*, **121**, 2065.

Haisch, Jr., K. E., Lada, E. A., & Lada, C. J. 2001b, *Astrophysical Journal*, **553**, L153.

Harker, D. E. & Desch, S. J. 2002, *Astrophysical Journal*, **565**, L109.

Hartigan, P., Edwards, S., & Ghandour, L. 1995, *Astrophysical Journal*, **452**, 736.

Hartmann, L., Calvet, N., Gullbring, E., & D'Alessio, P. 1998, *Astrophysical Journal*, **495**, 385.
Hartmann, L., Calvet, N., Watson, D. M., *et al.* 2005, *Astrophysical Journal*, **628**, L147.
Hayashi, C., Nakazawa, K., & Nakagawa, Y. 1985, in *Protostars and Planets II*, ed. D. C. Black & M. S. Matthews, University of Arizona Press, 1100–1153.
Herczeg, G. J., Najita, J. R., Hillenbrand, L. A., & Pascucci, I. 2007, *Astrophysical Journal*, **670**, 509.
Hernández, J., Calvet, N., Briceño, C., *et al.* 2007, *Astrophysical Journal*, **671**, 1784.
Hernández, J., Calvet, N., Hartmann, L., *et al.* 2005, *Astronomical Journal*, **129**, 856.
Hernández, J., Hartmann, L., Calvet, N., *et al.* 2008, *Astrophysical Journal*, **686**, 1195.
Hillenbrand, L. A. 2008, *Physica Scripta*, **T130**, 014024.
Hohenberg, C. M., Nichols, Jr., R. H., Olinger, C. T., & Goswami, J. N. 1990, *Geochimica et Cosmochimica Acta*, **54**, 2133.
Hollenbach, D., Gorti, U., Meyer, M., *et al.* 2005, *Astrophysical Journal*, **631**, 1180.
Hollenbach, D., Johnstone, D., Lizano, S., & Shu, F. 1994, *Astrophysical Journal*, **428**, 654.
Hollenbach, D. J., Yorke, H. W., & Johnstone, D. 2000, *Protostars and Planets IV*, ed. V. Mannings, A. P. Boss, & S. S. Russell, University of Arizona Press, 401–428.
Hood, L. L. & Horanyi, M. 1991, *Icarus*, **93**, 259.
Hood, L. L. & Horanyi, M. 1993, *Icarus*, **106**, 179.
Huélamo, N., Figueira, P., Bonfils, X., *et al.* 2008, *Astronomy and Astrophysics*, **489**, L9.
Hughes, A. M., Wilner, D. J., Calvet, N., *et al.* 2007, *Astrophysical Journal*, **664**, 536.
Huss, G. R. & Alexander, E. C. J. 1987, *Journal of Geophysic Research*, **92**, 710.
Huss, G. R., Weisberg, M. K., & Wasserburg, G. J. 1996, *Lunar and Planetary Institute Conference Abstracts*, **27**, 575.
Iida, A., Nakamoto, T., Susa, H., & Nakagawa, Y. 2001, *Icarus*, **153**, 430.
Ireland, M. J. & Kraus, A. L. 2008, *Astrophysical Journal*, **678**, L59.
Itoh, S., Kojima, H., & Yurimoto, H. 2004, *Geochimica et Cosmochimica Acta*, **68**, 183.
Itoh, S. & Yurimoto, H. 2003, *Nature*, **423**, 728.
Jacobsen, S. B. 2005, *Annual Review of Earth and Planetary Sciences*, **33**, 531.
Jones, R. H., Lee, T., Connolly, Jr., H. C., Love, S. G., & Shang, H. 2000, *Protostars and Planets IV*, ed. V. Mannings, A. P. Boss, & S. S. Russell, University of Arizona Press, 927–962.
Kamp, I. & Bertoldi, F. 2000, *Astronomy and Astrophysics*, **353**, 276.
Kamp, I., Freudling, W., & Chengalur, J. N. 2007, *Astrophysical Journal*, **660**, 469.
Kita, N. T., Nagahara, H., Togashi, S., & Morishita, Y. 2000, *Geochimica et Cosmochimica Acta*, **64**, 3913.
Kita, N. T., Tomomura, S., Tachibana, S., *et al.* 2005a, *Lunar and Planetary Institute Conference Abstracts* **36**, 1750.
Kita, N. T., Huss, G. R., Tachibana, S., *et al.* 2005b, *Astronomical Society of the Pacific Conference Series*, **341**, 558.
Kleine, T., Münker, C., Mezger, K., & Palme, H. 2002, *Nature*, **418**, 952.
Kleine, T., Palme, H., Mezger, K., & Halliday, A. N. 2005, *Science*, **310**, 1671.
Kominami, J. & Ida, S. 2004, *Icarus*, **167**, 231.
Konigl, A. & Pudritz, R. E. 2000, *Protostars and Planets IV*, ed. V. Mannings, A. P. Boss, & S. S. Russell, University of Arizona Press, 759–787.
Kraus, A., Ireland, M., Martinache, F., Lloyd, J. C., & Hillenbraud, L. C. 2008, Proceedings of the 5th Spitzer Conference, New Light on Young Stars, Pasadena, October 26–30.
Krot, A. N., Amelin, Y., Cassen, P., & Meibom, A. 2005a, *Nature*, **436**, 989.
Krot, A. N., Yurimoto, H., Hutcheon, I. D., & MacPherson, G. J. 2005b, *Nature*, **434**, 998.

Kunihiro, T., Rubin, A. E., McKeegan, K. D., & Wasson, J. T. 2004, *Geochimica et Cosmochimica Acta*, **68**, 3599.

Kurahashi, E., Kita, N. T., Nagahara, H., & Morishita, Y. 2008, *Geochimica et Cosmochimica Acta*, **72**, 3865.

Lada, C. J. & Lada, E. A. 2003, *Annual Review of Astronomy and Astrophysics*, **41**, 57.

Lada, C. J., Muench, A. A., Luhman, K. L., *et al.* 2006, *Astronomical Journal*, **131**, 1574.

Lahuis, F., van Dishoeck, E. F., Blake, G. A., *et al.* 2007, *Astrophysical Journal*, **665**, 492.

Lauretta, D. S. & Buseck, P. R. 2003, *Meteoritics and Planetary Science*, **38**, 59.

Lauretta, D. S., Buseck, P. R., & Zega, T. J. 2001, *Geochimica et Cosmochimica Acta*, **65**, 1337.

Lauretta, D. S., Nagahara, H., & Alexander, C. M. O. 2006, in *Meteorites and the Early Solar System II*, ed. D. S. Lauretta & H. Y. McSween, University of Arizona Press, 431–459.

Lawson, W. A., Lyo, A.-R., & Muzerolle, J. 2004, *Monthly Notices of the Royal Astronomical Society*, **351**, L39.

Lecavelier des Etangs, A., Vidal-Madjar, A., Roberge, A., *et al.* 2001, *Nature*, **412**, 706.

Liffman, K. 2003, *Publications of the Astronomical Society of Australia*, **20**, 337.

Lissauer, J. J. & Stevenson, D. J. 2007, in *Protostars and Planets V*, ed. B. Reipurth, D. Jewitt, & K. Keil, University of Arizona Press, 591–606.

Lisse, C. M., VanCleve, J., Adams, A. C., *et al.* 2006, *Science*, **313**, 635.

Lodders, K. 2003, *Astrophysical Journal*, **591**, 1220.

Malhotra, R. 1993, *Nature*, **365**, 819.

Markowski, A., Quitté, G., Kleine, T., *et al.* 2007, *Earth and Planetary Science Letters*, **262**, 214.

Mathieu, R. D., Ghez, A. M., Jensen, E. L. N., & Simon, M. 2000, *Protostars and Planets IV*, ed. V. Mannings, A. P. Boss, & S. S. Russell, University of Arizona Press, 703–730.

McKeegan, K. D., Aléon, J., Bradley, J., *et al.* 2006, *Science*, **314**, 1724.

Megeath, S. T., Hartmann, L., Luhman, K. L., & Fazio, G. G. 2005, *Astrophysical Journal*, **634**, L113.

Meibom, A., Desch, S. J., Krot, A. N., *et al.* 2000, *Science*, **288**, 839.

Merin, B., Oliveira, I., Brown, J., *et al.* 2008, Proceedings of the 5th Spitzer Conference, New Light on Young Stars, Pasadena.

Meyer, M. R., Backman, D. E., Weinberger, A. J., & Wyatt, M. C. 2007, in *Protostars and Planets V*, ed. B. Reipurth, D. Jewitt, & K. Keil, University of Arizona Press, 573–588.

Meyer, M. R. & Beckwith, S. V. W. 2000, in *ISO Survey of a Dusty Universe*, ed. D. Lemke, M. Stickel, & K. Wilke, Lecture Notes in Physics, Springer Verlag, vol. 548, 341–352.

Meyer, M. R., Carpenter, J. M., Mamajek, E. E., *et al.* 2008, *Astrophysical Journal*, **673**, L181.

Meyer, M. R., Hillenbrand, L. A., Backman, D., *et al.* 2006, *Publications of the Astronomical Society of the Pacific*, **118**, 1690.

Miura, H., Nakamoto, T., & Susa, H. 2002, *Icarus*, **160**, 258.

Mostefaoui, S., Kita, N. T., Togashi, S., *et al.* 2002, *Meteoritics and Planetary Science*, **37**, 421.

Mostefaoui, S., Lugmair, G. W., & Hoppe, P. 2005, *Astrophysical Journal*, **625**, 271.

Muzerolle, J., Calvet, N., Briceño, C., Hartmann, L., & Hillenbrand, L. 2000, *Astrophysical Journal*, **535**, L47.

Nagahara, H. & Ozawa, K. 2000, *Geochimica et Cosmochimica Acta*, **169**, 45.

Najita, J. R., Carr, J. S., Glassgold, A. E., & Valenti, J. A. 2007a, in *Protostars and Planets V*, ed. B. Reipurth, D. Jewitt, & K. Keil, University of Arizona Press, 507–522.

Najita, J. R., Strom, S. E., & Muzerolle, J. 2007b, *Monthly Notices of the Royal Astronomical Society*, **378**, 369.

Nakamura, T. 1999, *Geochimica et Cosmochimica Acta*, **63**, 241.

Natta, A., Testi, L., Calvet, N., *et al.* 2007, in *Protostars and Planets V*, ed. B. Reipurth, D. Jewitt, & K. Keil, University of Arizona Press, 767–781.

Nittler, L. R. 2003, *Earth and Planetary Science Letters*, **209**, 259.

O'dell, C. R. 2001, *Annual Review of Astronomy and Astrophysics*, **39**, 99.

Okazaki, R., Takaoka, N., Nagao, K., Sekiya, M., & Nakamura, T. 2001, *Nature*, **412**, 795.

Ormel, C. W., Cuzzi, J. N., & Tielens, A. G. G. M. 2008, *Astrophysical Journal*, **679**, 1588.

Ozawa, K. & Nagahara, H. 2001, *Geochimica et Cosmochimica Acta*, **65**, 2171.

Padgett, D. L., Cieza, L., Stapelfeldt, K. R., *et al.* 2006, *Astrophysical Journal*, **645**, 1283.

Palme, H. & Jones, A. 2003, in *Meteorites, Comets and Planets*, ed. A. M. Davis, Treatise on Geochemistry, vol. 1, Elsevier, 41–61.

Paque, J. M. & Cuzzi, J. N. 1997, *Lunar and Planetary Institute Conference Abstracts*, **28**, 1071.

Pascucci, I., Apai, D., Hardegree-Ullman, E. E., *et al.* 2008, *Astrophysical Journal*, **673**, 477.

Pascucci, I., Gorti, U., Hollenbach, D., *et al.* 2006, *Astrophysical Journal*, **651**, 1177.

Pascucci, I., Hollenbach, D., Najita, J., *et al.* 2007, *Astrophysical Journal*, **663**, 383.

Pascucci, I. & Sterzik, M. 2009, *Astrophysical Journal*, **702**, 724.

Pepin, R. O. 1991, *Icarus*, **92**, 2.

Polnau, E. & Lugmair, G. W. 2001, *Lunar and Planetary Institute Conference Abstracts*, **32**, 1527.

Posch, T., Mutschke, H., Trieloff, M., & Henning, T. 2007, *Astrophysical Journal*, **656**, 615.

Prato, L., Huerta, M., Johns-Krull, C. M., *et al.* 2008, *Astrophysical Journal*, **687**, L103.

Pudritz, R. E., Ouyed, R., Fendt, C., & Brandenburg, A. 2007, in *Protostars and Planets V*, ed. B. Reipurth, D. Jewitt, & K. Keil, University of Arizona Press, 277–294.

Quillen, A. C., Blackman, E. G., Frank, A., & Varnière, P. 2004, *Astrophysical Journal*, **612**, L137.

Ramsay Howat, S. K. & Greaves, J. S. 2007, *Monthly Notices of the Royal Astronomical Society*, **379**, 1658.

Riaz, B. & Gizis, J. E. 2008, *Astrophysical Journal*, **681**, 1584.

Richter, M. J., Jaffe, D. T., Blake, G. A., & Lacy, J. H. 2002, *Astrophysical Journal*, **572**, L161.

Rieke, G. H., Young, E. T., Engelbracht, C. W., *et al.* 2004, *Astrophysical Journal Supplement Series*, **154**, 25.

Roberge, A., Feldman, P. D., Weinberger, A. J., Deleuil, M., & Bouret, J.-C. 2006, *Nature*, **441**, 724.

Roberge, A., Weinberger, A. J., Redfield, S., & Feldman, P. D. 2005, *Astrophysical Journal*, **626**, L105.

Rubin, A. E., Kallemeyn, G. W., Wasson, J. T., *et al.* 2003, *Geochimica et Cosmochimica Acta*, **67**, 3283.

Sako, S., Yamashita, T., Kataza, H., *et al.* 2005, *Astrophysical Journal*, **620**, 347.

Scott, E. R. D. 2006, *Icarus*, **185**, 72.

Scott, E. R. D. & Krot, A. N. 2003, in *Meteorites, Comets and Planets*, ed. A. M. Davis, Treatise on Geochemistry, vol. 1, Elsevier, 1–72.

Setiawan, J., Henning, T., Launhardt, R., *et al.* 2008, *Nature*, **451**, 38.

Shakura, N. I. & Syunyaev, R. A. 1973, *Astronomy and Astrophysics*, **24**, 337.

Sheret, I., Ramsay Howat, S. K., & Dent, W. R. F. 2003, *Monthly Notices of the Royal Astronomical Society*, **343**, L65.

Sicilia-Aguilar, A., Hartmann, L., Calvet, N., *et al.* 2006, *Astrophysical Journal*, **638**, 897.

Sicilia–Aguilar, A., Hartmann, L. W., Hernández, J., Briceño, C., & Calvet, N. 2005, *Astronomical Journal*, **130**, 188.

Silverstone, M. D., Meyer, M. R., Mamajek, E. E., *et al.* 2006, *Astrophysical Journal*, **639**, 1138.

Skrutskie, M. F., Dutkevitch, D., Strom, S. E., *et al.* 1990, *Astronomical Journal*, **99**, 1187.

Spangler, C., Sargent, A. I., Silverstone, M. D., Becklin, E. E., & Zuckerman, B. 2001, *Astrophysical Journal*, **555**, 932.

Sterzik, M. F., Pascucci, I., Apai, D., van der Bliek, N., & Dullemond, C. P. 2004, *Astronomy and Astrophysics*, **427**, 245.

Strom, K. M., Strom, S. E., Edwards, S., Cabrit, S., & Skrutskie, M. F. 1989, *Astronomical Journal*, **97**, 1451.

Strom, S. E., Edwards, S., & Skrutskie, M. F. 1993, in *Protostars and Planets III*, ed. E. H. Levy & J. I. Lunine, University of Arizona Press, 837–866.

Tachibana, S. & Huss, G. R. 2003, *Astrophysical Journal*, **588**, L41.

Tachibana, S. & Huss, G. R. 2005, *Geochimica et Cosmochimica Acta*, **69**, 3075.

Tachibana, S., Huss, G. R., Kita, N. T., Shimoda, G., & Morishita, Y. 2006, *Astrophysical Journal*, **639**, L87.

Testi, L., Natta, A., Shepherd, D. S., & Wilner, D. J. 2003, *Astronomy and Astrophysics*, **403**, 323.

Thi, W. F., Blake, G. A., van Dishoeck, E. F., *et al.* 2001a, *Nature*, **409**, 60.

Thi, W. F., van Dishoeck, E. F., Blake, G. A., *et al.* 2001b, *Astrophysical Journal*, **561**, 1074.

Touboul, M., Kleine, T., Bourdon, B., Palme, H., & Wieler, R. 2007, *Nature*, **450**, 1206.

Trieloff, M., Kunz, J., & Allègre, C. J. 2002, *Earth and Planetary Science Letters*, **200**, 297.

Trieloff, M., Kunz, J., Clague, D. A., Harrison, D., & Allègre, C. J. 2000, *Science*, **288**, 1036.

Tsiganis, K., Gomes, R., Morbidelli, A., & Levison, H. F. 2005, *Nature*, **435**, 459.

Wadhwa, M., Srinivasan, G., & Carlson, R. W. 2006, in *Meteorites and the Early Solar System II*, ed. D. S. Lauretta and H. Y. McSween, University of Arizona Press, 715–731.

Weidenschilling, S. J. 1977, *Astrophysics and Space Science*, **51**, 153.

Weinberger, A. J., Becklin, E. E., Zuckerman, B., & Song, I. 2004, *Astronomical Journal*, **127**, 2246.

Weisberg, M. K., McCoy, T. J., & Krot, A. N. 2006, in *Meteorites and the Early Solar System II*, ed. D. S. Lauretta & H. Y. McSween, University of Arizona Press, 19–52.

White, R. J. & Hillenbrand, L. A. 2005, *Astrophysical Journal*, **621**, L65.

Wieler, R. 2002, *Reviews in Mineralogy and Geochemistry*, **47**, 22.

Wieler, R., Busemann, H., & Franchi, I. A. 2006, in *Meteorites and the Early Solar System II*, ed. D. S. Lauretta & H. Y. McSween, University of Arizona Press, 499–521.

Wood, J. A. & Hashimoto, A. 1993, *Geochimica et Cosmochimica Acta*, **57**, 2377.

Wyatt, M. C. 2008, *Annual Review of Astronomy and Astrophysics*, **46**, 339.

Yin, Q., Jacobsen, S. B., Yamashita, K., *et al.* 2002, *Nature*, **418**, 949.

Yoneda, S. & Grossman, L. 1995, *Geochimica et Cosmochimica Acta*, **59**, 3413.

Yu, Y., Hewins, R. H., Alexander, C. M. O., & Wang, J. 2003, *Geochimica et Cosmochimica Acta*, **67**, 773.

Yurimoto, H., Ito, M., & Nagasawa, H. 1998, *Science*, **282**, 1874.

Zanda, B., Bourot-Denise, M., & Hewins, R. H. 1995, *Meteoritics*, **30**, 605.

Zinner, E. K. 2005, in *Meteorites, Comets and Planets*, ed. A. M. Davis, Treatise on Geochemistry, vol. 1 Elsevier, 1–33.

Zuckerman, B., Forveille, T., & Kastner, J. H. 1995, *Nature*, **373**, 494.

10

Accretion of planetesimals and the formation of rocky planets

John E. Chambers, David P. O'Brien, and Andrew M. Davis

Abstract Here we describe the formation of rocky planets and asteroids in the context of the planetesimal hypothesis. Small dust grains in protoplanetary disks readily stick together forming millimeter-to-centimeter-sized aggregates, many of which experience brief heating episodes causing melting. Growth to kilometer-sized planetesimals might proceed via continued pairwise sticking, turbulent concentration, or gravitational instability of a thin particle layer. Gravitational interactions between planetesimals lead to rapid runaway and oligarchic growth forming lunar to Mars-sized protoplanets in 10^5 to 10^6 years. Giant impacts between protoplanets form Earth-mass planets in 10^7 to 10^8 years, and occasionally lead to the formation of large satellites. Protoplanets may migrate far from their formation locations due to tidal interactions with the surrounding disk. Radioactive decay and impact heating cause melting and differentiation of planetesimals and protoplanets, forming iron-rich cores and silicate mantles, and leading to some loss of volatiles. Dynamical perturbations from giant planets eject most planetesimals and protoplanets from regions near orbital resonances, leading to Asteroid Belt formation. Some of this scattered material will collide with growing terrestrial planets, altering their composition as a result. Numerical simulations and radioisotope dating indicate that the terrestrial planets of the Solar System were essentially fully formed in 100–200 million years.

The formation of rocky planets marks the last stage in the evolution of a protoplanetary disk, extending beyond the dissipation of the gas disk itself. The seeds of planet formation are micrometer-sized dust grains that make up roughly 1% of the mass of a typical protoplanetary disk. In the Solar System, some 10^{40} of these grains evolved into a handful of rocky planets in the space of 10–100 million years – a remarkable transformation indeed. The early stages of this process and the formation of particles up to about 1 mm in size are described in Chapter 7 of this volume. In this chapter, we consider the growth of millimeter-sized particles into fully formed planets.

10.1 Observational constraints on rocky-planet formation

10.1.1 Physical properties of rocky bodies

The current arrangement of mass in the Solar System provides some constraints for any theory of planet formation. The terrestrial planets contain roughly $2M_\oplus$ of material in total, while the giant planets contain $450M_\oplus$, including $\sim80M_\oplus$ in elements heavier than He (Saumon & Guillot 2004). The terrestrial planets fall into two categories: Earth and Venus are similar in size while Mercury and Mars are less massive by an order of magnitude. The Asteroid Belt contains $\sim5 \times 10^{-4}M_\oplus$ of material, which is three to four orders of magnitude less than the amount of solid mass that once existed in this region (Weidenschilling 1977b). Main-belt asteroids with diameters <120 km have a size distribution that suggests they formed as a result of collisional erosion. Larger asteroids have a different size distribution that is probably primordial (Bottke *et al.* 2005a).

The terrestrial planets and the Moon are differentiated, with dense iron-rich cores and rocky mantles. The uncompressed densities of Earth and Venus are similar. Mercury has a high density which suggests it has relatively large core. Conversely, the Moon has a low density, indicating a very small core. There is little observational evidence that asteroids are differentiated except for Vesta and Ceres (Thomas *et al.* 2005). However, iron meteorites from the cores of differentiated asteroids are quite common, and the irons found to date come from several dozen different parent bodies (Meibom & Clark 1999). Most meteorites come from asteroids that never differentiated. These "chondritic" meteorites consist of intimate mixtures of heterogeneous material: millimeter-sized rounded particles that were once molten, called chondrules, similarly sized calcium–aluminum-rich inclusions (CAIs), and micrometer-sized matrix grains.

The spectral properties of main-belt asteroids vary with distance from the Sun (Gradie & Tedesco 1982), suggesting a radial gradient in either composition or degree of thermal processing. The inner belt (<2.5 AU) contains many S-type asteroids (see Fig. 10.1) which appear to be dry. The middle belt (2.5–3.2 AU) is dominated by C-type asteroids that commonly contain hydrated minerals (Rivkin *et al.* 2003). A handful of cometary objects have been found in the outer half of the Asteroid Belt (Hsieh & Jewitt 2006), indicating that at least some objects in this region contain water ice. Jupiter's Trojans are predominantly D-type asteroids (Gradie & Tedesco 1982).

10.1.2 Chemical and isotopic compositions of the inner planets and asteroids

The CI class of chondritic meteorites are thought to have a similar elemental composition to that of the solar nebula (see Chapter 6 of this volume). All elements

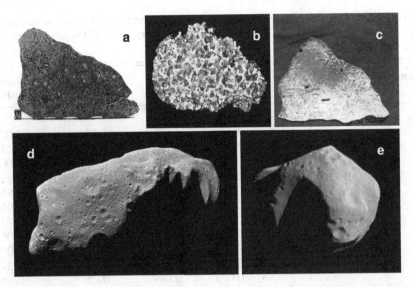

Figure 10.1 (a) a chondritic meteorite; (b) a pallasite stony-iron meteorite; (c) an iron meteorite; (d) S-type asteroid 243 Ida; (e) C-type asteroid 253 Mathilde. Credit: NASA.

are present in CI chondrites and the solar photosphere in roughly the same ratios, except for volatile H, C, N, O, and the noble gases, and elements such as Li that have been depleted in the Sun by nuclear reactions (Anders & Grevesse 1989; Lodders 2003; Palme & Jones 2003). All other types of meteorites and the bulk terrestrial planets are depleted in some elements relative to CI chondrites (see Table 10.1). The depletion patterns differ for different objects and meteorite classes, but the depletion generally increases with an element's volatility (Alexander *et al.* 2001; Davis 2006). Igneous differentiation has modified the elemental abundance patterns in the terrestrial planets and differentiated meteorites, but this effect can be accounted for by comparing elements such as K and U that have similar mineral/melt partitioning behavior but different volatility. There appears to be a progression in the degree of volatile-element depletion from primitive carbonaceous chondrites to ordinary chondrites to the terrestrial planets to some differentiated meteorites (Halliday *et al.* 2001; Davis 2006).

With the exception of oxygen, the Solar System has a remarkably uniform isotopic composition, with variations of less than one part in 1000 for all objects ≥ 1 cm in size. Larger isotopic anomalies are seen only in a few submillimeter-sized CAIs, at the level of a few percent for some elements (e.g. MacPherson 2007), and in micrometer-sized presolar, circumstellar grains, which show large isotopic variations in all elements (Zinner 2007). Oxygen is a notable exception to this rule, with non-mass-dependent variations of up to 20% among Solar System materials

Table 10.1 *Elemental abundances for several objects in the Solar System, normalized to Si and CI chondrites. Following Halliday et al. 2001, Lodders 2003. Siderophile elements are indicated by* *.

Element	50% Condensation temperature (K)	Earth mantle	Mars mantle	CV chondrite	Solar photosphere
W*	1789	0.13	0.6		
Al	1653	1.33	1.00	1.37	1.02
Ca	1517	1.38	1.00	1.37	1.11
Ni*	1353	0.096	0.019	0.86	1.00
Mg	1336	1.17	1.00	1.04	0.96
Fe*	1334	0.16	0.39	0.83	0.94
Cr	1296	0.55	1.00	0.91	0.96
Mn	1158	0.27	1.00	0.50	0.77
K	1006	0.22	0.30	0.38	1.06
Na	958	0.28	0.38	0.46	1.00
F	734	0.18	0.31	0.37	1.24
Zn	726	0.073	0.099	0.25	0.96
S	664			0.23	1.04
O	180[a]	0.47	0.47	0.54	1.87
C	40[b]				9.20

[a] Water ice.
[b] Methane clathrate.

(Kobayashi *et al.* 2003; Sakamoto *et al.* 2007; Seto *et al.* 2008), and a few tenths of a percent among planets and bulk meteorites (although Earth and the Moon have essentially identical O isotopic compositions). These variations are used to probe a wide variety of Solar System processes and provide one of the criteria used to classify meteorites. It seems likely that isotopic variations in O arose owing to photochemical effects in the solar nebula (e.g. Clayton 2002; Yurimoto & Kuramoto 2004; Lyons & Young 2005).

Solar System bodies exhibit a wide range of oxidation states. This is especially true of chondritic meteorites, which come from parent bodies that never melted and may provide a good reflection of the bulk nebular conditions under which they formed. The CI and CM carbonaceous chondrites are highly oxidized and contain little or no metal. At the opposite extreme, enstatite chondrites are highly reduced with abundant metal and little oxidized Fe present in silicates. The individual components of chondrites also exhibit a range of redox states. Type-II chondrules contain significant amounts of oxidized Fe, indicating formation under conditions much more oxidizing than the Solar System as a whole. The CAIs appear to have formed under reducing conditions typical of the bulk Solar System (see MacPherson *et al.* 2008).

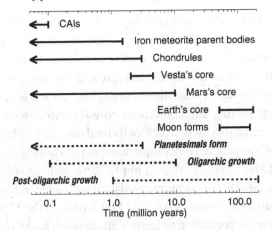

Figure 10.2 The timing of events early in the Solar System based on isotopic dating and numerical models. Data from Kleine *et al.* 2002, Halliday 2004, Kleine *et al.* 2005, Nimmo & Kleine 2007, Scott 2007, Touboul *et al.* 2007, 2008.

10.1.3 Cosmochemical timescale constraints

The early Solar System contained roughly a dozen short-lived isotopes that were probably injected into the solar nebula from several stellar sources (Meyer & Zinner 2006). These radionuclides, together with the U–Pb system, have contributed greatly to our knowledge of early Solar System chronology, and there is a growing consensus on the timing of events (see Fig. 10.2). The CAIs are the oldest objects known to have formed in the Solar System. Their absolute age is 4567.18 ± 0.11 Ma, determined using the U–Pb system (Amelin 2006). Lead-isotope dating (Amelin *et al.* 2002) and the $^{26}Al–^{26}Mg$ system both indicate there was a delay of up to 3 Myr between the formation of CAIs and chondrules (e.g. Kita *et al.* 2005). It has long been assumed that the parent bodies of iron and differentiated-silicate meteorites formed after CAIs and chondrules. However, recent work using the $^{182}Hf–^{182}W$ system suggests these bodies formed within 1 Myr of CAIs (Kleine *et al.* 2005), possibly pre-dating most chondrules. Some recent work on the $^{60}Fe–^{60}Ni$ system suggested that differentiated meteorites might have formed before ^{60}Fe was injected into the Solar System (Bizzarro *et al.* 2007), but more recent measurements demonstrate that this happened at an early stage and was mixed to the point at which heterogeneity was $<10\%$ (Dauphas *et al.* 2008). The $^{53}Mn–^{53}Cr$, $^{26}Al–^{26}Mg$, and $^{182}Hf–^{182}W$ systems indicate that the parent body of the HED basaltic achondrites, probably Vesta, differentiated quite early, within only a few million years of CAI formation (Lugmair & Shukolyukov 1998; Srinivasan *et al.* 1999; McKeegan & Davis 2007; Touboul *et al.* 2008).

The differentiation histories of Earth, Mars, and the Moon are constrained by short-lived isotope systematics and the U–Pb isotope system. The $^{182}Hf–^{182}W$

system is widely used since the parent isotope is lithophile (partitions into a planet's silicate mantle) while the daughter isotope is siderophile (tends to enter the core). Earth's mantle contains a small ^{182}W excess compared to chondritic meteorites (Kleine *et al.* 2002; Yin *et al.* 2002). Interpreting this excess requires assumptions about how well material was mixed during impacts while Earth was growing. Assuming complete mixing and continuous core formation with an exponentially declining impact rate, Earth reached 63% of its final mass in 11 Myr (Yin *et al.* 2002). In practice, mixing was probably incomplete, especially during large impacts. Interpreted in this light, the ^{182}Hf–^{182}W data imply Earth took longer to form (Halliday 2004; Kleine *et al.* 2004). This is more consistent with a mean core formation time of ~100 Myr (equivalent to an age of 4.45 Ga) based on the U–Pb system, and a similar timescale for the formation of Earth's atmosphere based on the I–Xe system (Allègre *et al.* 1995; Halliday 2004). The U–Pb dates must also be treated with some caution since Pb is relatively volatile and may have been depleted during Earth's formation (Halliday 2004). The oldest rocks on Earth's surface are only ~4.0 Ga. However, zircon crystals from the Jack Hills in Western Australia are as old as 4.4 Ga (Wilde *et al.* 2001), providing a firm lower limit on Earth's age.

The early history of Earth is greatly influenced by the probable impact of a Mars-sized body to form the Moon. Core-formation models suggest both Earth and the impactor were already differentiated by the time of the impact (Tonks & Melosh 1992). The lack of a clear ^{182}W excess in uncontaminated lunar samples implies that the Moon-forming impact took place >50 Myr after the start of the Solar System (Touboul *et al.* 2007). The oldest known lunar samples are ~150 Myr younger than CAIs, based on Sm–Nd dating (Touboul *et al.* 2007), which provides a lower limit on the Moon's age.

The differentiation history of Mars is constrained by ^{182}Hf–^{182}W and ^{147,146}Sm–^{143,142}Nd systematics. One model suggests core segregation occurred at 4556 ± 1 Ma and silicate differentiation (crust formation) occurred at 4526 ± 21 Ma (Borg & Drake 2005), equivalent to roughly 11 and 41 Myr after CAI formation, respectively. However, the Hf/W ratio of Mars is sufficiently uncertain that the ^{182}Hf–^{182}W data are consistent with core formation any time within the first 10 Myr of the Solar System (Nimmo & Kleine 2007). The unknown level of mixing during impacts adds to this uncertainty. The lack of samples from Venus and Mercury make it impossible to determine the differentiation chronologies of these planets at present.

10.2 Planetesimal formation

Many of the observations described in the previous section can be explained by the planetesimal theory, in which rocky planets form by the aggregation of huge numbers of asteroid-sized building blocks called "planetesimals" (Safronov 1969). This will be the focus of the rest of the chapter. It is currently unclear how planetesimals form. The two leading models, pairwise sticking of dust grains and gravitational instability of a dense layer of dust grains, both encounter severe difficulties in the face of turbulence, as we will see in Sections 10.2.2 and 10.2.3. A third group of models, in which turbulence leads to the formation of gravitationally bound clumps, and ultimately planetesimals, has enjoyed a good deal of success lately, and these models will be described in Section 10.2.4.

10.2.1 Turbulence in the nebula

Most young stars with optically thick disks appear to be accreting gas, which suggests that gas is continually flowing inwards through the disk and onto the star (Calvet *et al.* 2004). This inward viscous accretion is probably caused by turbulence in the differentially rotating disk. The existence of turbulence and the strength of the turbulent eddies can have a substantial impact on planetesimal formation, as we will see in Section 10.2.2.

The mechanism driving turbulence is unclear. Gas falling from the star's molecular cloud core onto the disk can generate turbulence, but this epoch probably lasts for only $\sim 10^5$ years. Gravitational torques can redistribute angular momentum in Keplerian disks, but these are probably only important in the cool, outer regions of massive disks (Laughlin & Bodenheimer 1994). Hydrodynamic turbulence, caused by convection and differential rotation in the disk, has long been proposed as a way to drive viscous accretion, and this remains an active area of research (Mukhopadhyay 2006). To date, laboratory experiments have failed to find sustained hydrodynamic turbulence, although these experiments have yet to probe regions of parameter space appropriate for protoplanetary disks (Cuzzi & Weidenschilling 2006).

Recently, attention has focused on turbulence generated by magneto-rotational instability (MRI). This arises when ionized material in the disk responds to both the differential rotation of the disk and the presence of an intrinsic magnetic field (Hawley *et al.* 1995). Collisions between ions and molecules mean that MRI also affects neutral gas which makes up most of the mass of the disk. Simulations show that MRI can be highly effective, but it is probably restricted to parts of the disk where the ionization level exceeds a minimum threshhold. This will be the case close to the star due to thermal ionization, and also in the outer disk and surface layers owing to ionization by cosmic rays and X-rays (see Chapters 3 and 4). Ionization will be lowest near the mid-plane a few astronomical units from the star where the disk is optically thick to ionizing radiation, and dust grains rapidly mop up charged particles. Turbulence levels may be substantially lower here, giving rise to a layered disk with viscous accretion mostly occurring in the surface layers (Matsumura & Pudritz 2006).

The turbulent viscosity ν is often parameterized using $\nu = \alpha c_s H$, where it is assumed that turbulent eddies will be smaller than the scale height H of the gas and will rotate more slowly than the sound speed c_s. The observed rates of gas accretion onto young stars suggest that $10^{-2} \lesssim \alpha \lesssim 10^{-4}$ (Hueso & Guillot 2005). Most studies assume that α is a constant, although in practice it probably varies with time and location.

10.2.2 The meter-size barrier

Laboratory experiments show that micrometer-sized dust grains, like those in proto-planetary disks, readily stick together during collisions, forming porous aggregates up to $\sim$1 cm in size (Poppe *et al.* 2000; Marshall & Cuzzi 2001; Krause & Blum 2004). The outcome of subsequent collisions between these aggregates depends on their internal structure and the impact velocity. Highly porous, fractal aggregates stick together at low collision velocities, and become compacted and fragmented at velocities $>$1 m s^{-1} (Blum & Wurm 2000). Moderately porous aggregates stick at low collision speeds, rebound and undergo some fragmentation at intermediate speeds, and can embed themselves in one another at speeds $>$10 m s^{-1} (Wurm *et al.* 2005). Compact objects like chondrules rebound during low-speed collisions and undergo some fragmentation at speeds $>$20 m s^{-1} (Ueda *et al.* 2001).

Growth is likely to become more difficult as bodies approach 1 m in size. The presence of gas affects the relative velocities of small particles and the outcome of collisions. Grains and aggregates settle vertically towards the disk mid-plane at different rates determined by their terminal velocities. A pressure gradient causes gas to orbit the star at a slightly different rate than a solid particle, and the resulting headwind/tailwind removes or adds angular momentum to the particle's orbit. As a result, particles drift radially towards the star, or the nearest local pressure maximum, at rates that depend on the size of the particle. If the gas is turbulent, particles become coupled to turbulent eddies via gas drag. Particles will experience a range of collision speeds even with other objects of the same size. All of these effects become progressively more important with increasing particle size up to $\sim$1 m.

Radial drift rates are highest for meter-sized objects, which typically move inwards by 1 AU every few hundred years (Weidenschilling 1977a). Rapid radial drift implies either that objects grow quickly from centimeter size to $\gg$1 m size, or that planetesimals at 1 AU begin life much further out in the disk. The rapid drifting of meter-sized bodies means they experience a headwind $\sim$50 m s^{-1} and undergo high-speed collisions with small particles entrained in the gas, possibly leading to erosion rather than accretion.

In the absence of turbulence, meter-sized objects collide with one another at low speeds, leading to rebound or merger rather than fragmentation. Collisions between large and small objects may often lead to sticking. Numerical coagulation models show that micrometer-sized grains will grow into kilometer-sized planetesimals in 10^3–10^5 orbital periods for plausible sticking and fragmentation probabilities (Weidenschilling 1997; Dullemond & Dominik 2005; Ciesla 2007a). Growth may also proceed readily in layered accretion disks. Models in which the mid-plane is initially laminar predict the formation of ~100 m bodies in 5×10^4 years at 1 AU (Ciesla 2007a). These objects are large enough to survive the subsequent onset of global turbulence when the mid-plane dust opacity falls to the level at which MRI begins to operate. The snowline may provide an ideal location for planetesimal formation in a layered disk since this can correspond to a pressure maximum in the disk (Kretke & Lin 2007) where radial drift ceases. This allows the growth of kilometer-sized bodies in only ~1000 orbits (Brauer *et al.* 2008).

A common aspect of many of these growth models is that a good deal of second-generation dust is produced during collisions between larger aggregates. This partially replenishes the dust initially present in the disk, which would otherwise be rapidly consumed via coagulation. This may help explain why some older protoplanetary disks still contain substantial amounts of dust. However, the short growth times predicted by these models may be hard to reconcile with U–Pb and ^{26}Al–^{26}Mg age data, which imply that some meteorite parent bodies required 2–3 Myr to form (Scott 2007).

Turbulence increases the collision speeds of all particles that are small enough to be dragged along by turbulent eddies. Meter-sized objects couple to the largest eddies and experience the greatest collision velocities as a result (Cuzzi & Weidenschilling 2006). The stochastic nature of turbulence means that identically sized particles can collide at high speeds, which is not the case in a laminar disk. Collisions between meter-sized particles are particularly energetic, so the question of whether growth stalls at this size depends sensitively on the strength and number of these objects.

Coagulation models that use relatively strong particles (disruption energy per unit mass $Q^\star = 5 \times 10^5$ erg g^{-1}) find that growth stalls in a turbulent disk when particles are ~1 cm in size (Weidenschilling 1984). At this point aggregration and fragmentation are roughly in equilibrium. Simulations using much weaker particles ($Q^\star = 10^3$ erg g^{-1}) yield a very different size distribution so that a few objects grow to ~10 m in size (Dullemond & Dominik 2005). Models for the growth of millimeter-sized chondrules in the presence of fine dust grains find that chondrules accrete porous dust rims that become compacted during subsequent collisions, increasing the chance that chondrules will stick together (Ormel *et al.* 2008). However, these compound objects generally do not grow beyond ~1 m in radius in the presence of turbulence.

The short drift lifetimes and high collision velocities experienced by meter-sized particles have led many researchers to conclude that dust aggregation stalls in this size range, creating a "meter-size barrier" to growth. Moving past this meter-size barrier via direct coagulation may require that the nebula is non-turbulent or that particles have very low strengths.

If a significant portion of the solid material is contained in meter-size bodies, substantial radial mixing is inevitable. Disk temperatures decrease with distance from the star, so meter-sized objects will drift inwards until they become hot enough to evaporate. Volatile ices such as CO, CO_2, and CH_4 evaporate in the outer disk. Water ice evaporates near the "snowline," which lies several AU from a star like the Sun (Kennedy & Kenyon 2008). Refractory materials such as silicates evaporate closer to the star. Vapor is initially concentrated near each evaporation front, gradually diffusing across the disk if the gas is turbulent. Radial drifting of ice-rich bodies probably changes the chemical composition and oxygen fugacity of the disk over time and at different locations. Numerical models suggest the inner disk becomes more oxidizing at early times, and progressively more reducing later when most H_2O in the outer disk is incorporated into large bodies (Ciesla & Cuzzi 2006).

Millimeter-sized particles such as chondrules and CAIs have drift lifetimes $\sim 10^5$ years at 1 AU. Both types of particle are present in most chondritic meteorites, even though chondrules typically formed 1–3 Myr after CAIs (Amelin et al. 2002; Kita et al. 2005). The CAIs could have survived by becoming incorporated into an early generation of planetesimals and later released during collisions. However, these planetesimals must have been large enough (>1 km) to avoid being lost by radial drift, but small enough (<10 km) to avoid substantial heating and melting by short-lived radioactive isotopes (Hevey & Sanders 2006). Melting would have been especially likely in planetesimals containing many CAIs due to the large amount of ^{26}Al that would be present. It seems unlikely that planetary growth would stall in this narrow size range for several million years. Instead, CAIs and chondrules were probably present in the solar nebula for extended periods of time. The fact that they were not quickly swept up into larger bodies argues that planetesimal formation was inefficient, at least in the Asteroid Belt.

Several factors prolong the lifetime of millimeter-sized particles such as CAIs. Rapid viscous spreading of a disk shortly after its formation will transport thermally processed material from the inner regions to larger radii (Dullemond et al. 2006). In a turbulent disk, some CAIs diffuse outwards against the overall inward flow, allowing a significant fraction to survive for at least 10^6 years (Cuzzi et al. 2003). Survival rates are especially enhanced when both vertical and radial motions in a disk are considered (Ciesla 2007b). At early times, millimeter-sized particles may become entrained in winds originating from the inner edge of the disk (the "X-wind"), traveling out of the disk plane and returning to the disk several astronomical units from the star (Shu *et al.* 2001). Later, when the inner disk becomes optically thin, CAIs will be pushed outwards by photophoresis – the asymmetric rebounding of gas molecules off a particle that is hotter on one side than the other (Wurm & Krauss 2006).

10.2.3 Planetesimal formation via gravitational instability

Gravitational instability (GI) may provide a way to form planetesimals directly, bypassing the meter-size barrier (Safronov 1969; Goldreich & Ward 1973). If small particles sediment to a thin layer at the disk mid-plane, they will form gravitationally bound clusters, provided that the velocity dispersion v_{rel} of the particles satisfies the Toomre instability criterion $v_{rel} \lesssim \pi G \Sigma_{solid}/\Omega$, where Ω is the orbital frequency and Σ_{solid} is the local surface density of particles. In the absence of other forces, a gravitationally bound clump will contract over time to form a planetesimal.

The velocity dispersion of the particles depends on the thickness of the particle layer, so this is the key factor determining whether GI will take place. Particles settle to the disk mid-plane owing to the gravitational pull of the star and the disk itself. Settling is opposed by turbulence, which causes particles to diffuse vertically away from the mid-plane. As particles become concentrated near the mid-plane, they begin to drag along the gas so that it orbits the star at the same speed as a solid particle, given by Kepler's law. Gas above and below the mid-plane typically orbits at sub-Keplerian speeds due to the outward pressure gradient in the disk. This vertical velocity shear generates turbulence, even in an otherwise laminar disk (Weidenschilling 1980), except in rare locations where there is no radial pressure gradient.

The thickness of the particle layer is determined by a balance between gravity and turbulent mixing. Small particles form a thicker layer since they are more strongly coupled to the gas. For particles <1 cm in size, at 1 AU, strong vertical density gradients stabilize the particle layer against turbulence (Sekiya 1998). In this case the thickness of the particle layer no longer depends on particle size, and particles become somewhat more concentrated near the mid-plane.

Calculations that consider settling and turbulence find that gravitationally bound clumps are unlikely to form unless solid material is concentrated radially as well as vertically until $\Sigma_{solid}/\Sigma_{gas} \gtrsim 1$ (Garaud & Lin 2004; Gómez & Ostriker 2005). This is two orders of magnitude greater than the ratio for the solar nebula as a whole, so another mechanism is required to raise $\Sigma_{solid}/\Sigma_{gas}$ before GI can occur. This may happen in several ways. In a massive disk, where the gas itself is close to being gravitationally unstable, spiral density waves develop. These waves correspond to gas pressure maxima, so particles drift towards them (Haghighipour & Boss 2003;

Rice *et al.* 2006). Small particles (<1 cm at 1 AU) drift inwards in the Epstein drag regime (see Chapter 3). Epstein drift rates increase with radial distance, so particles tend to accumulate in the inner disk, increasing $\Sigma_{solid}/\Sigma_{gas}$ as a result (Youdin & Shu 2002). At late times, $\Sigma_{solid}/\Sigma_{gas}$ increases as gas begins to disperse (see Chapter 9 of this volume). Meter-sized particles can also become concentrated in the center of long-lived anticyclonic vortices if these are present near the mid-plane of the disk (Klahr & Bodenheimer 2006).

Until now, we have considered turbulence generated by the presence of a particle layer. However, any source of turbulence, such as MRI, can potentially frustrate gravitational instability. Cuzzi & Weidenschilling (2006) have estimated that global turbulence levels equivalent to $\alpha \geq 2 \times 10^{-4}$ and $\alpha \geq 10^{-7}$ will prevent meter-sized and millimeter-sized particles, respectively, from settling to a layer thin enough to undergo GI. This suggests GI will be restricted to laminar regions of the disk such as Dead Zones, unless bodies can first grow significantly beyond 1 m in size as a result of pairwise sticking.

10.2.4 Particle concentration

Intrinsic turbulence in the disk poses severe problems for planetesimal formation by either pairwise sticking or gravitational instability. This has led to the development of a third class of model that embraces turbulence as a necessary ingredient for planetesimal formation.

Turbulence can concentrate meter-sized bodies as they rapidly drift to locations where the gas pressure is temporarily at a maximum. Further concentration occurs due to the "streaming instability"—regions containing many particles tend to drag gas along with them so the particles experience a reduced headwind and drift inwards more slowly as a result. Particles further out in the disk drift inwards at the usual rate and so objects accumulate in regions where the density of particles is already high (Johansen & Youdin 2007). If a large fraction of solid material exists in meter-sized bodies, these processes, combined with collisional damping, can rapidly form bodies >100 km in size (Johansen *et al.* 2007). However, it is unclear how much solid material will exist in meter-sized bodies since these objects are rapidly lost as a result of inward drift and collisional fragmentation.

A more promising model involves the concentration of millimeter-sized particles in stagnant regions between the smallest eddies, forming gravitationally bound clumps (Cuzzi *et al.* 2001). Calculations show that turbulence can increase the local solid/gas density up to ~100 (i.e. 10^4 times that in the protoplanetary disk as a whole), at which point the high particle density tends to shut off the turbulence (Cuzzi *et al.* 2008).

If a gravitationally bound clump forms, the relative velocities of the particles within the clump must be damped before the clump can collapse to form a solid planetesimal. Gas drag and particle–particle collisions will do this, with damping being more effective for small particles. The collapse rate is limited by the pressure of gas within the clump – as particles move inwards, the gas becomes compressed, opposing further collapse (Cuzzi & Weidenschilling 2006). The rate of collapse is set by the time required for particles to settle to the center, which may be hundreds of orbital periods for millimeter-sized particles (Cuzzi & Weidenschilling 2006).

Low-density clumps orbit the star at roughly Keplerian speeds, while gas in the surrounding disk typically orbits at $\sim 50\,\mathrm{m\,s^{-1}}$ more slowly. As a result, clumps are vulnerable to disruption by the ram pressure of the gas. However, numerical simulations suggest that the mutual gravitational attraction of solid particles within a clump is sufficient to keep them largely intact during collapse, provided the initial clump is $\gtrsim 10^3$ km in diameter (Cuzzi *et al.* 2008; see Figure 10.3).

In this model, large gravitationally bound clumps form only occasionally, which would explain why planetesimal formation in the Asteroid Belt continued for several million years. This mechanism also reproduces the narrow size distribution of chondrules seen in chondritic meteorites (Cuzzi *et al.* 2001). While the mean chondrule size differs from one meteorite type to another, the size distributions closely

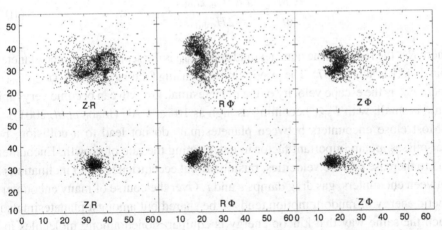

Figure 10.3 The effect of ram pressure acting on a clump of particles in a gas disk. The six panels show the positions of the particles after 1.75 orbital periods. Each point in the plots is a "superparticle" representing a large number of millimeter-sized objects. The clump is initially 10 model units in diameter, equivalent to $\sim 10^4$ km. Upper panels (front, above, and side views): without any cohesion, the clump is soon shredded. Lower panels: when gravitational forces between particles are included, the clump remains mostly intact and becomes more compact. From Cuzzi *et al.* (2007).

match the distribution expected under turbulent concentration provided that the turbulence level is somewhat different in each case (Cuzzi *et al.* 2001).

10.3 Growth of rocky planets

10.3.1 Runaway growth

Gravity plays an increasingly important role in the growth of objects once they grow larger than 1 km in size. Energetic collisions can shatter large planetesimals, but gravity is often sufficient to hold the fragments together and retain some material from the impactor as well, resulting in net growth. Numerical simulations show that impacts onto a 1 km radius planetesimal at 20 m s^{-1} often lead to erosion unless the impactor is <one-fifth of the mass of the target (Leinhardt & Richardson 2002). However, impacts at the same speed onto a 10 km radius planetesimal result in net growth, even when the impactor is as large as the target (Benz 2000).

The probability of a collision between two objects depends on their relative velocity v_{rel}. When two bodies are moving slowly with respect to each other, there is time for gravity to pull them closer together, focusing their trajectories towards each other. The increase in a planetesimal's mass M as it accretes smaller bodies is roughly

$$\frac{dM}{dt} \simeq \frac{\pi R^2 v_{rel} \Sigma_{solid}}{2H_{solid}} F_{grav} \qquad (10.1)$$

where R is the radius of the planetesimal and H_{solid} is the scale height of the planetesimal disk (Lissauer 1987). The gravitational focusing factor $F_{grav} \simeq 1 + (v_{esc}/v_{rel})^2$, where v_{esc} is the escape velocity of the planetesimal. Growth can become very rapid if v_{rel} is small, with F_{grav} becoming as high as 10^4 at 1 AU (Inaba *et al.* 2001).

Most close encounters between planetesimals do not lead to a collision, but near misses are an important factor in determining the rate of growth. Encounters tend to increase relative velocities, raising orbital eccentricites e and inclinations i. Between encounters, gas drag damps e and i. Over the course of many encounters, kinetic energy of random motion tends to be shared out amongst planetesimals in much the same way that kinetic energy is equipartitioned among molecules in a gas. This "dynamical friction" means that large bodies tend to have small relative velocities and vice versa.

A combination of gravitational focusing and dynamical friction leads to runaway growth, where the largest few planetesimals grow the fastest, while most objects grow slowly (Wetherill & Stewart 1993). At 1 AU, runaway growth generates bodies >100 km in radius only ~10^4 years after planetesimals appear in large numbers (Wetherill & Stewart 1993).

10.3.2 Oligarchic growth

During runaway growth, the relative velocities of planetesimals are determined by the strength of encounters between planetesimals, and the growth rate of the largest bodies depends on the mean planetesimal mass m. Eventually, some objects grow large enough that they begin to control the velocity distribution of the smaller planetesimals. This happens when $2M\Sigma_M > m\Sigma_m$, where M is the mass of the largest bodies, and Σ_M and Σ_m denote the surface densities of the large bodies and smaller planetesimals, respectively (Ida & Makino 1993). For a mean planetesimal radius of 10 km, this transition occurs when the largest bodies have masses of about $10^{-6}M_\oplus$ (Thommes *et al.* 2003).

The evolution now enters a slower phase called "oligarchic growth." As the largest bodies increase in mass, their perturbations grow larger, so v_{rel} increases and growth slows as a result. The largest objects, called "planetary embryos," still grow more quickly than most planetesimals, leading to a roughly bimodal mass distribution. Close encounters between embryos typically push their orbits apart, while subsequent dynamical friction with smaller planetesimals circularizes the orbits of the embryos leaving them widely separated (Kokubo & Ida 1995). As a result, each annular region of the disk becomes dominated by a single embryo (Kokubo & Ida 1998). When neighboring embryos do collide, they typically do so at low speeds, comparable to their mutual escape velocity, leading to merger with relatively little mass escaping as fragments (Agnor & Asphaug 2004).

Embryos accrete most of their mass from a "feeding zone" centered on their orbit, with a full width ~10 Hill radii – roughly 0.01 AU for a $10^{-3}M_\oplus$ body at 1 AU. Because embryos mostly accrete local material, objects in different regions of the disk are likely to have different compositions. However, the inward drift of planetesimals and collision fragments means that some radial mixing occurs at this stage.

During oligarchic growth, the relative velocities of planetesimals are much higher than their escape speed. As a result, collisions between planetesimals tend to be disruptive, generating large quantities of dust and collision fragments. The orbits of these fragments are rapidly circularized by gas drag, allowing the embryos to sweep up many of the fragments and speeding up growth (Wetherill & Stewart 1993). However, a substantial fraction of the solid mass is probably lost as fragments rapidly drift inwards and evaporate before they can be accreted (Inaba *et al.* 2003; Chambers 2008). Dust production near 1 AU peaks ~10^5 years after runaway growth begins, when the largest bodies have masses comparable to the Moon, and declines substantially after 10^6 years (Kenyon & Bromley 2004).

Embryos larger than Mars acquire significant atmospheres of captured nebular gas. Planetesimals passing through these atmospheres are slowed, increasing the probability of capture and speeding up growth (Inaba & Ikoma 2003). The

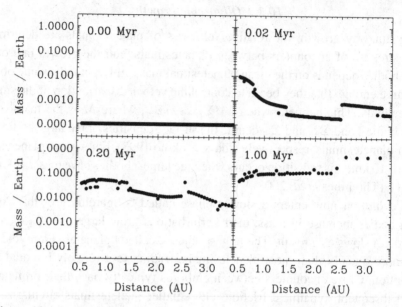

Figure 10.4 A simulation of oligarchic growth in the inner region of a proto-planetary disk around a solar-mass star. In the inner disk, embryos grow to ~0.1 Earth-masses in $\lesssim 10^6$ years. Growth then slows dramatically. Embryos continue to grow larger beyond the snowline at 2.5 AU. The simulation uses the semi-analytic model of Chambers (2008) with $\Sigma_{\text{solid}} \propto 1/a$.

importance of this effect depends on the density gradient within the atmosphere. This in turn depends on the size distribution of dust that is present since this determines the opacity of the atmosphere and the rate at which energy released by impacting planetesimals can be radiated away.

In the absence of planetary migration, oligarchic growth forms lunar to Mars-sized bodies in the inner few astronomical units of a disk in ~10^6 years (Weidenschilling *et al.* 1997; Kokubo & Ida 2002; Chambers 2006; see Fig. 10.4).

10.3.3 Post-oligarchic growth

Oligarchic growth continues until planetary embryos contain about half of the total solid mass (Kenyon & Bromley 2006). At this point, perturbations between neighboring embryos become stronger than damping due to dynamical friction with the remaining planetesimals. The orbits of neighboring embryos begin to cross one another and their eccentricities and inclinations increase rapidly. Gravitational focusing becomes much weaker, and the growth rate drops by roughly two orders of magnitude.

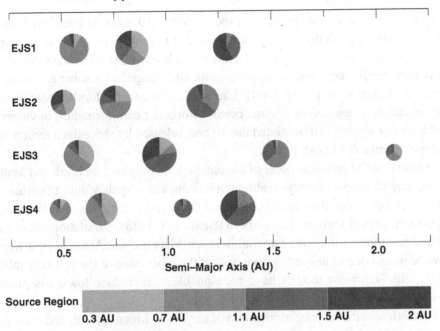

Figure 10.5 The results of four simulations of post-oligarchic accretion in the inner Solar System. Each row of symbols shows planets formed in one simulation, with symbol radius proportional to the cube root of the planetary mass. The differently shaded segments of each symbol indicate the fraction of material coming from different parts of the disk. From O'Brien *et al.* (2006).

Numerical simulations suggest that Earth took $\sim 10^8$ years to grow to its current mass, with the growth rate declining roughly exponentially over time (Chambers 2001; Raymond *et al.* 2005; O'Brien *et al.* 2006). According to these simulations, Earth reached half its current mass in 10–30 Myr. The final stage of planet formation is highly stochastic since few planetary embryos are involved and each undergoes many close encounters for every collision. Small uncertainties in the initial conditions can lead to major differences in the outcome, altering the number of terrestrial planets that form and their orbits and masses (Chambers 2001; O'Brien *et al.* 2006). The protracted nature of the last stage of accretion means that considerable radial mixing takes place. Each terrestrial planet in the Solar System probably acquired material from throughout the inner disk. However, the fractions would have been different in each case (see Fig. 10.5) so heterogeneities established during the oligarchic growth phase were not completely erased. This is reflected in the slightly different oxygen isotope ratios found on Earth and Mars (Clayton & Mayeda 1996). The numerical simulations suggest that a wide range of terrestrial-planet systems will be found in orbit around other stars, but within each system the compositions will vary with distance.

Earth and Venus each probably experienced ~ 10 giant impacts with other embryos during post-oligarchic accretion. Head-on collisions result in a merger with little mass loss. Conversely, when two embryos of similar mass collide at an oblique angle, they tend to separate again and escape, with some exchange of material (Agnor & Asphaug 2004). Large impacts cause extensive heating and the formation of temporary magma oceans, initiating core formation in embryos that have not already differentiated due to heat released by short-lived radioactive isotopes (Tonks & Melosh 1992).

Mercury and Mars are an order of magnitude less massive than Earth and Venus. These may be leftover embryos, although it is unclear why they didn't continue to grow. Mercury's high density suggests it may have lost much of its mantle during a high-speed impact with another embryo (Benz *et al.* 1988). Simulations show that around one third of the solid material between Mars and the Asteroid Belt would have been lost due to unstable resonances at the inner edge of the belt (Chambers 2001), although most models have been unable to reproduce low-mass planets similar to Mars.

The high specific angular momentum of the Earth–Moon system, and the small size of the Moon's core, suggest the Moon formed during an oblique impact between the Earth and another differentiated embryo (Cameron & Ward 1976). Hydrodynamical simulations show that an impact by a Mars-mass body onto Earth at slightly above escape velocity generates a debris disk in orbit, while the cores of the two bodies rapidly coalesce (Canup & Asphaug 2001; Canup 2004). Temperatures in the debris disk are high, so material is initially a mixture of liquid and vapor. As the disk radiates away energy, the material solidifies and accretes into a single large satellite in $\sim 10^3$ years (Kokubo *et al.* 2000; Pahlevan & Stevenson 2007). Substantial mixing of material across the proto-lunar disk and between the disk and Earth is likely, which probably explains why Earth and the Moon have identical oxygen isotope compositions (Pahlevan & Stevenson 2007).

The low Fe abundance in the lunar mantle suggests the Moon-forming impact happened late in Earth's accretion (Canup & Asphaug 2001). It may have been the last collision with another embryo. Simulations of terrestrial-planet formation find that low-velocity, oblique impacts are common (Agnor *et al.* 1999), so that planets like Earth and Venus are likely to experience at least one such impact during their formation. This suggests large satellites may be a common outcome of terrestrial-planet formation.

10.3.4 Migration of rocky planets

The runaway, oligarchic, and early post-oligarchic stages of planet formation take place while gas is present in the disk. Gravitational interactions between a planet and the gas can alter a planet's orbit, causing the planet to migrate radially on timescales shorter than the disk lifetime. At present there is much uncertainty about the strength and the direction of migration, and whether it was important during the formation of the Solar System. However, it seems likely that migration played an important role in the formation of some extrasolar planetary systems (Trilling *et al.* 2002).

A planet embedded in a gas disk launches spiral density waves in the gas interior and exterior to the planet's orbit. The gas in these waves is denser than in the surrounding regions, and the non-uniform distribution of mass exerts gravitational torques on the planet, especially at Lindblad resonances where the orbital period of the gas is $(1 \pm 1/m)P$, where P is the orbital period of the planet and m is an integer. Gas moving with the same average angular speed as the planet (the co-rotation resonance) also exerts a torque on the planet, although this is typically weak for low-mass planets.

The torques acting on a planet are generally not in equilibrium since the outer Lindblad resonances lie closer to the planet than the inner ones, especially if the disk has a steep outward pressure gradient (Ward 1997). Planets with a mass $< 10 M_\oplus$ lose angular momentum to the disk and move inwards, undergoing "Type-I migration." A second regime (Type-II migration) affects giant planets that are massive enough to clear a gap in the disk.

Analytical and numerical calculations find that Type-I migration is rapid, increasing linearly with the mass of a planet. An Earth-mass planet at 1 AU has a lifetime of $\sim 10^5$ years (Tanaka *et al.* 2002). This is significantly shorter than both the time required to form such a planet and the lifetime of a typical protoplanetary disk (Haisch *et al.* 2001). As the gas disperses, migration slows down. At some point, planets migrate slowly enough to survive for the remaining life of the disk. Simulations suggest planetary accretion is inefficient as a result: 50–75% of solid mass may be lost from the inner disk due to Type-I migration (McNeil *et al.* 2005). Planet formation may be most effective in low-mass disks where planets migrate less rapidly.

Several factors may alter Type-I migration rates. A planet changes the distribution of gas in the vertical direction in the disk, affecting the illumination of the gas by the star. This can reduce migration rates for large terrestrial planets by a factor of two (Jang-Condell & Sasselov 2005). Radiative transfer in the region around a planet can also slow or reverse migration in the inner regions of a disk (Paardekooper & Mellema 2006). When planets grow to $\sim 5 M_\oplus$, the co-rotation torque becomes increasingly important, slowing or possibly reversing Type-I migration (Masset *et al.* 2006).

In the presence of a significant toroidal magnetic field, Type-I migration will be slowed or reversed due to additional resonances associated with the magnetic field (Terquem 2003). If the disk is turbulent due to MRI, overdense regions will exert torques on a planet's orbit that may be at least as large as those generated in a laminar disk (Laughlin *et al.* 2004). These torques will fluctuate over time, causing the orbital radius of a low-mass planet to undergo a random walk. In most cases, this reduces the dynamical lifetime, but a small fraction of planets can survive for much longer times, possibly exceeding the lifetime of the disk (Johnson *et al.* 2006).

Disk torques also alter e and i, typically reducing these quantities for terrestrial-mass planets. Here, the inner and outer Lindblad resonances operate in the same sense, so e and i are damped on a timescale that is 10^2–10^3 times shorter than the migration time (Tanaka & Ward 2004). Disk–torque damping may play a signifi-cant role in circularizing the orbits of terrestrial planets (Kominami & Ida 2002), although damping will no longer operate after $\sim$10 Myr due to dispersal of the disk (Pascucci *et al.* 2006).

10.3.5 Chemical evolution of rocky bodies

The previous sections considered the growth of dust grains into planets from a dynamical point of view. Here we discuss aspects of the chemical evolution that occur at each stage of growth.

Meteorites provide perhaps the best record of the chemical evolution of small bod-ies in the Solar System, and this record is supplemented by asteroidal spectroscopy. Meteorites show progressive degrees of thermal processing on their parent aster-oids, from primitive carbonaceous chondrites that contain percent-level quantities of water, through ordinary chondrites that show a wide range of degree of thermal metamorphism, to the achondrites that have been melted and differentiated.

The various degrees of thermal processing in meteorites are matched by differ-ences in the depletion of volatile elements and water (see Davis 2006 for a detailed review of volatile element loss in meteorites and planets). The CI chondrites, which have similar elemental composition to the solar photosphere except for H, C, N, O, and the noble gases, contain about 10% water in the form of hydrated silicates. The CM chondrites have about half as much H_2O and show evidence for exposure to liquid water. They are also somewhat depleted in volatile trace elements, but this could be a nebular signature unrelated to water loss. The CV chondrites contain lit-tle water and are somewhat more depleted in volatile trace elements, but it has been argued that liquid water was once present and altered minerals within CV chon-drites before being lost (Krot *et al.* 1995). The ordinary chondrites are essentially dry and are comparable to CV chondrites in the degree of volatile-element deple-tion. The depletions in the ordinary chondrites show little relation to metamorphic grade of the meteorite. There can be some offset between the volatile-element deple-tions seen in siderophile and lithophile elements in ordinary chondrites: lithophile elements with 50% condensation temperatures below $\sim$1200 K are depleted, but siderophiles are not depleted until temperatures of 700–800 K (see Figure 2 of Davis

2006). This indicates that the volatility fractionations affecting silicates and metal occurred under different conditions. Since chondrites are sedimentary rocks consisting largely of chondrules that were melted and rapidly cooled in zero gravity, the volatility depletions in bulk chondrites almost certainly arose in the nebula, likely during the thermal processing events that melted chondrules (see Chapters 7 and 8).

A number of achondritic meteorite groups, including silicate achondrites, stony-irons, and irons, are more strongly depleted in volatile elements than chondrites or even Mars, Earth, and the Moon. The HED basaltic achondrites, which likely come from Vesta, and the angrites are quite dry and crystallized from essentially water-free melts; they are significantly depleted in volatile K and Rb relative to the terrestrial planets (see Figure 3 of Davis 2006). Perhaps the most extreme case of fractionation is the depletion of some groups of iron meteorites in Ga/Ni and Ge/Ni by factors of up to 10^3 and 10^4, respectively, compared to the Solar System ratio. About 85% of iron meteorites can be classified into one of 13 groups based on trace-element abundances, each of which likely represents a parent body. The remaining 15% are ungrouped and likely come from many parent bodies. Many of the ungrouped irons are strongly depleted in Ga and Ge (see Figure 4 of Davis 2006). Evaporation of volatile elements at plausible nebular pressures leads to substantial isotopic mass fractionation effects (see Davis & Richter 2007 for a recent review), yet depletions in volatile elements like K in chondrites and achondrites (Humayun & Clayton 1995), K in chondrules (Alexander *et al.* 2000), and Ge in iron meteorites (Luais 2007) are not accompanied by isotopic fractionation effects. It appears that the depletions arose due to high-temperature processing in the solar nebula prior to asteroid formation, under conditions where gas–solid or gas–liquid exchange could occur during evaporation. Thermal metamorphism of asteroids is likely to have led to the loss of water and carbon (probably as CO). For early-formed bodies, a likely heat source for this process was the decay of short-lived radionuclides, principally ^{26}Al; impact heating must be invoked for later loss.

Infrared spectroscopy indicates a significant variation in asteroid properties with distance from the Sun, but little unambiguous evidence for melting and differentiation. Only Vesta and the Vestoids, which appear to be collisional fragments of Vesta, have a signature matching the basaltic achondrites. The spectra of S-type asteroids show olivine and pyroxene with a substantial metal component. These objects could be the parent bodies of pallasites, mesosiderites, and other stony-iron achondrites. However, they could also be related to ordinary chondrites, which are the most common type of meteorite to fall on Earth and clearly did not come from differentiated asteroids. The abundant C-type asteroids often show signatures of water-bearing phyllosilicates (Rivkin *et al.* 2003). The D-type asteroids, which are most abundant in the outer Asteroid Belt, are inferred to have anhydrous silicates and organics, but the presence of ice remains an open question. Spacecraft observations of asteroids show they have irregular shapes and high porosities, suggesting that they are rubble piles that have experienced considerable collisional evolution (Britt *et al.* 2006), but it remains unclear how much collisions have affected the volatile content of asteroids.

Siderophile trace elements, such as Ir, are strongly partitioned into the metal phase during core formation, but Earth's mantle is richer in these elements than metal–silicate partitioning experiments suggest it should be. This has led to the hypothesis that the Earth accreted a "late veneer" of chondritic material, amounting to ~1% of its mass, after the planet's core formed (Drake & Righter 2002). The isotopic ratios of Mo and Ru exhibit small variations between different meteorite classes, but the variations in ^{92}Mo and ^{100}Ru are anti correlated, suggesting that there were variations in the ratio of *s*- to *r*- and *p*-process isotopes among different meteorite parent bodies (Dauphas *et al.* 2004). Most Mo in the Earth's mantle arrived before core formation was complete, while almost all the Ru was delivered in the late veneer. Nonetheless, the ^{92}Mo and ^{100}Ru in Earth's mantle follows the same correlation seen in meteorites, indicating that the material in the late veneer came from the same part of the Solar System as the material that makes up the rest of the Earth (Dauphas *et al.* 2004).

Earth and Mars clearly contain H_2O. Venus's atmosphere is very dry, and composed mainly of CO_2, but the high D/H ratio of the small amount of water present suggests Venus was once much wetter than today (Zahnle 1998). Mercury is perhaps too small and too close to the Sun to have acquired and retained water. Water may have been present in much of the material that accreted to form the Earth. Small amounts of water may have been adsorbed onto dust grains at 1 AU by physisorption or chemisorption (Drake 2005). Once Jupiter formed, substantial amounts of water could have been delivered to the growing Earth in the form of planetesimals and planetary embryos from the Asteroid Belt (Morbidelli *et al.* 2000). It is also possible that Earth lay beyond the snowline at some point during the evolution of the solar nebula (Chiang *et al.* 2001) so that local planetesimals contained ice.

Water and other volatiles could have been supplied to Earth by comets and asteroids as part of the late veneer. The arguments for and against this hypothesis have recently been reviewed by Drake (2005). The D/H ratio measured in three comets to date is ~2× higher than on Earth, suggesting that comets could not have supplied more than ~50% of Earth's water (Robert 2001). However, these comets may not be representative of objects colliding with the early Earth. If the Ar/H_2O ratio measured in comet Hale–Bopp is typical, comets would have delivered 2×10^4 times more Ar than is presently found in Earth's atmosphere if they were the main source of Earth's water (Swindle & Kring 2001). Consideration of the abundances of noble metals and noble gases led Dauphas & Marty (2002) to estimate that comets contributed <1% of the Earth's water. It is unlikely that carbonaceous chondrites supplied most of the late veneer since these objects have different Os isotope ratios than Earth's mantle, and almost all the Os in the mantle arrived during the late veneer (Walker *et al.* 2001; Drake & Righter 2002; but see also Dauphas *et al.* 2002). A carbonaceous chondrite source may also be hard to reconcile with the observed Xe isotopic distribution and the Xe/Kr ratio of Earth (Dauphas 2003). Ordinary chondrites are the only primitive meteorites that match the Os isotopic composition of the mantle (Drake & Righter 2002). These come from parent bodies that are currently anhydrous, although they once may have contained water (Grossman *et al.* 2000). Arguments based on Os isotopes do not constrain the amount of water delivered before core formation ended, and simulations of terrestrial-planet formation typically find that substantial amounts of water-rich material can be incorporated before core formation is complete (Morbidelli *et al.* 2000; Raymond *et al.* 2004; O'Brien *et al.* 2006).

10.4 The effect of the giant planets and the formation of the Asteroid Belt

The giant planets, especially Jupiter and Saturn, significantly influenced accretion in the inner Solar System, with important consequences for the properties of the terrestrial planets, described in Section 10.4.1. The influence of the giant planets is especially strong in the Asteroid Belt. Given that meteorites are our primary samples of primitive Solar System material, understanding the role of dynamical and collisional processes in the formation and evolution of the Asteroid Belt is of fundamental importance for theories of planet formation (Section 10.4.2).

10.4.1 Effect of giant planets on inner Solar System accretion

The most widely accepted model for the formation of giant planets is the "core-accretion" hypothesis. In this model, large planetary embryos formed in the outer Solar System via oligarchic growth on a timescale of $\sim$1 Myr. Embryos grew larger than those in the terrestrial-planet region because more solid mass was present beyond the snowline, and because feeding-zone widths increase with distance from the Sun (Lissauer 1987). Once an embryo grows to a critical mass $M_{crit} \sim 10\,M_\oplus$, gravity becomes too strong to maintain a static atmosphere and runaway gas accretion begins, forming a giant planet with a solid "core" at the center of a gas-rich envelope. Simulations show that Saturn- to Jupiter-mass planets can form this way in $\sim$1$-$10 Myr, with the main source of uncertainty arising from the unknown size distribution of dust grains in the envelope (Pollack *et al.* 1996; Hubickyj *et al.* 2005). This timescale is broadly consistent with the observed lifetimes of protoplanetary disks (Chapter 9; Strom *et al.* 1993; Kenyon & Hartmann 1995; Zuckerman *et al.* 1995; Haisch *et al.* 2001). If the core-accretion model is correct, oligarchic growth would have finished in the terrestrial-planet region, and the final stage of accretion would already be underway, by the time Jupiter and Saturn reached their current masses.

Some models for Jupiter's interior structure find that its core is smaller than M_{crit} (e.g. Chabrier *et al.* 1992; Saumon & Guillot 2004). This led Boss (1997) to propose that giant planets form via gravitational instability instead. In this model, portions of the protoplanetary disk become gravitationally unstable and collapse, forming planets in $\lesssim 10^3$ yr. This is the same mechanism that may form planetesimals but applied to the gas component of the disk. If giant planets form this way, they may have been present during the runaway and oligarchic growth stages of terrestrial-planet formation.

If Jupiter and Saturn formed quickly, their perturbations would raise the relative velocities among planetesimals elsewhere in the Solar System. Calculations by Kortenkamp & Wetherill (2000), incorporating perturbations by Jupiter and Saturn and gas drag, suggest that runaway growth could be nearly halted in the Asteroid Belt region. In this case, the largest bodies to form would be comparable to the largest asteroids observed today (this was also suggested earlier by Wetherill 1989 and Wetherill & Stewart 1989). In the terrestrial-planet region, Kortenkamp & Wetherill found that runaway growth would be slowed, but not necessarily prevented.

Under either model for giant-planet formation, the growth of the terrestrial planets is likely to proceed through the stages described in Sections 10.3.1 to 10.3.3. Numerical simulations and isotopic dating suggest the final stage of terrestrial-planet growth took place after the solar nebula dispersed. Jupiter and Saturn must have been almost fully formed by this time. As a result, the terrestrial planets were subject to perturbations from the giant planets during the final stage of growth.

Perturbations are especially important at mean-motion and secular resonances. A mean-motion resonance (MMR) occurs when the orbital period of a body is a fraction of that of the perturbing planet. For example, the 3:1 MMR with Jupiter occurs at a semi-major axis $a \simeq 2.5$ AU, where the orbital period is one third that of Jupiter. A secular resonance occurs when the precession rate of a body's longitude of perihelion ϖ or of its longitude of ascending node Ω equals one of the eigenfrequencies of the linearized secular equations for the Solar System. The most important secular resonances in the inner Solar System are the ν_5 and ν_6, which correspond to eigenfrequencies dominated by Jupiter and Saturn, respectively, when ϖ lies close to the precession rate for one of these planets. A body trapped in a strong resonance can experience a rapid increase in eccentricity to values near unity (Farinella *et al.* 1994), causing the object to collide with the Sun or become ejected from the Solar System following a close approach to Jupiter.

The strength of a resonance depends on the eccentricities of the planets involved (Morbidelli & Henrard 1991; Moons & Morbidelli 1993). Thus, the eccentricities of Jupiter (e_J) and Saturn (e_S) can have a significant effect on the growth of terrestrial planets. Simulations of terrestrial-planet formation often find that e_J and e_S decrease over time, due to the ejection of material from the system (e.g. Petit *et al.* 2001;

Chambers & Cassen 2002; O'Brien *et al.* 2006). If no other mechanisms are invoked to enhance e_J and e_S, they must have once been $\sim 2\times$ larger than at present. When e_J and e_S are large, the amount of material from the Asteroid Belt that is incorporated into the terrestrial planets is reduced because the main-belt resonances become stronger, and material is cleared from the Asteroid Belt more rapidly (e.g. Chambers & Cassen 2002; Raymond *et al.* 2004; O'Brien *et al.* 2006).

Numerical models suggest that the orbits of the outer planets migrated by up to a few astronomical units early in the Solar System as the planets scattered and ejected residual planetesimals (Fernandez & Ip 1984; Malhotra 1993). In this case, e_J and e_S may have been small initially, but would have increased substantially if migration drove the orbits of Jupiter and Saturn across their mutual 2:1 MMR (Tsiganis *et al.* 2005). This event would have destabilized the orbits of many main-belt asteroids, and may be linked to the increased impact rate on the inner planets ~ 3.9 Gyr ago, during the lunar Late Heavy Bombardment (Gomes *et al.* 2005). If $e_J, e_S \simeq 0$ when the inner planets were forming, the ejection rate for asteroids in the main belt would have been relatively low, allowing more material to diffuse into the inner Solar System. This is consistent with the hypothesis that the Asteroid Belt was an important source of Earth's water (e.g. Morbidelli *et al.* 2000; Raymond *et al.* 2004). However, simulations with $e_J, e_S \sim 0$ find that the orbits and masses of the terrestrial planets typically agree less well with the observed values than in simulations where $e_J, e_S \gtrsim 0.05$ (O'Brien *et al.* 2006).

As gas in the solar nebula dissipates, the gravitational potential of the disk changes, causing secular resonances to "sweep" radially through the Solar System. The sweeping of secular resonances through the terrestrial-planet region, especially the ν_5 resonance, could excite the eccentricities of planetary embryos and force rapid accretion of the terrestrial planets (Nagasawa *et al.* 2005). Subsequent tidal interactions with the nebular gas may circularize the orbits of the inner planets leading to their current low eccentricities. This scenario may depend sensitively on the timing of nebular gas dissipation and the initial orbits of Jupiter and Saturn.

The extrasolar planetary systems observed to date often have significantly different giant-planet configurations than the Solar System, including giant planets with highly eccentric orbits and "hot Jupiters" orbiting within 0.1 AU of the central star. Understanding the kind of terrestrial planets that might exist in these systems is a difficult task, as it is not possible to detect and characterize Earth-mass extrasolar planets with current technology. Thus, much of the work on this issue has been done using numerical simulations of planet formation. Levison & Agnor (2003), Raymond (2006), and Raymond *et al.* (2006a) studied a wide range of real and hypothetical giant-planet systems and demonstrated that different giant-planet configurations can lead to substantially different terrestrial-planet systems, in terms of their masses, orbital configurations, water contents, and the likelihood that planets will form in a star's habitable zone (Kasting *et al.* 1993).

Raymond *et al.* (2006b), Fogg & Nelson (2007), and Mandell *et al.* (2007) have examined the effect of giant-planet migration on terrestrial-planet formation. These authors find that a giant planet migrating through the inner part of a protoplanetary disk will accrete or scatter away a substantial fraction of the embryos in this region, while other embryos are trapped in resonances and migrate inwards with the planet. However, the surviving embryos, together with others scattered inwards from the outer disk, contain enough mass to subsequently form terrestrial planets exterior to the giant planet's orbit. A substantial fraction of the mass incorporated in these planets comes from the outer disk, so these objects are likely to be very water-rich compared to the terrestrial planets of the Solar System.

10.4.2 Formation of the Asteroid Belt

The current mass of the Asteroid Belt is $\sim 5 \times 10^{-4}$ $M_{\oplus}$, with most of this mass contained in the few largest asteroids. The growth rate of asteroids in the solar nebula is proportional to the surface density of solid material. The formation of Vesta in only a few million years, inferred from the HED meteorites, implies that the Asteroid Belt once contained $\gtrsim 100$ times more mass than is present today (Wetherill 1989). If the surface density followed a relatively smooth profile, several $M_{\oplus}$ of solids would have been present in this region (e.g. Lecar & Franklin 1973; Weidenschilling 1977b; Safronov 1979; Wetherill 1989). One might expect that runaway, oligarchic, and late-stage growth would have proceeded until all surviving solid material in the Asteroid Belt resided in one or two planets. However, the physical and chemical makeup of meteorites and their oxygen isotope variations show that they come from at least 100 different parent bodies (Meibom & Clark 1999). The 2–3 Myr spread in the ages of chondrules and CAIs in chondrites, argues that planetary growth was actually quite slow in the Asteroid Belt, perhaps because planetesimal formation was inefficient (see Section 10.2.4).

The orbits of the asteroids are very excited dynamically (see Fig. 10.6). Asteroids with diameters $\gtrsim 50$ km have mean proper (time-averaged) orbits with $e = 0.135$ and $i = 10.9°$ (from the catalog of Knežević & Milani 2003). These values of e and i are substantially higher than would be expected owing to simple perturbations from Jupiter and Saturn alone. Additionally, the Asteroid Belt shows evidence for significant radial mixing. The different taxonomic spectral types (S-type, C-type, etc.) are each spread over a range of semi-major axes (on a scale ~ 1 AU), partially overlapping one another, rather than confined to narrow radial zones (Gradie & Tedesco 1982).

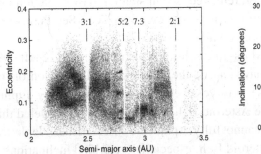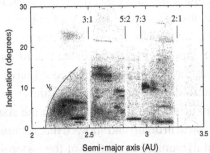

Figure 10.6 Proper orbital elements of numbered asteroids, from the catalog of Knežević & Milani (2003). Also shown are the locations of the main mean-motion and secular resonances.

It was first proposed that runaway growth of large bodies never occurred in the Asteroid Belt and that most of the primordial mass was removed by collisional erosion (e.g. Chapman & Davis 1975). This requires that most of the original mass was contained in small bodies that were easily disrupted during collisions, which would be the case if Jupiter and Saturn formed early, as discussed in Section 10.4.1. Bodies comparable in size to the largest existing asteroids, such as Ceres and Vesta, would be essentially impossible to catastrophically disrupt, so these objects must have been rare. The existence of a basaltic crust on Vesta argues against this hypothesis. This crust formed in the first few million years of the Solar System (Ghosh & McSween 1998; Lugmair & Shukolyukov 1998; Srinivasan *et al.* 1999), but has remained relatively intact, apart from one large impact basin. If the mass of the primordial Asteroid Belt was significantly larger than today, and collisions were the only process removing mass from the belt, it is likely that Vesta's basaltic crust would have been eroded away over the age of the Solar System (Davis *et al.* 1985). Furthermore, it is difficult to explain the dynamical excitation and radial mixing of the Asteroid Belt in a purely collisional scenario.

Instead, several dynamical mechanisms have been proposed to explain the characteristics of the Asteroid Belt. Heppenheimer (1980) and Ward (1981) suggested that secular resonances would have swept through the Asteroid Belt as the solar nebula was dissipating. Numerous authors have studied this mechanism in more detail (e.g. Lemaitre & Dubru 1991; Lecar & Franklin 1997; Nagasawa *et al.* 2000). These studies find that a combination of resonant eccentricity excitation and orbital decay due to gas drag can remove most of the original solid mass from the Asteroid Belt, and generate the eccentric orbits seen in the belt today. The observed inclinations of the asteroids can only be reproduced if the nebula interior to Jupiter's orbit disperses first, followed by the slow dissipation (e-folding time $\gtrsim 3$ Myr) of the outer nebula (Nagasawa *et al.* 2002). In this case, the resonances sweep across the main belt after gas there has been removed, so substantial mass depletion and the observed radial mixing of spectral types does not take place. (See Petit *et al.* 2002 and O'Brien *et al.* 2007 for more detailed reviews.)

Safronov (1979) suggested that Mars- to Earth-mass planetary embryos left over from the accretion process could have caused the excitation and depletion of the Asteroid Belt. In this model, these embryos were scattered inwards by Jupiter onto orbits that traveled through the Asteroid Belt. Petit *et al.* (1999) modeled this scenario in detail and showed that it cannot fully explain the observed mass depletion and dynamical excitation of the Asteroid Belt, especially the orbital inclinations.

Wetherill (1992) proposed instead that runaway and oligarchic growth were able to proceed in the Asteroid Belt, generating planetary embryos *in situ*. His model, and subsequent integrations of planetary embryos in the Asteroid Belt and terrestrial-planet region (e.g. Chambers & Wetherill 1998, 2001), showed that these embryos would be perturbed by one another and by Jovian and Saturnian resonances, becoming excited and often being driven out of the Asteroid Belt. In the majority of these simulations, all embryos are removed from the belt. Figure 10.7 shows a representative example.

Petit *et al.* (2001) modeled the effects of planetary embryos on a population of massless test particles representing asteroids. These authors found that, for essentially any initial distribution of planetary embryos in the main-belt region, $\sim$99% of the asteroids would be dynamically removed with a median lifetime of only a few million years, while the remaining objects would have orbital a, e, and i distributions roughly similar to the observed asteroids. The radial mixing of different asteroid taxonomic types was also reproduced in these simulations. The rate of depletion of the asteroid population depends on the eccentricities and orbital configurations of Jupiter and Saturn (Chambers & Wetherill 2001), and there may be additional effects on the asteroid population from dynamical processes occurring during the Late Heavy Bombardment. In general, though, it appears that this is the most promising model for explaining the origin and early evolution of the Asteroid Belt. These issues are discussed in further detail in O'Brien *et al.* (2007).

An interesting consequence of this model is that bodies from the terrestrial-planet region may be scattered outwards while the planets are forming and implanted in the Asteroid Belt. With dynamical and collisional modeling of this process, Bottke *et al.* (2006) find that this may be the origin of most iron meteorites. This would explain the diversity of iron-meteorite types, why there is little observational or meteoritical evidence of mantle material from differentiated bodies in the Asteroid Belt, and the fact that most iron-meteorite parent bodies appear to have formed $\gtrsim$1 Myr before the parent bodies of the chondritic meteorites (Kleine *et al.* 2005).

Figure 10.8 shows observational estimates of the main-belt size distribution for objects $\gtrsim$1 km in diameter (Jedicke *et al.* 2002; Bottke *et al.* 2005a).

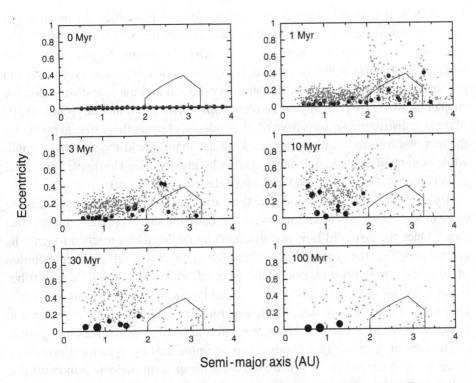

Figure 10.7 Snapshots from a simulation of accretion in the inner Solar System, from O'Brien *et al.* (2006). Jupiter and Saturn are present at $t = 0$ on their current orbits, black particles are embryos and gray particles are planetesimals. By 30 Myr all embryos and many of the planetesimals are cleared from the Asteroid Belt region (shown as the black line). The remaining asteroids are dynamically excited and have experienced significant radial displacement from their initial locations.

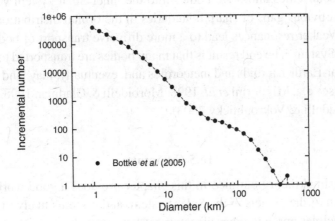

Figure 10.8 Observational estimate of the main-belt size distribution from Bottke *et al.* (2005a), which is based on the debiasing of the cataloged asteroid population and Sloan Digital Sky Survey data from Jedicke *et al.* (2002).

A collisionally evolved population in which asteroid properties are independent of size would have a simple power-law size distribution (Dohnanyi 1969). The observed size distribution is actually wavy, with peaks around 10 and 100 km. The bump at $\sim$100 km is probably a remnant of the size distribution that evolved during the accretion of the asteroids. The bump at $\sim$10 km and the size distribution for smaller asteroids are probably the result of collisional evolution. The specific shape of the size distribution results from the dependence of strength on size, in particular the fact that asteroids with radius <100 m are mainly held together by strength while larger bodies are mainly held together by gravity. (See Holsapple *et al.* 2002 and O'Brien & Greenberg 2003 for more detailed discussions.)

Bottke *et al.* (2005a,b) found that the current asteroid size distribution arose early in its history, when the total mass and collision rate were much higher than today. Once the Asteroid Belt was dynamically depleted and reached roughly its current mass (via the processes described above), there was little further evolution of the size distribution, and hence it has been referred to as a "fossil" size distribution. Collisions still occur, albeit at a reduced rate, and large collisions lead to the formation of asteroid families, which are groups of asteroids that are clustered in orbital-element (a, e, i) space. Numerous asteroid families can be seen in Fig. 10.6.

The current Asteroid Belt continues to be influenced by dynamical processes. Asteroids with diameters $\lesssim$10 km undergo significant semi-major axis mobility due to the Yarkovsky effect, a radiation force resulting from the asymmetry between absorbed and re-emitted radiation by an orbiting, rotating body (see Bottke *et al.* 2002). The Yarkovsky effect can drive bodies into resonances, either strong ones such as the 3:1 MMR with Jupiter and the ν_6, or weaker, higher-order resonances that may involve interactions with Mars as well as Jupiter and Saturn. Strong resonances increase eccentricities and drive bodies into the inner Solar system very quickly, and lead to obvious gaps or sharp boundaries in the orbital distribution as seen in Fig. 10.6. Weaker resonances lead to a more diffusive transport of bodies into the inner Solar System. The end result is that many bodies are transported to near-Earth space as near-Earth asteroids and meteoroids and, eventually, may land on Earth as meteorites (see e.g. Migliorini *et al.* 1998; Morbidelli & Gladman 1998; Morbidelli 1999; Morbidelli & Vokrouhlický 2003).

10.5 Summary

The planetesimal theory outlined in this chapter provides a good working model for the origin of the planets in the Solar System, and it seems likely that the same general principles apply in other planetary systems as well.

Experiments show that micrometer-sized dust grains seen in disks readily stick together building up millimeter- to centimeter-sized aggregates. It is currently unclear how 1–100 km planetesimals form. These objects may emerge from continued particle sticking, or by the sedimentation of aggregates to a thin layer followed by gravitational instability. Both routes are probably restricted to laminar regions. In a turbulent disk, turbulence may concentrate particles into gravitationally bound clumps which then shrink to form planetesimals.

Gravitational interactions between planetesimals and embryos give the largest embryos low relative velocities favorable to further rapid growth. Planetary embryos grow to roughly Mars size in 10^5–10^6 years after planetesimals form. Growth slows substantially when embryos contain most of the mass. Occasional impacts between embryos and the sweep-up of remaining planetesimals complete the formation of terrestrial planets. The Moon probably formed during a giant impact near the end of Earth's growth. While the gas disk is still present, gravitational interactions can cause radial migration of an embryo's orbit, although the rate and direction of migration are poorly constrained at present.

Energy released during impacts causes large embryos and planets to melt and differentiate, forming iron-rich cores and silicate mantles. Earth acquired most of its water before its core finished forming, possibly from the Asteroid Belt. It gained the last 1% of its mass in the form of non-fractionated material after core formation was complete.

Planetary-mass bodies probably formed in the Asteroid Belt and were responsible for its dynamical excitation, radial mixing, and mass depletion. The orbits of these bodies became unstable once Jupiter and Saturn formed. These objects and most remaining planetesimals fell into the Sun or were ejected from the Solar System. The Asteroid Belt may have been further depleted when the giant planets passed through a resonance before reaching their current orbits. The Asteroid Belt has lost relatively little mass due to collisional erosion, and most asteroids $\gtrsim 100$ km in diameter are probably primordial.

Cosmochemical analysis of meteorites provides important constraints on the timing of planet formation. The first differentiated bodies formed $\lesssim 1.5$ Myr after the start of the Solar System. However, processing of dust grains in the nebula and the formation of planetesimals continued for at least the first 2–3 Myr. Vesta formed within 5 Myr, while Mars was fully grown and differentiated by 10–20 Myr. Earth took longer to grow, and it was not until 50–150 Myr that the planet was fully formed and the Moon was present.

Dust plays an important role in many aspects of planet formation. It provides the initial reservoir of material for rocky planets and the cores of giant planets, and dust probably affects the level of turbulence in a protoplanetary disk. Interactions between dust grains and the gas in the disk determine the rate at which planetesimals form. Later, during the runaway and oligarchic growth stages, substantial amounts of second-generation dust are generated as a result of collisions.

Dust affects the structure of early planetary atmospheres and is probably an important factor controlling the rate at which giant planets grow. Long after the planets are fully formed, dust continues to be produced by collisions between asteroids in regions where giant-planet perturbations prevented the growth of planets.

References

Agnor, C. & Asphaug, E. 2004, *Astrophysical Journal*, **613**, L157.

Agnor, C. B., Canup, R. M., & Levison, H. F. 1999, *Icarus*, **142**, 219.

Alexander, C. M. O., Grossman, J. N., Wang, J., *et al.* 2000, *Meteoritics and Planetary Science*, **35**, 859.

Alexander, C. M. O., Boss, A. P., & Carlson, R. W. 2001, *Science*, **293**, 64.

Allègre, C. J., Manhès, G., & Göpel, C. 1995, *Geochimica et Cosmochimica Acta*, **59**, 1445.

Amelin, Y. 2006, *Meteoritics and Planetary Science*, **41**, 7.

Amelin, Y., Krot, A. N., Hutcheon, I. D., & Ulyanov, A. A. 2002, *Science*, **297**, 1678.

Anders, E. & Grevesse, N. 1989, *Geochimica et Cosmochimica Acta*, **53**, 197.

Benz, W. 2000, *Space Science Reviews*, **92**, 279.

Benz, W., Slattery, W. L., & Cameron, A. G. W. 1988, *Icarus*, **74**, 516.

Bizzarro, M., Ulfbeck, D., Trinquier, A., *et al.* 2007, *Science*, **316**, 1178.

Blum, J. & Wurm, G. 2000, *Icarus*, **143**, 138.

Borg, L. & Drake, M. J. 2005, *Journal of Geophysical Research – Planets*, **110**, 12.

Boss, A. P. 1997, *Science*, **276**, 1836.

Bottke, W. F., Durda, D. D., Nesvorný, D., *et al.* 2005a, *Icarus*, **175**, 111.

Bottke, W. F., Durda, D. D., Nesvorný, D., *et al.* 2005b, *Icarus*, **179**, 63.

Bottke, W. F., Nesvorny, D., Grimm, R. F., Morbidelli, A., & O'Brien, D. P. 2006, *Nature*, **439**, 821.

Bottke, W. F., Vokrouhlický, D., Rubincam, D. P., & Brož, M. 2002, in *Asteroids III*, ed. W. F. Bottke, A. Cellino, P. Paolicchi, & R. P. Binzel, University of Arizona Press, 395–408.

Brauer, F., Henning, T., & Dullemond, C. P. 2008, *Astronomy and Astrophysics* **487**, L1.

Britt, D. T., Consolmagno, G., & Lebofsky, L. 2006, in *Encyclopedia of the Solar System*, 2nd edn., ed. L.-A. McFadden, P. Weissman, T. Johnson, & L. Versteeg-Buschman, Elsevier, 349–364.

Calvet, N., Muzerolle, J., Briceño, C., *et al.* 2004, *Astronomical Journal*, **128**, 1294.

Cameron, A. G. W. & Ward, W. R. 1976, *Lunar and Planetary Science*, **7**, 120.

Canup, R. M. 2004, *Icarus*, **168**, 433.

Canup, R. M. & Asphaug, E. 2001, *Nature*, **412**, 708.

Chabrier, G., Saumon, D., Hubbard, W. B., & Lunine, J. I. 1992, *Astrophysical Journal*, **391**, 817.

Chambers, J. 2006, *Icarus*, **180**, 496.

Chambers, J. 2008, *Icarus*, **198**, 256.

Chambers, J. E. 2001, *Icarus*, **152**, 205.

Chambers, J. E. & Wetherill, G. W. 1998, *Icarus*, **136**, 304.

Chambers, J. E. & Wetherill, G. W. 2001, *Meteoritics and Planetary Science*, **36**, 381.

Chambers, J. E. & Cassen, P. 2002, *Meteoritics and Planetary Science*, **37**, 1523.

Chapman, C. R. & Davis, D. R. 1975, *Science*, **190**, 553.

Chiang, E. I., Joung, M. K., Creech-Eakman, M. J., *et al.* 2001, *Astrophysical Journal*, **547**, 1077.

Ciesla, F. J. 2007a, *Astrophysical Journal*, **654**, L159.

Ciesla, F. J. 2007b, *Science*, **318**, 613.

Ciesla, F. J. & Cuzzi, J. N. 2006, *Icarus*, **181**, 178.

Clayton, R. N. 2002, *Nature*, **415**, 860.

Clayton, R. N. & Mayeda, T. K. 1996, *Geochimica et Cosmochimica Acta*, **60**, 1999.

Cuzzi, J. N., Davis, S. S., & Dobrovolskis, A. R. 2003, *Icarus*, **166**, 385.

Cuzzi, J. N., Hogan, R. C., Paque, J. M., & Dobrovolskis, A. R. 2001, *Astrophysical Journal*, **546**, 496.

Cuzzi, J. N., Hogan, R. C., & Shariff, K. 2007, *Lunar and Planetary Institute Conference Abstracts*, **38**, 1439.

Cuzzi, J. N., Hogan, R. C., & Shariff, K. 2008, *Astrophysical Journal*, **687**, 1432.

Cuzzi, J. N. & Weidenschilling, S. J. 2006, in *Meteorites and the Early Solar System II*, ed. D. S. Lauretta & H. Y. McSween, Jr. University of Arizona Press, 353–381.

Dauphas, N. 2003, *Icarus*, **165**, 326.

Dauphas, N., Cook, D. L., Sacarabany, A., *et al.* 2008, *Astrophysical Journal*, **686**, 560.

Dauphas, N., Davis, A. M., Marty, B., & Reisberg, L. 2004, *Earth and Planetary Science Letters*, **226**, 465.

Dauphas, N. & Marty, B. 2002, *Journal of Geophysical Research – Planets*, **107**, 5129.

Dauphas, N., Reisberg, L., & Marty, B. 2002, *Geochemistry Journal*, **36**, 409.

Davis, A. M. 2006, in *Meteorites and the Early Solar System II*, ed. D. S. Lauretta & H. Y. McSween, Jr., University of Arizona Press, 295–307.

Davis, A. M. & Richter, F. M. 2007, in *Meteorites, Comets, and Planets, 2nd edn.*, ed. A. M. Davis, Elsevier, Treatise on Geochemistry, vol. 1, see http://www.sciencedirect.com/science/referenceworks/9780080437514.

Davis, D. R., Chapman, C. R., Weidenschilling, S. J., & Greenberg, R. 1985, *Icarus*, **63**, 30.

Dohnanyi, J. S. 1969, *Journal of Geophysical Research*, **74**, 2431.

Drake, M. J. 2005, *Meteoritics and Planetary Science*, **40**, 519.

Drake, M. J. & Righter, K. 2002, *Nature*, **416**, 39.

Dullemond, C. P., Apai, D., & Walch, S. 2006, *Astrophysical Journal*, **640**, L67.

Dullemond, C. P. & Dominik, C. 2005, *Astronomy and Astrophysics*, **434**, 971.

Farinella, P., Froeschle, Ch., Froeschle, Cl., *et al.* 1994, *Nature*, **371**, 315.

Fernandez, J. A. & Ip, W.-H. 1984, *Icarus*, **58**, 109.

Fogg, M. J. & Nelson, R. P. 2007, *Astronomy and Astrophysics*, **461**, 1195.

Garaud, P. & Lin, D. N. C. 2004, *Astrophysical Journal*, **608**, 1050.

Ghosh, A. & McSween, H. Y. 1998, *Icarus*, **134**, 187.

Goldreich, P. & Ward, W. R. 1973, *Astrophysical Journal*, **183**, 1051.

Gomes, R., Levison, H. F., Tsiganis, K., & Morbidelli, A. 2005, *Nature*, **435**, 466.

Gómez, G. C. & Ostriker, E. C. 2005, *Astrophysical Journal*, **630**, 1093.

Gradie, J. & Tedesco, E. 1982, *Science*, **216**, 1405.

Grossman, J. N., Alexander, C. M. O., Wang, J., & Brearley, A. J. 2000, *Meteoritics and Planetary Science*, **35**, 467.

Haghighipour, N. & Boss, A. P. 2003, *Astrophysical Journal*, **598**, 1301.

Haisch, Jr., K. E., Lada, E. A., & Lada, C. J. 2001, *Astrophysical Journal*, **553**, L153.

Halliday, A. N. 2004, *Nature*, **427**, 505.

Halliday, A. N., Wänke, H., Birck, J.-L., & Clayton, R. N. 2001, *Space Science Reviews*, **96**, 197.

Hawley, J. F., Gammie, C. F., & Balbus, S. A. 1995, *Astrophysical Journal*, **440**, 742.

Heppenheimer, T. A. 1980, *Icarus*, **41**, 76.

Hevey, P. J. & Sanders, I. S. 2006, *Meteoritics and Planetary Science*, **41**, 95.

Holsapple, K., Giblin, I., Housen, K., Nakamura, A., & Ryan, E. 2002, in *Asteroids III*, ed. W. F. Bottke, A. Cellino, P. Paolicchi, & R. P. Binzel, University of Arizona Press, 443–462.

Hsieh, H. H. & Jewitt, D. 2006, *Science*, **312**, 561.

Hubickyj, O., Bodenheimer, P., & Lissauer, J. J. 2005, *Icarus*, **179**, 415.

Hueso, R. & Guillot, T. 2005, *Astronomy and Astrophysics*, **442**, 703.

Humayun, M. & Clayton, R. N. 1995, *Geochimica et Cosmochimica Acta*, **59**, 2131.

Ida, S. & Makino, J. 1993, *Icarus*, **106**, 210.

Inaba, S. & Ikoma, M. 2003, *Astronomy and Astrophysics*, **410**, 711.

Inaba, S., Tanaka, H., Nakazawa, K., Wetherill, G. W., & Kokubo, E. 2001, *Icarus*, **149**, 235.

Inaba, S., Wetherill, G. W., & Ikoma, M. 2003, *Icarus*, **166**, 46.

Jang-Condell, H. & Sasselov, D. D. 2005, *Astrophysical Journal*, **619**, 1123.

Jedicke, R., Larsen, J., & Spahr, T. 2002, *Asteroids III*, 71, ed. W. F. Bottke, A. Cellino, P. Paolicchi, & R. P. Binzel, University of Arizona Press.

Johansen, A., Oishi, J. S., Mac Low, M.-M., *et al.* 2007, *Nature*, **448**, 1022.

Johansen, A. & Youdin, A. 2007, *Astrophysical Journal*, **662**, 627.

Johnson, E. T., Goodman, J., & Menou, K. 2006, *Astrophysical Journal*, **647**, 1413.

Kasting, J. F., Whitmire, D. P., & Reynolds, R. T. 1993, *Icarus*, **101**, 108.

Kennedy, G. M. & Kenyon, S. J. 2008, *Astrophysical Journal*, **673**, 502.

Kenyon, S. J. & Bromley, B. C. 2004, *Astrophysical Journal*, **602**, L133.

Kenyon, S. J. & Bromley, B. C. 2006, *Astronomical Journal*, **131**, 1837.

Kenyon, S. J. & Hartmann, L. 1995, *Astrophysical Journal Supplement Series*, **101**, 117.

Kita, N. T., Huss, G. R., Tachibana, S., *et al.* 2005, in *Astronomical Society of the Pacific Conference Series*, **341**, 558.

Klahr, H. & Bodenheimer, P. 2006, *Astrophysical Journal*, **639**, 432.

Kleine, T., Mezger, K., Palme, H., & Münker, C. 2004, *Earth and Planetary Science Letters*, **228**, 109.

Kleine, T., Mezger, K., Palme, H., Scherer, E., & Munker, C. 2005, *Geochimica et Cosmochimica Acta*, **69**, 5805.

Kleine, T., Münker, C., Mezger, K., & Palme, H. 2002, *Nature*, **418**, 952.

Knežević, Z. & Milani, A. 2003, *Astronomy and Astrophysics*, **403**, 1165.

Kobayashi, S., Imai, H., & Yurimoto, H. 2003, *Geochemical Journal*, **37**, 663.

Kokubo, E. & Ida, S. 1995, *Icarus*, **114**, 247.

Kokubo, E. & Ida, S. 1998, *Icarus*, **131**, 171.

Kokubo, E. & Ida, S. 2002, *Astrophysical Journal*, **581**, 666.

Kokubo, E., Ida, S., & Makino, J. 2000, *Icarus*, **148**, 419.

Kominami, J. & Ida, S. 2002, *Icarus*, **157**, 43.

Kortenkamp, S. J. & Wetherill, G. W. 2000, *Icarus*, **143**, 60.

Krause, M. & Blum, J. 2004, *Physical Review Letters*, **93**, 021103.

Kretke, K. A. & Lin, D. N. C. 2007, *Astrophysical Journal*, **664**, L55.

Krot, A. N., Scott, E. R. D., & Zolensky, M. E. 1995, *Meteoritics and Planetary Science*, **30**, 748.

Laughlin, G. & Bodenheimer, P. 1994, *Astrophysical Journal*, **436**, 335.

Laughlin, G., Steinacker, A., & Adams, F. C. 2004, *Astrophysical Journal*, **608**, 489.

Lecar, M. & Franklin, F. 1997, *Icarus*, **129**, 134.

Lecar, M. & Franklin, F. A. 1973, *Icarus*, **20**, 422.

Leinhardt, Z. M. & Richardson, D. C. 2002, *Icarus*, **159**, 306.

Lemaitre, A. & Dubru, P. 1991, *Celestial Mechanics and Dynamical Astronomy*, **52**, 57.

Levison, H. F. & Agnor, C. 2003, *Astronomical Journal*, **125**, 2692.

Lissauer, J. J. 1987, *Icarus*, **69**, 249.

Lodders, K. 2003, *Astrophysical Journal*, **591**, 1220.

Luais, B. 2007, *Earth and Planetary Science Letters*, **262**, 21.

Lugmair, G. W. & Shukolyukov, A. 1998, *Geochimica et Cosmochimica Acta*, **62**, 2863.

Lyons, J. R. & Young, E. D. 2005, *Nature*, **435**, 317.

MacPherson, G. J. 2007, in *Meteorites, Comets, and Planets*, 2nd edn., ed. A. M. Davis, Elsevier, Treatise on Geochemistry, vol. 1; see; http://www.sciencedirect.com/ science/referenceworks/9780080437514.

MacPherson, G. J., Mittlefehldt, D. W., & Jones, J. H. 2008, *Reviews in Mineralogy and Geochemistry*, 68.

Malhotra, R. 1993, *Nature*, **365**, 819.

Mandell, A. M., Raymond, S. N., & Sigurdsson, S. 2007, *Astrophysical Journal*, **660**, 823.

Marshall, J. & Cuzzi, J. 2001, *Lunar and Planetary Science*, **32**, 1262.

Masset, F. S., D'Angelo, G., & Kley, W. 2006, *Astrophysical Journal*, **652**, 730.

Matsumura, S. & Pudritz, R. E. 2006, *Monthly Notices of the Royal Astronomical Society*, **365**, 572.

McKeegan, K. D. & Davis, A. M. 2007, in *Meteorites, Comets, and Planets*, 2nd edn., ed. A. M. Davis, Elsevier, Treatise on Geochemistry, vol. 1; see http://www.sciencedirect. com/ science/referenceworks/9780080437514.

McNeil, D., Duncan, M., & Levison, H. F. 2005, *Astronomical Journal*, **130**, 2884.

Meibom, A. & Clark, B. E. 1999, *Meteoritics and Planetary Science*, **34**, 7.

Meyer, B. S. & Zinner, E. 2006, in *Meteorites and the Early Solar System II*, ed. D. S. Lauretta & H. Y. McSween, Jr. Univ. Arizona Press, 69–108.

Migliorini, F., Michel, P., Morbidelli, A., Nesvorny, D., & Zappala, V. 1998, *Science*, **281**, 2022.

Moons, M. & Morbidelli, A. 1993, *Celestial Mechanics and Dynamical Astronomy*, **57**, 99.

Morbidelli, A. 1999, *Celestial Mechanics and Dynamical Astronomy*, **73**, 39.

Morbidelli, A., Chambers, J., Lunine, J. I., *et al.* 2000, *Meteoritics and Planetary Science*, **35**, 1309.

Morbidelli, A. & Gladman, B. 1998, *Meteoritics and Planetary Science*, **33**, 999.

Morbidelli, A. & Henrard, J. 1991, *Celestial Mechanics and Dynamical Astronomy*, **51**, 169.

Morbidelli, A. & Vokrouhlický, D. 2003, *Icarus*, **163**, 120.

Mukhopadhyay, B. 2006, *Astrophysical Journal*, **653**, 503.

Nagasawa, M., Ida, S., & Tanaka, H. 2002, *Icarus*, **159**, 322.

Nagasawa, M., Lin, D. N. C., & Thommes, E. 2005, *Astrophysical Journal*, **635**, 578.

Nagasawa, M., Tanaka, H., & Ida, S. 2000, *Astronomical Journal*, **119**, 1480.

Nimmo, F. & Kleine, T. 2007, *Icarus*, **191**, 497.

O'Brien, D. P. & Greenberg, R. 2003, *Icarus*, **164**, 334.

O'Brien, D. P., Morbidelli, A., & Bottke, W. F. 2007, *Icarus*, **191**, 434.

O'Brien, D. P., Morbidelli, A., & Levison, H. F. 2006, *Icarus*, **184**, 39.

Ormel, C. W., Cuzzi, J. N., & Tielens, A. G. G. M. 2008, *Astrophysical Journal*, **679**, 1588.

Paardekooper, S.-J. & Mellema, G. 2006, *Astronomy and Astrophysics*, **459**, L17.

Pahlevan, K. & Stevenson, D. J. 2007, *Earth and Planetary Science Letters*, **262**, 438.

Palme, H. & Jones, A. 2003, in *Meteorites, Comets, and Planets*, ed. A. M. Davis, Elsevier, Treatise on Geochemistry, vol. 1, 41.

Pascucci, I., Gorti, U., Hollenbach, D., *et al.* 2006, *Astrophysical Journal*, **651**, 1177.

Petit, J., Chambers, J., Franklin, F., & Nagasawa, M. 2002, in *Asteroids III*, ed. W. F. Bottke, A. Cellino, P. Paolicchi, & R. P. Binzel, University of Arizona Press, 711–738.

Petit, J., Morbidelli, A., & Chambers, J. 2001, *Icarus*, **153**, 338.

Petit, J., Morbidelli, A., & Valsecchi, G. B. 1999, *Icarus*, **141**, 367.

Pollack, J. B., Hubickyj, O., Bodenheimer, P., *et al.* 1996, *Icarus*, **124**, 62.

Poppe, T., Blum, J., & Henning, T. 2000, *Astrophysical Journal*, **533**, 454.

Raymond, S. N. 2006, *Astrophysical Journal*, **643**, L131.

Raymond, S. N., Barnes, R., & Kaib, N. A. 2006a, *Astrophysical Journal*, **644**, 1223.

Raymond, S. N., Mandell, A. M., & Sigurdsson, S. 2006b, *Science*, **313**, 1413.

Raymond, S. N., Quinn, T., & Lunine, J. I. 2004, *Icarus*, **168**, 1.

Raymond, S. N., Quinn, T., & Lunine, J. I. 2005, *Astrophysical Journal*, **632**, 670.

Rice, W. K. M., Lodato, G., Pringle, J. E., Armitage, P. J., & Bonnell, I. A. 2006,
    Monthly Notices of the Royal Astronomical Society, **372**, L9.

Rivkin, A. S., Davies, J. K., Johnson, J. R., *et al.* 2003, *Meteoritics and Planetary Science*,
    38, 1383.

Robert, F. 2001, *Science*, **293**, 1056.

Safronov, V. S. 1969, *Evolution of the Protoplanetary Cloud and Formation of the
    Earth and Planets*, Nauka Press, Moscow (English Translation: NASA TTF-677).

Safronov, V. S. 1979, in *Asteroids*, ed. T. Gehrels, University of Arizona Press, 975–991.

Sakamoto, N., Seto, Y., Itoh, S., *et al.* 2007, *Science*, **317**, 231.

Saumon, D. & Guillot, T. 2004, *Astrophysical Journal*, **609**, 1170.

Scott, E. R. D. 2007, *Annual Review of Earth and Planetary Science*, **35**, 577.

Sekiya, M. 1998, *Icarus*, **133**, 298.

Seto, Y., Sakamoto, N., Fujino, K., *et al.* 2008, *Geochimica et Cosmochimica Acta*, **72**, 2723.

Shu, F. H., Shang, H., Gounelle, M., Glassgold, A. E., & Lee, T. 2001, *Astrophysical
    Journal*, **548**, 1029.

Srinivasan, G., Goswami, J. N., & Bhandari, N. 1999, *Science*, **284**, 1348.

Strom, S. E., Edwards, S., & Skrutskie, M. F. 1993, in *Protostars and Planets III*, ed.
    E. H. Levy & J. I. Lunine, University of Arizona Press, 837–866.

Swindle, T. D. & Kring, D. A. 2001, Eleventh Annual V. M. Goldschmidt Conference,
    Abstract 3785.

Tanaka, H., Takeuchi, T., & Ward, W. R. 2002, *Astrophysical Journal*, **565**, 1257.

Tanaka, H. & Ward, W. R. 2004, *Astrophysical Journal*, **602**, 388.

Terquem, C. E. J. M. L. J. 2003, *Monthly Notices of the Royal Astronomical Society*,
    341, 1157.

Thomas, P. C., Parker, J. W., McFadden, L. A., *et al.* 2005, *Nature*, **437**, 224.

Thommes, E. W., Duncan, M. J., & Levison, H. F. 2003, *Icarus*, **161**, 431.

Tonks, W. B. & Melosh, H. J. 1992, *Icarus*, **100**, 326.

Touboul, M., Kleine, T., & Bourdon, B. 2008, in *Lunar and Planetary Institute
    Conference Abstracts*, **39**, 2336.

Touboul, M., Kleine, T., Bourdon, B., Palme, H., & Wieler, R. 2007, *Nature*, **450**, 1206.

Trilling, D. E., Lunine, J. I., & Benz, W. 2002, *Astronomy and Astrophysics*, **394**, 241.

Tsiganis, K., Gomes, R., Morbidelli, A., & Levison, H. F. 2005, *Nature*, **435**, 459.

Ueda, T., Murakami, Y., Ishitsu, N., *et al.* 2001, *Earth, Planets, and Space*, **53**, 927.

Walker, R. J., Horan, M. F., Morgan, J. W., & Meisel, T. 2001, in *Lunar and Planetary
    Institute Conference Abstracts*, **32**, 1152.

Ward, W. R. 1981, *Icarus*, **47**, 234.

Ward, W. R. 1997, *Icarus*, **126**, 261.

Weidenschilling, S. J. 1977a, *Monthly Notices of the Royal Astronomical Society*, **180**, 57.

Weidenschilling, S. J. 1977b, *Astrophysics and Space Science*, **51**, 153.

Weidenschilling, S. J. 1980, *Icarus*, **44**, 172.

Weidenschilling, S. J. 1984, *Icarus*, **60**, 553.

Weidenschilling, S. J. 1997, *Icarus*, **127**, 290.

Weidenschilling, S. J., Spaute, D., Davis, D. R., Marzari, F., & Ohtsuki, K. 1997, *Icarus*, **128**, 429.

Wetherill, G. W. 1989, in *Asteroids II*, ed. R. P. Binzel, T. Gehrels, & M. S. Matthews, University of Arizona Press, 661–680.

Wetherill, G. W. 1992, *Icarus*, **100**, 307.

Wetherill, G. W. & Stewart, G. R. 1989, *Icarus*, **77**, 330.

Wetherill, G. W. & Stewart, G. R. 1993, *Icarus*, **106**, 190.

Wilde, S. A., Valley, J. W., Peck, W. H., & Graham, C. M. 2001, *Nature*, **409**, 175.

Wurm, G. & Krauss, O. 2006, *Icarus*, **180**, 487.

Wurm, G., Paraskov, G., & Krauss, O. 2005, *Icarus*, **178**, 253.

Yin, Q., Jacobsen, S. B., Yamashita, K., *et al.* 2002, *Nature*, **418**, 949.

Youdin, A. N. & Shu, F. H. 2002, *Astrophysical Journal*, **580**, 494.

Yurimoto, H. & Kuramoto, K. 2004, *Science*, **305**, 1763.

Zahnle, K. 1998, in *Astronomical Society of the Pacific Conference Series*, **148**, 364.

Zinner, E. 2007, in *Meteorites, Comets, and Planets*, 2nd edition, ed. A. M. Davis, Elsevier, Treatise on Geochemistry, vol. 1;
see http://www.sciencedirect.com/science/referenceworks/9780080437514.

Zuckerman, B., Forveille, T., & Kastner, J. H. 1995, *Nature*, **373**, 494.

Appendix 1

Common minerals in the Solar System

Dante S. Lauretta

A1.1 Mineralogy of chondrite components

A1.1.1 Calcium–aluminum-rich inclusions

Our description of meteorite mineralogy starts with the minerals characteristic of the calcium–aluminum-rich inclusions (CAIs). The mineralogy of CAIs varies systematically with their composition. The most Al-rich CAIs contain spinel, hibonite, and/or grossite. More rarely, corundum or calcium mono-aluminate is present. As the bulk composition becomes more Si-rich, the melilite solid solution becomes important. With additional Mg and Si in the bulk composition, fassaite and anorthite are present.

Inclusions that are predominantly melilite with minor spinel, perovskite, and hibonite are referred to as Type A. Most Type-A CAIs have a porous structure and are called fluffy Type-A CAIs. Some Type-A CAIs have a compact form and generally rounded shapes. These are referred to as "compact" Type-A CAIs. Type-B1 CAIs are characterized by coarse-grained, melilite-rich mantles surrounding cores composed of melilite, spinel, fassaite, and anorthite. Type-B2 inclusions have the same mineralogy, but lack the melilite-rich mantle. Type-B3 inclusions contain significant amounts of forsterite in addition to melilite. Type-C inclusions are similar to Type B2s, but anorthite is more abundant than melilite. All Type-B and Type-C inclusions have compact morphologies.

A1.1.2 Chondrules

Aluminum-rich chondrules

Aluminum-rich chondrules are a broad class of objects with compositions intermediate between those of CAIs and the more common ferromagnesian chondrules. Their bulk compositions are generally Mg-, Si-rich and Ca-, Al-poor relative to most CAIs. The most abundant minerals are usually olivine, orthopyroxene, and plagioclase. Pigeonite and other clinopyroxenes in the diopside–hedenbergite solid solution may be present, but melilite is absent.

There are three basic types of Al-rich chondrules: (1) olivine phyric, Al-rich chondrules, (2) plagioclase phyric, Al-rich chondrules, and (3) glassy Al-rich chondrules. The term phyric is a general textural term referring to the presence of large crystals in an igneous rock set in a finer-grained groundmass or glass.

Ferromagnesian chondrules

Ferromagnesian chondrules are primarily composed of the minerals olivine, orthopyroxene, and clinopyroxene. Olivine incorporates only minor amounts of elements other than oxygen, silicon, magnesium, and iron. Manganese and nickel commonly are the additional elements present in highest concentrations. Compositions of olivine are commonly expressed as molar percentages of forsterite (Fo) and fayalite (Fa) (e.g. $Fo_{70}Fa_{30}$). Likewise, pyroxene compositions are expressed as molar percentages of enstatite (En), ferrosilite (Fs), and, if appropriate, wollastonite (Wo). These objects also contain a silicate–glass mesostasis rich in Na, Al, Si, Ca, and K, relative to chondrite bulk composition. Oxides such as chromite and magnetite and phosphates, such as whitlockite and apatite, are present in varying abundances.

The most commonly used classification scheme divides chondrules into two broad categories: Type I and Type II, based on bulk FeO contents. Type-I chondrules are characterized by the presence of FeO-poor olivine and pyroxene (Fo and En > 90). Type-II chondrules contain FeO-rich olivine and pyroxene (Fo and En < 90). The textural characteristics of both the Type-I and Type-II series are gradational. These two main categories are further subdivided into Subtypes A and B based on their abundances of olivine and pyroxene. Type-IA and IIA chondrules contain abundant olivine (> 80 vol%), Type-IB and IIB chondrules contain abundant pyroxene (> 80 vol%), and chondrules with intermediate abundances of olivine and pyroxene are classified as Type IAB or IIAB.

Metals and sulfides

Many chondrules contain minor amounts of metals, sulfides, and oxides. These phases also occur as distinct grains and assemblages embedded in the chondrite matrix. The metallic mineral kamacite is a common chondrule component that contains significant amounts of minor elements such as cobalt, chromium, and phosphorus. Taenite is another alloy of iron and nickel. Sulfide minerals such as troilite, pyrrhotite, and pentlandite are also abundant in many chondrules.

The enstatite chondrites formed under extremely reducing conditions, reflected in their strange and unique mineralogy. In addition to the more common sulfide phases, minerals such as niningerite, alabandite, oldhamite, daubreelite, sphalerite, caswellsilverite, djerfisherite, and other minor phases typify these samples. Other reduced minerals include perryite, graphite, and cohenite.

Chondrite matrix

Matrix is the fine-grained groundmass in which the chondrules, inclusions, and larger mineral grains occur. When viewed through a petrographic microscope, matrix is opaque, black, and relatively featureless. The grain sizes of its constituent minerals range from 5 μm down to grains as small as a few nanometers. Its composition and mineralogy varies widely across the different chondrite types. In the most primitive meteorites it mainly consists of amorphous materials and the anhydrous phases olivine, pyroxene, and metals, together with lesser amounts of oxides, sulfides, and a variety of other minor minerals.

The matrix of some meteorites (the CI, CM, and CR chondrites) contains significant amounts of phyllosilicates, generally either serpentines or smectites, and other hydrated phases, such as tochilinite. The serpentine group describes a group of common rock-forming hydrous magnesium–iron phyllosilicate minerals. The smectite group of minerals includes dioctahedral smectites, such as montmorillonite and nontronite, and trioctahedral smectites, such as saponite. Tochilinite is a mixed sulfide–hydroxide phase.

Chondrite matrices are also the carriers of presolar grains. These phases cover a broad mineralogy and include diamonds and graphite, silicon carbide (SiC), titanium carbide (TiC), silicon nitride (Si_3N_4), corundum, spinel, silicates, and even rare metal grains.

A1.2 Composition of common minerals in the Solar System

The composition of common minerals in the Solar System is given in Table A1.1.

Table A1.1 *List of the most common minerals in the Solar System*

Common CAI components	Corundum	Al_2O_3
	Hibonite	$CaAl_{12}O_{19}$
	Grossite	$CaAl_4O_7$
	Calcium monoaluminate	$CaAl_2O_4$
	Spinel	$MgAl_2O_4$
	Fassaite	$Ca(Mg, Al, Ti^{3+}, Ti^{4+})(Al, Si)_2O_6$
	Melilite solid solution	
	gehlenite	$Ca_2Al_2SiO_7$
	åkermanite	$Ca_2MgSi_2O_7$
	Plagioclase solid solution	
	anorthite	$CaAl_2Si_2O_8$
	albite	$NaAlSi_3O_8$

Table A1.1 *(cont.)*

Major silicates	Olivine solid solution	
	forsterite	Mg_2SiO_4
	fayalite	Fe_2SiO_4
	Orthopyroxene solid solution	
	enstatite	$MgSiO_3$
	ferrosilite	$FeSiO_3$
	Clinopyroxene solid solution	
	diopside	$CaMgSi_2O_6$
	hedenbergite	$CaFeSi_2O_6$
	Pigeonite	$(Ca,Mg,Fe)(Mg,Fe)Si_2O_6$
	Augite	$(Ca,Na)(Mg,Fe,Al,Ti)(Si, Al)_2O_6$
Oxides	Chromite	$FeCr_2O_4$
	Magnetite	Fe_3O_4
Phosphates	Apatite	$Ca_5(PO_4)_3(F,Cl,OH)$
	Whitlockite	$Ca_3(PO_4)_2$
Metals and	Kamacite	α-Fe,Ni(Ni < 7.5%)
sulfides	Taenite	γ-Fe,Ni (8–55% Ni)
	Schreibersite	$(Fe, Ni)_3P$
	Perryite	$(Ni, Fe)_x(Si,P)$
	Troilite	FeS
	Pyrrhotite	Fe_7S_8
	Pentlandite	$(Fe, Ni)_9S_8$
	Niningerite	$(Fe,Mg,Mn)S$
	Alabandite	$(Mn,Fe)S$
	Oldhamite	CaS
	Daubreelite	$FeCr_2S_4$
	Sphalerite	$(Zn,Fe)S$
	Caswellsilverite	$NaCrS_2$
	Dierfisherite	$K_3(Na, Cu)(Fe, Ni)_{12}S_{14}$
Carbides,	Graphite	C
nitrides,	Cohenite	Fe_3C
carbonates,	Silicon carbide	SiC
and sulfates	Titanium carbide	TiC
	Silicon nitride	Si_3N_4
	Calcite	$CaCO_3$
	Anhydrite	$CaSO_4$
Hydrated	Serpentine solid solution	
phases	chrysotile	$Mg_3Si_2O_5(OH)_4$
	greenalite	$Fe_3Si_2O_5(OH)_4$
	Cronstedtite	$(Mg, Fe^{2+})_2(Al, Si, Fe^{3+})O_5(OH)_4$
	Montmorillonite	$(Na, Ca)_{0.33}(Al, Mg)_2(Si_4O_{10})(OH)_2 \cdot$ nH_2O
	Nontronite	$Ca_{0.5}(Si_7Al_{0.8}Fe_{0.2})(Fe_{3.5}Al_{0.4}Mg_{0.1})$ $O_{20}(OH)_4$
	Saponite	$(Ca_{0.5}Na)_{0.33}(Mg, Fe^{2+})_3(Si, Al)_4$ $O_{10}(OH)_2 \cdot 4H_2O$
	Tochilinite	$2[(Fe, Cu, Ni)S] \cdot 1.7[(Fe, Ni)(OH)_2]$

Appendix 2

Mass spectrometry

Peter Hoppe

A2.1 Secondary ion mass spectrometry

Secondary ion mass spectrometry (SIMS) is a widely used analytical technique in fields such as microelectronics, metallurgy, biology, geochemistry, and cosmochemistry. Major SIMS applications in cosmochemistry are measurements of the isotopic compositions of the light- to intermediate-mass elements and of minor and trace element abundances of nanometer- to micrometer-sized samples. In the context of this book, the major application of SIMS is the study of presolar dust and organics found in primitive Solar System materials. The basic principle of SIMS can be described as follows: the sample of interest is bombarded with primary ions (several keV energy), mostly oxygen or cesium. This triggers a collisional cascade in the target and secondary particles (atomic and molecular ions, neutrals) are emitted from the uppermost layers. The information depth, i.e. from where the secondary particles originate, is typically 5–20 nm and depends on parameters such as primary particle energy, angle of incidence, and target composition. Typically, some permil or percent of the sputtered particles are ionized and can be analyzed in a mass spectrometer.

Secondary ion mass spectrometry is a powerful technique, which has several advantages: detection limits are ppm for most elements and ppb for favorable elements, all elements (except the noble gases) are detectable, isotopes can be distinguished, and a high lateral resolution, ranging from ≈ 50 nm to several µm, depending on the type of instrument and application (see below), can be achieved. Disadvantages of SIMS are its destructive nature and the fact that secondary ion yields vary by more than six orders of magnitude which makes isotope studies of certain elements very difficult or impossible.

Oxygen and cesium are commonly used as primary ions. In the oxygen ion source (Duoplasmatron) ions (mostly O_2^+, O^-) are produced in an arc discharge. Usually O^- is used as it reduces sample charging. Since O has the second highest electronegativity of all elements the bombardment of samples with O leads to the preferred formation of positive secondary ions. This can favorably be utilized for isotope studies of, e.g. Mg, Ca, and Ti. Because primary O ions have a large spread in initial energy, they are difficult to focus into a small spot and the achievable lateral resolution in imaging applications (see below) is generally worse than if Cs^+ primary ions are used. The latter are produced by surface ionization when passing through a hot tungsten membrane which leads to essentially monoenergetic ions. Since Cs has the lowest electronegativity of all elements its use leads to the preferred formation of negative secondary ions. The Cs source is generally used for isotope studies of H, C, N (which is measured as CN since N does not form negative secondary ions), O, and Si.

The SIMS instruments are also called ion microprobes. Ion microprobes consist of three major parts. These include the production and extraction of ions, the mass separation (either by a magnetic sector or by time of flight), and the ion detection (electron multiplier, Faraday cup, microchannel-plate). Two types of ion micro-probes have been mostly used for the study of presolar grains (see Chapter 2): the Cameca IMSxf series ($x = 3$–6) and the latest generation instrument, the Cameca NanoSIMS 50. Both are magnetic sector-type instruments. The IMSxf series instrument, which has been used since the 1980s, can be operated in two different modes, the so-called microscope and microprobe modes. It is a single collector instrument, i.e. the different isotopes must be measured one after another. In the microscope mode, mass-filtered secondary ion images, produced by a defocused primary ion beam (typically 100 μm) can be directly projected on a microchannel-plate (MCP) and visualized on a fluorescent screen (FS). The lateral resolution is determined by the secondary ion optics and is typically > 1 μm. Digitized isotope images are acquired with a high-performance CCD camera attached to the MCP/FS detector. This mode cannot be used for precise isotope measurements, but has been successfully used to locate specific types of presolar grains such as the oxides and the supernova SiC X grains. In the microprobe mode a focused primary ion beam is used which permits precise isotope measurements. Isotope images can be created by rastering the focused primary beam over the sample; in this mode the lateral resolution is limited by the primary beam size (typically ≈ 1 μm for Cs^+ ions, several μm for O^- ions). The NanoSIMS instrument was introduced into the field of cosmochemistry in 2001. It can be operated only in the microprobe mode but is superior to the IMSxf series in several respects: the primary beam can be focused into much smaller spots with sizes of ≈ 50 nm (Cs^+) and ≈ 300 nm (O^-), respectively. The NanoSIMS has a significantly higher detection efficiency for secondary ions at high mass resolution (30 × higher than in the IMS3f for conditions used for O isotope measurements) and simultaneous detection of up to seven isotopes is possible. Ion imaging with the NanoSIMS has been an important tool for the *in situ* detection of presolar grains (specifically silicates) and organics in primitive Solar System materials.

A2.1.1 Resonance ionization mass spectrometry

Resonance ionization mass spectrometry (RIMS) has become an important tool in the study of presolar grains as it permits the measurement of the isotopic compositions of rare, heavy elements in individual, micrometer-sized grains. The most advanced instrument in the field of cosmochemistry is CHARISMA (Chicago-Argonne Resonant Ionization Spectrometer for Mass Analysis). Use of RIMS combines laser ablation, laser ionization, and time-of-flight (TOF) mass spectrometry. It is superior to SIMS for the isotopic study of heavy trace elements and important results were obtained for Sr, Zr, Mo, Ru, and Ba isotopic compositions of micrometer-sized presolar SiC and graphite grains.

The principle of CHARISMA is as follows: the sample is ablated with a pulsed Nd:YAG laser, focused into a spot of micrometer size. Neutral species drift up while ions are suppressed. Two lasers (Ti:Sapphire), tuned to resonantly ionize the element of interest, are fired into the ablated, neutral material. The ions are extracted and accelerated and analyzed with a TOF mass spectrometer where mass separation occurs due to different flight times of species with different mass. The useful yield (detected ions/sputtered atoms) is around 1% in the current set-up. A planned modification is aiming at a useful yield of around 30% which would open up new possibilities for the isotope study of trace elements (e.g. the rare earth elements) in presolar grains. With the new set-up it will also be possible to extend the analyses to sub-micrometer-sized grains, which are much more representative of presolar grains than the currently studied micrometer-sized grains.

Appendix 3

Basics of light absorption and scattering theory

Hans-Peter Gail

The most important application of the theory of absorption and scattering by dust particles is the limiting case, when the radius a of the particles, always assumed to be of spherical shape, is small compared to the wavelength λ of the light that interacts with the particle. The essential size parameter is $x = 2\pi a/\lambda$.

The small particle limit $x \ll 1$. In electromagnetic theory it is shown that the absorption and scattering cross-sections of small dielectric particles are (e.g. Bohren & Huffman 1983):

$$\sigma_\lambda^{\text{abs}} = \pi a^2\, Q_\lambda^{\text{abs}}, \quad \sigma_\lambda^{\text{sc}} = \pi a^2\, Q_\lambda^{\text{sc}}, \tag{A3.1}$$

where the absorption and scattering efficiencies Q_λ^{abs} and Q_λ^{sc} are

$$Q_\lambda^{\text{abs}} = 4x\, \text{Im}\,\alpha, \quad Q_\lambda^{\text{sc}} = \frac{8x^4}{3}\alpha\alpha^*, \quad \text{with} \quad \alpha = \frac{\epsilon - \epsilon_{\text{m}}}{\epsilon + 2\epsilon_{\text{m}}}. \tag{A3.2}$$

Here $\epsilon = \epsilon_r + i \cdot \epsilon_i$ is the complex dielectric function, that describes in electromagnetic theory the reaction of matter on an external electric field and α the polarizability[1] of a homogeneous spherical particle per unit volume. It is assumed that the particle is embedded in a material with real part ϵ_{m} of its dielectric function and vanishing imaginary part. For particles embedded in vacuum ($=$ ISM) we have $\epsilon_{\text{m}} = 1$. The complex dielectric function has to be determined by laboratory measurements for the materials of interest.

The small particle limit usually applies to interstellar dust ($a \lesssim 0.1\,\mu\text{m}$) for wavelengths from the optical to the IR spectral region. If this limit applies, scattering is inefficient compared to absorption (cf. Fig. A3.1) and can be neglected in most applications.

---

[1] $\text{Im}\,\alpha = $ imaginary part of α, $\alpha^* = $ complex conjugate of α.

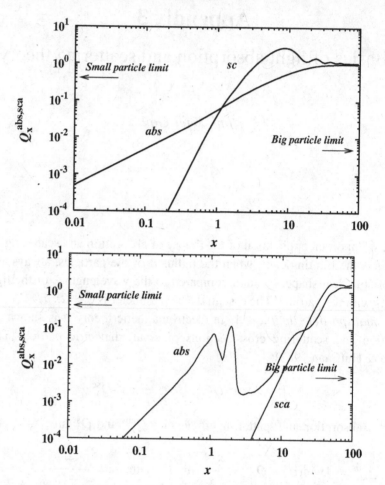

Figure A3.1 Variation of absorption and scattering efficiency Q_x^{abs} and Q_x^{sc}, respectively, with size parameter x, calculated from Mie theory for spherical particles. *Top:* for an assumed dielectric function $\epsilon = 1.5 + i \cdot 0.5$. The scattering efficiency shows for $x > 1$ the oscillations due to interference effects typical for particles with diameters of the order of a few wavelengths. *Bottom:* for optical constants of amorphous MgFeSiO₄ (from the Jena databases) dust grains with $a = 0.1\,\mu m$. The absorption efficiency shows the two strong silicate features around $\lambda = 9.7\,\mu m$ and $\lambda = 18\,\mu m$.

The absorption coefficient of a swarm of particles with particle density n_p and size distribution $f(a)$ is

$$\kappa_\lambda^{abs} = n_p \int da\, f(a)\sigma_\lambda^{abs}(a) = \frac{3}{4\pi\rho_c}\hat{Q}_\lambda^{abs}\varrho_d\,, \qquad (A3.3)$$

where $\hat{Q}^{abs} = Q^{abs}/a$ is independent of particle size, ρ_c is the mass density of the solid material, and ϱ_d the mass density of the ensemble of dust grains.

Core-mantle grains. If particles consist of a core with dielectric coefficient ϵ_1 and radius r, and a mantle with dielectric coefficient ϵ_2 and thickness Δ, then

$$\alpha = \frac{(\epsilon_2 - 1)(\epsilon_1 + 2\epsilon_2) + f\,(2\epsilon_2 + 1)(\epsilon_1 - \epsilon_2)}{(\epsilon_2 + 2)(\epsilon_1 + 2\epsilon_2) + f\,(2\epsilon_2 - 2)(\epsilon_1 - \epsilon_2)}, \quad f = \left(\frac{r}{r + \Delta}\right)^3. \quad \text{(A3.4)}$$

This is relevant for calculating extinction of dust grains coated by an ice mantle, e.g. for dust in molecular clouds or in the cool ($T < 150\,\text{K}$) parts of accretion disks.

The big particle limit $x \gg 1$, *Mie theory.* The absorption and scattering efficiencies tend to (geometric optics limit)

$$\lim_{\lambda \to 0} Q_\lambda^{\text{abs,sc}} = 1, \quad \text{(A3.5)}$$

i.e. the cross-sections $\sigma_\lambda^{\text{abs,sc}}$ tend to the geometric cross-section and optical properties of grains in that regime do not depend on wavelength and material properties (grey absorption). The transition to the big particle limit is important, e.g. if dust in accretion disks starts to grow to bigger sizes and characteristic absorption features (e.g. the infrared features of silicate dust, Fig. A3.1, lower part) become less visible and finally disappear (see Chapter 6).

In the transition regime $x \approx 1 \ldots 10$ one has to calculate optical properties from more elaborate theories. For spherical particles an exact solution exists, the Mie theory, that is treated in detail, e.g. in Bohren & Huffman (1983) and Krügel (2002). Figure A3.1 shows an example for the transition from the small to the big particle limit.

Inhomogeneous particles. For particles composed of a matrix and inclusions one approach for calculating optical properties is to assume an average dielectric coefficient $\langle \epsilon \rangle$ for the composed particle. A number of so-called mixing rules have been proposed; a frequently used one is the Maxwell–Garnett mixing rule (cf. Bohren & Huffman 1983). For a matrix with dielectric coefficient ϵ_{m}, and a number of different kinds of inclusions with dielectric coefficients ϵ_j and volume fractions f_j one uses

$$\langle \epsilon \rangle = \frac{(1 - f)\epsilon_{\text{m}} + \sum_j f_j \beta_j \epsilon_j}{1 - f + \sum_j f_j \beta_j}, \quad \beta_j = \frac{3\epsilon_{\text{m}}}{\epsilon_j + 2\epsilon_{\text{m}}} \quad \text{(A3.6)}$$

with $f = \sum_j f_j$ = volume fraction filled by inclusions. The matrix may also be vacuum ($\epsilon = 1$), in which case optical properties of fluffy particles can be calculated; this is necessary for dust particles in accretion disks that are agglomerates of a big number of very small subgrains (see Fig. 2.1 and Chapters 3 and 7). Using mixing rules may, however, yield results with only moderate accuracy (cf. e.g. Voshchinnikov *et al.* 2007).

Discrete dipole approximation. For particles with complex shape and/or complex composition, presently the only viable method for calculating optical properties is the discrete dipole approximation (DDA). This decomposes a grain in a very big number of cubes that are ascribed the polarizability α according to the dielectric function of the dust material at the mid-point of a cube. The mutual polarization of the cubes by the external field and the induced dipoles of all other dipoles is calculated from a linear equations system and the absorption and scattering efficiencies are derived from this. The method is computationally demanding. The theoretical background and the application of the method are described in Draine (1988) and Draine & Flatau (1994).

Laboratory determination of the dielectric function. The dielectric function $\epsilon(\omega)$ is measured in the laboratory for the materials of interest mainly in two different ways:

(1) The material of interest is powdered and pressed with some suited carrier material (e.g. KBr) with known dielectric function ϵ_m into pellets. The transmission

$$T_\lambda = I_{out}/I_{in} = \exp\left(-l\kappa_\lambda^{abs}\right) \tag{A3.7}$$

of a transparent slab of thickness l is measured. The absorption coefficient κ_λ^{abs} is given by Eq. (A3.3).

(2) The reflectivity is measured from a flat surface for normal incidence of the incoming light. Electromagnetic theory shows that the reflectivity $R = I_{refl}/I_{in}$ is given in terms of the real and imaginary parts n and k, respectively, of the complex index of refraction (cf. Bohren & Huffman 1983),

$$n + ik = \epsilon^{1/2} \tag{A3.8}$$

as

$$R = \frac{(n-1)^2 + k^2}{(n+1)^2 + k^2}. \tag{A3.9}$$

This method is particularly suited for strongly opaque materials.

Data for almost all dust materials of interest for protoplanetary disks and for more general astrophysical applications have been determined by laboratory work. Tables for ϵ or for n and k can be obtained from the Jena–St. Petersburg database.[2]

Theory of the dielectric function. The discussion of absorption properties of astrophysically relevant solids is frequently based on the classical Lorentz model for dielectric materials. This assumes that the electrons and ions forming the solid matter are located at fixed equilibrium positions in the solid, determined by internal forces. An applied electromagnetic field shifts the charged particles, labeled by

[2] http://www.mpia-hd.mpg.de/HJPDOC/orgn.php

index j, off their equilibrium positions $\vec{x}_{0,j}$, by a displacement $\vec{x}_j$, while internal forces tend to drive them back to equilibrium. The equation of motion for the displacements of particles with masses m_j and charges e_j is

$$m_j \ddot{\vec{x}}_j + m_j \Gamma_j \dot{\vec{x}}_j - m_j \omega_{0,j}^2 \vec{x}_j = e_j \vec{E}_{\text{loc},j} \, .$$

Here Γ_j and $\omega_{0,j}$ are the damping constants and resonance frequencies of the oscillators, and $\vec{E}_{\text{loc},j}$ is the local field at location $x_{0,j}$, given by the external field and the fields of the induced dipoles with moment $\vec{\mu}_l = e_l \vec{x}_l$ of all other oscillators $l \neq j$. A Fourier decomposition yields for the spectral components of the displacements

$$\vec{x}_j(\omega) = \frac{e_j}{m_j} \frac{1}{\omega_{0,j}^2 - \omega^2 - i\Gamma_j \omega} \cdot \vec{E}_{\text{loc},j}(\omega) \, .$$

The polarization of matter is $\vec{P} = \sum_j e_j \vec{x}_j$, where the summation is over all dipoles in the unit volume, and the electric displacement is $\vec{D} = \vec{E} + 4\pi \vec{P}$. The dielectric function $\epsilon(\omega)$ is defined by $\vec{D} = \epsilon \vec{E}$, which also holds for the spectral components. Thus

$$\epsilon(\omega) = 1 + 4\pi \sum_j n_{\text{osc},j} e_j \frac{x_j(\omega)}{E_{\text{loc},j}(\omega)} \, .$$

Here, all identical oscillators in the solid are grouped together, and j now runs over the different kinds of oscillators and $n_{\text{osc},j}$ is their density per unit volume.

The relation between the local field and the external field depends on the symmetry of the lattice structure of the solid. In practical applications almost exclusively isotropy is assumed, in which case one has $\vec{E}_{\text{loc}} = (\epsilon + 2)/3 \cdot \vec{E}$. One can solve, then, for $\epsilon(\omega)$ with the result

$$\epsilon(\omega) = 1 + \sum_j \frac{\omega_{\text{p},j}^2}{\omega_j^2 - \omega^2 - i\gamma_j \omega} \, , \tag{A3.10}$$

where $\omega_{\text{p},j}^2 = (4\pi/3) n_{\text{osc},j} e_j / m_j$ is the so-called plasma frequency and ω_j the middle frequency of the oscillator profile. Equation (A3.10) is the Lorentz–Drude model for the dielectric function of condensed matter. It also holds if free charge carriers are present. In this case one (or more) of the frequencies ω_j equals zero and the corresponding term equals the result for the Drude model of electric conductivity.

By fitting a Lorentz–Drude model, e.g. by the method of least squares to a set of measured values for T_λ respectively R_λ, determined for a sufficiently big number of wavelengths, distributed over a wide wavelength range, a set of coefficients ω_j,

$\omega_{p,j}$, and Γ_j is obtained which allows $\epsilon(\omega)$ or $n + ik$ for each frequency of interest to be calculated.

In electromagnetic theory it is shown that the real and imaginary part of $\epsilon(\omega)$ are not independent of each other, but are connected by a pair of integral equations, the Kramers–Kronig relation (e.g. Bohren & Huffman 1983). Equation (A3.10) satisfies these relations, i.e. using a Lorentz–Drude model fitted to the laboratory data automatically guarantees that the optical data satisfy this basic physical requirement.

Glossary

accretion: growth by assimilation of material, e.g. formation of stars, planets, and planetesimals by the accumulation of gas or smaller objects in the protoplanetary disk.

accretion disk: a disk-like structure formed by material falling into a central mass. Due to the conservation of angular momentum any initial angular velocity of a gas parcel in a collapsing cloud will increase dramatically and prohibit the material from falling directly to the central object. The infalling gas parcels follow elliptical orbits; these orbits cross at the plane perpendicular to the angular momentum of the object. The collisions between the gas parcels lead to the formation of a disk. The disk, if viscous, allows the redistribution of the angular momentum through radial spreading, enabling the continued accretion of mass to the central object.

accretion shock: rapid deceleration, compression, and heating of gas resulting from collapse into the nebular disk.

achondrite: (differentiated stony meteorite): a stony meteorite of non-solar composition. Achondrites are formed by the melting and recrystallization on or within meteorite parent bodies and have distinct textures and mineralogy indicative of igneous processes.

adiabatic: a change in the pressure and temperature of a gas during which no heat is exchanged with its environment.

AGB star: asymptotic giant branch star, a cool and luminous red giant star. These evolved stars drive an intense wind that removes material from the surface at an increasing rate as the end of its lifecyle approaches. AGB stars are the proposed source of many types of presolar grains.

aliphatic: a class of saturated or unsaturated carbon compounds, in which the carbon atoms are joined in open chains, and which contain no aromatic rings.

349

alkylation: transformation to any of a series of univalent groups of the general formula C_nH_{2n+1} derived from aliphatic hydrocarbons.

ALMA: the Atacama Large Millimeter Array, a very large imaging interferometer currently under construction in Chile. ALMA will greatly increase the spatial resolution at millimeter and submillimeter wavelengths.

amino acid: nitrogenous organic compounds that serve as the structural units of proteins.

amoeboid: an object with an irregular shape.

AMS: accelerator mass spectrometry.

angrite: medium- to coarse-grained achondrite meteorites of basaltic composition that are depleted in alkali elements relative to solar composition and contain abundant Ca–Al–Ti-rich pyroxene, Ca-rich olivine, and anorthitic plagioclase.

angular momentum: a quantity obtained by multiplying the linear momentum (mass × velocity) vector of a body by its position vector. Angular momentum is a quantity that is conserved in closed systems.

AOA: amoeboid olivine aggregate.

aromatic: organic compounds characterized by cyclic ring structure.

asteroid: a class of minor bodies orbiting the Sun.

basalt: a dark, fine-grained, mafic extrusive igneous rock composed primarily of plagioclase and pyroxene.

bipolar outflow: a stream of matter in two opposing directions from a central object, usually a star. Bipolar outflows represent significant periods of mass loss in a star's life. They occur during the protostellar and pre-main-sequence stages and, again, during the red giant phase just before the formation of a planetary nebula.

brown dwarf: a star-like object whose mass is too low to sustain hydrogen fusion at its core. The stellar–substellar boundary depends slightly on the metallicity, but it is typically around 75–80 Jupiter-masses.

CAI: calcium–aluminum-rich inclusion.

carbonaceous chondrite: chondritic meteorites that have bulk refractory lithophile abundances, normalized to Si, that are equal to or greater than those in the solar photosphere. Despite their name, many carbonaceous chondrites are relatively poor in carbon.

carbonyl: functional group composed of a carbon atom double-bonded to an oxygen atom.

catalysis: a chemical reaction involving a material which promotes or increases the rate of the reaction without itself undergoing any permanent chemical change.

chondrite: originally defined as a meteorite that contained chondrules; now also implies a bulk chemical composition, for all but the most volatile elements, that is similar to that of the Sun.

chondrule: approximately spherical assemblages, characteristic of most chondrites, that existed independently prior to incorporation in the meteorite and that show evidence for partial or complete melting.

chromosphere: a normally transparent region lying between the photosphere and the corona of a main-sequence star such as the Sun, and of intermediate temperature.

circumstellar disk: dust and/or gas orbiting a star at a range of radii. Most common types are massive, gas-rich accretion disks, lower-mass protoplanetary disks, or gas-poor tenuous debris disks.

clathrate: a chemical substance consisting of a lattice of one type of molecule trapping and containing a second type of molecule.

clinopyroxene: a mineral of the pyroxene group that crystallizes in the monoclinic system.

comet: a class of small natural bodies that follow independent orbits about the Sun and which develop an extended coma when present in the inner Solar System.

condensation: phase transition from the gaseous to a solid or liquid phase.

cosmic rays: energetic particles of extraterrestrial origin. Cosmic-ray particles include electrons, protons, gamma rays, and atomic nuclei from a large region of the periodic table. The kinetic energies of these particles span over 14 orders of magnitude.

cosmogenic: that produced by interaction with cosmic radiation, e.g. ^{21}Ne is a cosmogenic nuclide, produced by spallation reactions.

crater: bowl-shaped hole on a surface made by a volcanic explosion or the impact of a body.

cryptocrystalline: a rock texture in which the individual crystallites are too small to be distinguished by conventional optical microscopy.

daughter isotope: a nuclide produced by decay of a radioactive parent isotope. The daughter nuclide may be stable or unstable (radioactive).

debris disk: a circumstellar disk in which the majority of the dust is not derived from the collapsing molecular cloud, but from the collisions of minor bodies in the disk. The typical masses and optical depths of these disks are several orders of magnitudes lower than those typical to accretion disks.

desorption: changing from an adsorbed state on a surface to a gaseous or liquid state.

deuterium: heavy stable H isotope that has a mass of 2 atomic mass units.

ejecta: materials ejected either from a crater by the action of volcanism or a meteoroid impact, or from a stellar object, such as a supernova.

electron microprobe: an analytical instrument that utilizes a finely focused beam of electrons to excite characteristic X-rays in the sample. These are then analyzed using either a crystal spectrometer or a solid-state, energy-dispersive detector.

endothermic reaction: a reaction that requires external energy.

enstatite: the Mg-rich end-member of the orthopyroxene solid solution series.

enstatite chondrite: highly reduced chondritic meteorite composed primarily of enstatite, metal, and sulfide.

Epstein drag: a law describing the drag of a medium (e.g. a gas) on an object moving through it. The Epstein drag law applies when the object is smaller than the mean free path of the constituent particles (or molecules) of the medium.

eucrite: a broad class of achondritic meteorite. Both basaltic and cumulate eucrites have been recognized. Basaltic eucrites are pigeonite–plagioclase rocks thought to represent either residual liquids or primary partial melts. Cumulate eucrites are coarse-grained gabbros composed of low-Ca pyroxene and plagioclase and are cumulates from a fractionally crystallizing mafic melt.

eutectic: the lowest temperature at which a mixture of two or more elements can be wholly or partly liquid. Crystallization under these conditions frequently leads to a petrologically distinctive texture.

evaporation: phase transition during which a liquid is transformed into a gas, a slow vaporization of a liquid.

exothermic reaction: a reaction that liberates heat energy.

extinct radionuclide: a radioactive nuclide which has now decayed below detection limits. While these nuclides are no longer detectable, their one-time presence is revealed by their decay products.

extrasolar: originating outside of our Solar System; e.g. an extrasolar planet is a planet which orbits a star other than the Sun.

fall: a meteorite that was seen to fall. Such meteorites are usually recovered soon after fall and are relatively free of terrestrial contamination and weathering effects.

fassaite: a variety of augite that has a very low iron content. It is not recognized as an official, separate mineral.

fayalite: the Fe-rich end-member of the olivine solid solution series.

feeding zone: the region of the protoplanetary disk from which a growing protoplanet can easily accrete material.

ferromagnesian: containing iron and magnesium.

find: a meteorite that was not seen to fall but was found and recognized subsequently.

Fischer–Tropsch synthesis: production of organic molecules by the hydrogenation of carbon monoxide in the presence of a suitable catalyst.

fractionation: the physical separation of one phase, element, or isotope from another.

fugacity: a measure of the chemical potential of a gaseous species; it is the equivalent for a non-ideal gas of the partial pressure of an ideal gas.

fullerene: a form of carbon having a large molecule consisting of an empty cage of sixty or more carbon atoms.

FU Orionis outbursts: the brightening of an FU Orionis star. Such outbursts are thought to occur when the mass accretion rate through the accretion disk increases by orders of magnitude.

FU Orionis star (FU Ori-type star or Fuor): a pre-main-sequence star that recently underwent an extreme brightening at visual wavelengths, typically of five magnitudes or more. They are named after their prototype FU Orionis. These stars rapidly brighten, then remain almost steady or slowly decline by a magnitude or two over a period of decades. It is proposed that the brightening is driven by a runaway disk accretion, a result of an instability.

galactic cosmic rays (GCRs): high-energy particles that flow into our Solar System from far away in the galaxy. GCRs are mostly protons, electrons, and atomic nuclei.

GEMS: glass with embedded metals and sulfides.

graphite: crystalline form of carbon having a laminar structure.

Hα emission: a red emission line at 6562.8 Å emitted by a hydrogen atom when its excited electron jumps from the $n = 3$ energy level to the $n = 2$ energy level. This transition line is a powerful and easily accessible probe of hot hydrogen gas.

Half-life: the time required for half of the atoms in a given sample of a radioactive nuclide to decay.

HED: Howardite–Eucrite–Diogenite clan of meteorites.

HED parent body: the parent asteroid of the HED meteorites, thought to be the asteroid 4 Vesta based on the spectral similarities between this object and the HED meteorites. A known dynamical mechanism exists that can deliver material from Vesta to Earth-crossing orbits, supporting this assumption.

Herbig Ae/Be star: intermediate-mass ($\sim$1.5–6 $M_\odot$) star with a circumstellar disk, typically younger than 5–7 Myr.

Herbig–Haro objects: emission-line nebulae which are produced by shock waves in the supersonic outflow of material from young stars; also referred to as Herbig–Haro nebulae.

Hertzsprung–Russell diagram: a diagram of bolometric magnitude against effective temperature for stars. The color–magnitude diagram is a similar plot, showing the absolute or apparent visual magnitude as a function of color index.

HST: see Hubble Space Telescope.

Hubble Space Telescope (HST): a 2.4 m mirror-diameter space telescope operating at ultraviolet, optical, and near-infrared wavelengths, operated by NASA and ESA. The HST has a versatile suite of instruments and has been serviced and upgraded several times by astronauts.

IDP: interplanetary dust particle.

igneous: a rock or mineral that solidified from molten or partly molten material.

induction: electromagnetic induction is the appearance of an electric potential difference in response to the presence of a time-varying electromagnetic field.

infrared excess emission: infrared emission exceeding the expected levels of photospheric emission from the star. In most cases it emerges from circumstellar dust grains, thus the presence of infrared excess emission is a commonly used marker for disks (young stars) or circumstellar shells/outflows (evolved stars).

IOM: insoluble organic matter.

ion microprobe: an analytical instrument that uses a finely focused ion beam to ionize atoms in the target and eject them into a mass spectrometer so that their precise mass distribution can be determined.

ion–molecule reaction: a reaction between an ionized atom or molecule and a neutral atom or molecule.

ionization: the process by which a neutral atom or molecule acquires a positive or negative charge.

IRAC: Infrared Array Camera, the short-wavelength (3–8 μm) imaging instrument on board the Spitzer Space Telescope.

IRAS: Infrared Astronomical Satellite, an all-sky infrared survey telescope that obtained infrared photometry at 12, 25, 60, and 160 μm wavelengths.

irradiation: the process by which an item is exposed to radiation.

IRS: Infrared Spectrograph, the spectrograph on board the Spitzer Space Telescope. The instrument had a low- and high-resolution mode, providing sensitive spectra from 5 to 40 μm.

ISM: interstellar medium.

ISO: Infrared Space Observatory, an ESA-led infrared space telescope that carried out targeted observations. ISO had both imaging and spectroscopic modes.

isochemical: without change in bulk chemical composition.

isochron: refers to points in the parameter space that have an identical age. An expression commonly used for denoting the curve on the Hertzsprung–Russell diagram connecting the stellar populations of a given age, or for a sample suite when the radiogenic daughter nuclide is plotted against the radioactive parent nuclide. In this case, the two nuclides are usually normalized to a non-radiogenic nuclide.

kamacite: Fe,Ni alloy of 7 wt% Ni or less with the body-centered-cubic structure. It occurs as large plates or single crystals in iron meteorites, abundant grains in chondrites and rare grains in most achondrites.

Kepler Space Telescope: a NASA-built optical space telescope that monitors 170 000 stars continuously for four years to search for transiting exoplanets. The goal of the Kepler mission is to determine the frequency of Earth-sized planets around other stars.

Keplerian rotation: an orbital velocity that matches that of a gravitationally bound object around the central object as described by Kepler's third law.

kerogen: insoluble macromolecular organic matter, operationally defined as the organic residue left after acid demineralization of a rock.

Kuiper Belt: a region in the outer Solar System beyond Neptune's orbit populated by small icy planetesimals or Kuiper Belt objects, and dwarf planets. Many short-period comets (possessing orbits of less than 200 years) are thrown into the Solar System from the Kuiper Belt.

liquidus: the maximum temperature where the solid and liquid phase of a material can co-exist. At temperatures exceeding the liquidus the material is fully liquid.

lithophile: a geochemical class of elements. Lithophile elements are those which tend to concentrate in the silicate phase, e.g. Si, Mg, Ca, Al, Na, K, and rare-earth elements.

luminosity: a measure of the energy output of a star.

magma ocean: a hypothetical stage in the evolution of a planetary object during which virtually the entire surface of the object is covered with molten lava.

magmatic: that associated with molten silicate material.

magneto-hydrodynamics: the study of the behavior of an electrically conducting fluid in the presence of a magnetic field.

mass fractionation: a process which causes the proportions of the isotopes or elements to change in a manner which is dependent on the differences in their mass.

mass spectrometer: an analytical instrument in which the sample is converted into a beam of ions which can be separated from each other on the basis of their mass-to-charge ratio, generally by a magnetic or electrostatic field, so that the relative proportions of entities of different mass, commonly isotopes, can be determined.

matrix: the fine-grained material that occupies the space in a rock, such as a meteorite, between the larger, well-characterized components such as chondrules, inclusions, etc.

mesosiderite: class of stony-iron meteorite consisting of subequal proportions of silicate material (related to eucrites and diogenites) and Fe–Ni metal.

mesostasis: interstitial, generally fine-grained material occupying the space between larger mineral grains in an igneous rock; generally, therefore, the last material to solidify from a melt.

metallicity: the proportion of matter in a star that is made up of chemical elements other than hydrogen and helium.

metamorphism: solid-state modification of a rock, e.g. recrystallization, caused by elevated temperature and pressure.

metastable: a state that is characterized by a potential energy minimum that is not, however, the ground state of the system.

meteorite: a natural object of extraterrestrial origin that survives passage through a planetary atmosphere.

micrometeorite: a small extraterrestrial particle that has survived entry into the Earth's atmosphere.

Miller–Urey synthesis: formation of organic molecules by the passage of an electric discharge, or energetic radiation, through a mixture of methane and ammonia over refluxing water. It is often used for analogous experiments.

MIPS: Multi-band Imaging Photometer for Spitzer provides imaging capabilities in broad bands at 24, 70, and 160 μm. In addition, the instrument has a low-resolution spectroscopic mode that operates between 55 and 95 μm.

molecular cloud: a cold, dense interstellar cloud which contains a high fraction of molecules. It is widely believed that the relatively high density of dust particles in these clouds plays an important role in the formation and protection of the molecules.

MRI: magneto-rotational instability.

neutron star: a stellar remnant, the end product of the evolution of some massive stars that undergo a supernova explosion. Neutron stars are primarily composed of neutrons.

noble gas: the gases He, Ar, Kr, Ne, Xe, and Rn which rarely undergo chemical reactions, also known as inert gases and rare gases.

nova: a star that exhibits a sudden surge of energy, temporarily increasing its luminosity by as much as 17 magnitudes or more (although 12 to 14 magnitudes is typical). Novae are old disk-population stars, and are all close binaries with one component, a main-sequence star filling its Roche lobe, and the other component a white dwarf. Unlike supernovae, novae retain their stellar form and most of their substance after the outburst.

nucleosynthesis: the process of creating new atomic nuclei either by nuclear fusion or nuclear fission.

olivine: a magnesium iron silicate with the formula $(Mg,Fe)_2SiO_4$ in which the magnesium-to-iron ratio varies between the two end-members of the series:

forsterite (Mg-rich) and fayalite (Fe-rich). It gives its name to the group of minerals with a related structure (the olivine group) which includes monticellite and kirschsteinite.

orbital resonance: occurs when two orbiting bodies have periods of revolution that are in a simple integer ratio so that they exert a regular gravitational influence on each other.

ordinary chondrite: collective name for the most common variety of chondritic meteorite, subdivided into H, L, and LL groups on the basis of Fe content and distribution.

orthopyroxene: a mineral of the pyroxene group that crystallizes in the orthorhombic form.

oxidation: the addition of oxygen, removal of hydrogen, or the removal of electrons from an element or compound.

oxidation state: the measure of oxidation: the hypothetical charge of an atom if all bonds to other atoms were ionic.

oxygen fugacity: a function expressing the molar free energy of O in a manner analogous to the way pressure measures free energy of an ideal gas. In practice, O fugacity is equivalent to the partial pressure of O.

p-process: a nucleosynthetic process thought to be responsible for the synthesis of the rare heavy proton-rich nuclei which are bypassed by the r- and s-processes.

PAH: polycyclic aromatic hydrocarbon.

pallasite: class of stony-iron meteorites in which the Fe–Ni metal forms a continuous framework enclosing nodules of olivine.

parent body or parent asteroid: the object on or in which a given meteorite or class of meteorites was located prior to ejection as approximately meter–sized objects.

parent nuclide or parent isotope: an unstable nuclide which changes spontaneously into another (daughter) nuclide.

parsec: a unit of distance commonly used in astronomy; 1 parsec is the distance at which 1 AU subtends an angle of 1 arcsecond, equivalent to 3.26 light years or 3.0857×10^{16} m.

perihelion: the point of an orbit that is closest to the Sun.

photochemistry: chemical reactions occurring when molecules or atoms are illuminated, often with energetic ultraviolet radiation.

photolysis: chemical decomposition brought about by the action of light.

photosphere: the region of the stars from which externally visible light originates. Most solar spectroscopic abundance data have been obtained for this region of the Sun.

phyllosilicate: a family of silicate minerals characterized by a structure that consists of sheets or layers, invariably hydrated.

plagioclase: an important group of igneous-rock-forming tectosilicate minerals. Plagioclase comprises a solid solution series: $NaAlSi_3O_8$ (albite) to $CaAl_2Si_2O_8$ (anorthite).

planetary nebula: an expanding envelope of hot ionized gas expelled by an evolved star during its mass loss.

planetesimals: small rocky or icy bodies formed in protoplanetary disks that may become planetary building blocks.

Population I stars: relatively young stars with relatively high metallicity typically found in the disk of a galaxy.

Population II stars: old stars with low metallicities, characteristic to the galactic halo or to globular clusters.

porphyritic: a texture found in many igneous rocks, characterized by relatively large crystals, known as phenocrysts, set in a fine-grained or glassy matrix.

Poynting–Robertson (PR) drag: a relativistic effect of the re-radiation of the starlight absorbed by a dust particle on orbit that causes deceleration and the loss of angular moments, leading to an inward spiraling and eventual loss of the particle. The PR drag occurs because the dust particle absorbs photons with momentum pointing radially away from the central star, but re-radiates the energy isotropically in its own orbiting frame of reference – i.e. anisotropically, as seen from the reference frame of the star.

ppm: parts per million, generally by weight.

prebiotic: the state of matter before life existed, but conducive to the formation of life.

presolar: grains that predate the Solar System.

protostar: the earliest phase of stellar evolution during which a large part of the stellar energy is provided by gravitational contraction, rather than fusion reactions.

protostellar disk: an accretion disk surrounding a protostar.

purine: a heterocyclic compound containing fused pyrimidine and imidazole rings.

pyrolysis: heating and decomposition under controlled conditions, typically in a vacuum or an inert atmosphere.

pyroxene: a group of rock-forming silicate minerals that share a common structure which is comprised of single chains of silica tetrahedra. They crystallize in the monoclinic and orthorhombic system.

quantum tunneling: penetration by a particle into a potential energy region which is classically forbidden.

r-process: a nucleosynthesis process enabled by the successive rapid capture of neutrons. Due to the rapid capture regions of great nuclear instability are bridged; this model can successfully explain the formation of all elements heavier than Bi as well as the neutron-rich isotopes heavier than Fe. The essential feature of the r-process is the production and consumption of great numbers of neutrons in a very short time (< 100 s). The presumed source for such a large flux of neutrons is a supernova, at the boundary between the neutron star and the ejected material.

radiogenic: that which is made by radioactive decay.

radiogenic nuclide: a nuclide produced by decay of a radioactive parent nuclide, e.g. ^{206}Pb produced from the decay of ^{238}U.

radionuclide: a nuclide that is unstable against radioactive decay.

Raman spectroscopy: an analysis technique that utilizes the small energy differences, i.e. wavelength shifts, of monochromatic photons when they inelastically scatter on the atoms of a crystal: the energy differences reflect the phonon spectrum of the crystal and can be used to characterize the lattice structure of the target.

reaction cross-section: proportionality constant which relates the abundance of a target nucleus to the rate at which a given nuclear reaction occurs.

red giant: a relatively short evolutionary stage of an evolved star. These late-type (K or M) luminous stars have very large radii. A star reaches this phase after it has exhausted the nuclear fuel in its core; their luminosity is supported by energy production in an H-burning shell. Within the lifetime of the galaxy, only main-sequence stars of Type F and earlier have had enough time to evolve to the red giant phase (or beyond). The red giant phase corresponds to the establishment of a deep convective envelope.

reduction: process involving the addition of H, the removal of O, or decreasing the valence of an element. The opposite of oxidation.

refractory: term describing the high-temperature stability of an element or phase, the opposite of volatile.

regolith: the fragmented layer found on the surface of many planetary or subplanetary objects. It is created from the local competent lithologies by meteoroid impact and subsequently comminuted and turned over by such impacts.

rubble pile: a hypothetical structural model for planetesimals, asteroids, and comets in which the body is constructed by components that are not held together via structural bonds, but only by gravity.

s-process: nucleosynthetic process, in which heavy, stable, neutron-rich nuclei are formed through the capture of neutrons and subsequent beta decay to a stable daughter isotope. Because the neutron capture occurs much slower than the decay to stable isotopes, the nucleosynthesis proceeds through the series of stable isotopes. This is a process of nucleosynthesis that is believed to take place in intershell regions during the red giant phase of evolution.

SEM: scanning electron microscope.

shock: abrupt and major change in pressure and temperature, e.g. as caused by an impact, accretion, or other dynamic event.

shock metamorphism: alteration of a rock by shock-induced mechanical deformation or phase transformation above or below the solidus.

shock wave: discontinuity in temperature and pressure propagating in a solid, liquid, or gas with supersonic velocity, caused by impact or explosion.

siderophile: a geochemical class of elements. Siderophile elements are those which tend to go into the metal phase, e.g., Ni, Co, Au, As, Ge, Ga, Ir, Os, Re.

SIMS: secondary ion mass spectrometry.

SMA: Submillimeter Array, large interferometer array operating on the top of Mauna Kea, Hawaii, operating at the millimeter wavelength range that allows high-resolution studies of protostellar environments.

SMOW: standard mean ocean water, the reference standard for O isotopes.

solar nebula: the disk of gas and dust around the proto-Sun from which the Solar System formed.

solar wind: high-velocity charged light particles, mostly electrons and protons, ejected from the solar corona.

solid solution: a substance in which two or more components, such as the atoms of two or more elements, are randomly mixed on such fine a scale that the resulting solid is homogeneous.

solidus: the temperature below which a material is completely solid and no liquid exists.

spallation: a nuclear reaction in which a bombarded nucleus breaks up into many particles. This process accounts for the relatively high abundance of Be, Li, and B in cosmic rays.

spectral energy distribution: the flux density or emitted energy of an object as a function of wavelength. Typical use of the spectral energy distribution is to identify and characterize the temperature and luminosity of multiple emitting components of a single system, such as in the case of star + disk systems.

Spitzer Space Telescope: a cryogenically cooled infrared space telescope on a thermally stable Earth-trailing orbit operated by NASA. The telescope has very sensitive imaging capabilities with the IRAC (3 to 8 μm) and MIPS (24–160 μm) cameras as well as spectroscopic capability with the IRS instrument (5–40 μm).

sputtering: expulsion of atoms or ions from a solid, caused by impact of energetic particles.

STEM: scanning transmission electron microscopy.

Strecker–cyanohydrin synthesis: the synthesis of amino acids from aldehydes, HCN and NH_3 in the presence of an aqueous fluid.

sublimation: a transition from solid to vapor phase directly without passing through the liquid phase.

supernova, Type Ia: in a white dwarf star system the white dwarf star can be able to accrete mass from its companion and gradually increase its mass. When the star approaches its Chandrasekhar limit, it collapses and the increasing temperature and pressure enables the fusion of a significant fraction of the white dwarf's mass, forming a Type-Ia supernova.

supernova, Type Ib and Ic: massive stars that burn up their hydrogen in their cores collapse and explode as supernovae. The key difference between Type Ib,c and Type II is that the Type Ib,c progenitors have lost their hydrogen envelopes prior to the SN explosion, probably due to intense stellar winds.

supernova, Type II: Type-II supernovae involve massive stars ($>9\,M_\odot$) that burn their gases out within a few million years. If the star is massive enough, it will continue to undergo nucleosynthesis after the core has turned to helium and then to carbon. Heavier elements such as phosphorus, aluminum, and sulfur are created in shorter and shorter periods of time until silicon results. It takes less than a day for the silicon to fuse into iron; the iron core gets hotter and hotter and in less than a second the core collapses. Electrons are forced into the nuclei of their atoms, forming neutrons and neutrinos, and the star explodes, throwing as much as 90% of its material into space at speeds exceeding 30 000 km s^{-1}. After the supernova explosion, there remains a small, hot neutron star, possibly visible as a pulsar, surrounded by an expanding cloud.

T Tauri stars: very young, low-mass stars, less than 10 million years old and under 3 solar-masses, that are still undergoing gravitational contraction. T Tauri stars represent an intermediate stage between a protostar and a low-mass main-sequence star like the Sun. The prototype for this class of stars is T Tau.

taenite: iron with a face-centered cubic structure that is stable at high temperatures and/or when alloyed with a suitable proportion of a face-centered metal such as Ni.

TDU: third dredge-up in an AGB star, a powerful convective process that enriches the outer layers of the star with nucleosynthesis products.

TEM: transmission electron microscope.

transition disk: protoplanetary disk with a large inner cavity that is devoid or depleted in small dust grains and/or in gas. It has been hypothesized that some of these disks represent the rapid transition from optically thick primordial disks to optically thin debris disks reached through inside-out dispersal of the disk.

turbulence: a flow regime characterized by low-momentum diffusion, high-momentum convection, and rapid variation of pressure and velocity in space and time. Flow that is not turbulent is called laminar flow. The Reynolds number characterizes whether flow conditions lead to laminar or turbulent flow.

van der Waals force: an attractive force between two atoms due to a dipole moment in one molecule inducing a dipole moment in the other molecule which then interact.

viscosity: the measure of a material's resistance to flow. Viscosity is a result of the internal friction of the material's molecules. Materials with a high viscosity do not flow readily; materials with a low viscosity are more fluid.

volatile: an element that condenses from a gas or evaporates from a solid at a relatively low temperature.

VLT: Very Large Telescope, the combination of four 8m diameter ground-based telescopes and a set of 1.8 m auxiliary telescopes at Paranal Observatory of the European Southern Observatory. The telescopes can be used individually or as an interferometer (VLTI) providing very high spatial resolution.

VSMOW: Vienna standard mean ocean water.

white dwarf: a dim, dense, planet-sized star that marks the evolutionary endpoint for all but the most massive stars. White dwarfs form from the collapse of stellar cores in which nuclear fusion has stopped. White dwarfs consist of electron-degenerate matter, which provides the pressure needed to prevent further collapse, providing that the mass of the dwarf doesn't exceed the Chandrasekhar mass.

Wolf–Rayet star: a hot (25 000 to 50 000 K), massive (more than 25 solar masses), luminous star in an advanced stage of evolution, which is losing mass in the form of a powerful stellar wind. Wolf–Rayets are believed to be O stars that have lost their hydrogen envelopes, leaving their helium cores exposed.

X-wind: a theoretical model in which wind from a central star and the inner edge of its accompanying accretion disk join together to form magnetized jets of gas capable of launching material above the disk at velocities of hundreds of kilometers per second.

XRF: X-ray fluorescence.

Yarkovsky effect: a thrust produced when small, rotating bodies absorb sunlight, heat up, and then re-radiate the energy after a short delay produced by thermal inertia. The Yarkovsky effect produces a slow but steady drift in the semi-major axis of an object's orbit.

zodiacal cloud: a disk composed of few micron-sized dust grains in the Solar System. The zodiacal cloud is thought to be replenished by asteroid collisions and by dust grains freed from evaporating comets.

Index

Note. Material in tables is indicated by the letter *t* after the page number. Figures are indicated by the letter *f* after the page number. Footnotes are indicated by n after the page number.

Printed in the United States
By Bookmasters